STRUCTURE AND BONDING is issued at irregular intervals, according to the material received. With the acceptance for publication of a manuscript, copyright of all countries is vested exclusively in the publisher. Only papers not previously published elsewhere should be submitted. Likewise, the author guarantees against subsequent publication elsewhere. The text should be as clear and concise as possible, the manuscript written on one side of the paper only. Illustrations should be limited to those actually necessary.

SPRINGER-VERLAG		SPRINGER-VERLAG NEW YORK INC.
D-6900 Heidelberg 1	D-1000 Berlin 33	
P. O. Box 1780	Heidelberger Platz 3	175, Fifth Avenue
Telephone (06221) 491 01	Telephone (0311) 822001	New York, N. Y. 10010
Telex 04-61 723	Telex 01-83319	Telephone 673-2660

STRUCTURE AND BONDING

Volume 9

With 33 Figures

Springer-Verlag
Berlin Heidelberg GmbH 1971

ISBN 978-3-540-05320-0 ISBN 978-3-540-36444-3 (eBook)
DOI 10.1007/978-3-540-36444-3

Contents

Absorption Spectra and Ligand Field Parameters of Tetragonal 3*d*-Transition Metal Fluorides

D. Oelkrug

Institut für Physikalische Chemie, Universität Tübingen

Table of Contents

I. Introduction

Ligand field parameters of transition metal complexes can be obtained from spectroscopic data with some reliability as empirical values. Their nonempirical calculation, however, is a difficult task which highlights the problem of formulating the interactions of the atoms in such a complex. However, calculation is still one of the best ways of learning something about the physical background of the splitting of terms, especially about the effects of structure and bonding properties in the complex. It has become evident during the last fifteen years that the electrostatic model of the ligand field is inadequate in many respects, the effects of covalent bonding being much more important. To show that this is the case for typical "ionic" complexes, too, is one of the main objects of this paper in which the ligand field transitions of tetragonal transition metal fluorides are examinex. It is restricted to fluorides and tetragonal ligand fields for the following reasons:

1. If the electrostatic model is applicable at all, it should give the best results for compounds in which the bonds are of highest ionic character. Indeed, semi-empirical point charge calculations seem to be successful in the case of fluorides (*64*). Point charge ligands, however, are far removed from reality. Therefore a rigorous electrostatic calculation should include the spatial extension of the ligand electrons.

2. If covalent bonding effects are responsible for the splitting of the d-orbitals, one can start from the assumption that there is only a small overlap between central ion and fluoride orbitals. Therefore the method of calculation may be simplified. The most important simplification has been initiated by *Jørgensen et al.* (*45*) who assumed the antibonding effect on the *d*-orbitals to be proportional to the square of the overlap integrals. No significant quantitative results can be expected from this procedure, but we might obtain relative orbital energy differences and define ligand field parameters in terms of overlap integrals (angular overlap model).

3. In tetragonal complexes the *d*-orbitals are split into four components, not just two, as in octahedral complexes. A model therefore must give three correct energy differences, not just one. Even if one is taken from experiments to adjust the energy scale, the two others can be used to test the quality of a model.

4. The crystal structures of many transition metal fluorides are well known. Such knowledge is necessary for the interpretation of absorption spectra, as it is obvious that the term splittings are strongly related to the surroundings of the metal ions.

II. Microsymmetry of Transition Metal Fluorides

The structural properties of transition metal fluorides have been reviewed, e.g. by *Babel* (*2*). If electronic transition energies are considered, however, one need not look at the total variety of structure types,but only at the nearest neighbours of a transition metal ion. As a first approximation, the unit $[Me^{n+} F_x]^{-x+n}$ seems to behave like an isolated "cluster", even in the solid state. This approximation, however, is no longer justified when the vibronic interactions or the oscillator strengths, here, in particular, of intercombination bands (*26,63*), are to be considered. In such cases the phonon modes of the crystal or the linkage of the metal ions must be taken into account.

The transition metal ions in fluorocomplexes are generally six-coordinated with an octahedral arrangement of the fluoride ions. The regular coordination predominates, but sometimes there appear distinct deviations from O_h symmetry which can be explained with the help of the Jahn-Teller (JT) theorem (*42*), according to which a non-linear molecule in a degenerate orbital state may distort for energy reasons in order to remove as much degeneracy as possible. Thus only the d^3, d^5, d^8 high-spin complexes and the d^6 low-spin complexes should form regular octahedra in any case. The rest of the configurations with doubly or triply degenerate ground states may show a JT distortion, but not necessarily, for the theorem does not give any idea of the magnitude of the effect.

It is, however, odvious that compounds with an octahedral E_g ground state practically always appear in at least tetragonally distorted octahedra, decause the σ-antibonding e_g orbitals are occupied by one or three electrons. The interaction between the electronic state an dthe e_g vibrational mode (the only one which is JT active) therefore results in a considerable energy gain for the D_{4h} symmetry compared with the O_h symmetry, so that the energy minimum is manifested in a static distortion.

Typical representatives of this class are the high-spin configurations d^4 (Cr^{2+}, Mn^{3+}) and d^9 (Cu^{2+}), see Tab. 1. No fluoride complexes of the low-spin configuration d^7 (Co^{2+}, Ni^{3+}) are known. The distortion of the MeF_6 octahedra is obvious in almost all cases, with the exception of $CsCuAlF_6$, the powder diagramm of which shows cubic symmetry (*3*). In $KCrF_3$ and Na_2CrF_4, the Cr—F distances are unknown. The neutron diffraction investigations of *Scatturin et al.* (*74*) suggest an even stronger distortion than the isostructural copper(II) fluorides show.

Unfortunately, the JT theorem in a case of E_g configuration says nothing about the direction of the distortion. The elongated octahedron (having two axial Me—F distances longer than the other four) as well as the compressed one (having four equatorial Me—F distances longer than the other two) results in a non-degenerate ground state (B_{1g} or A_{1g}). The theoretical calculations have not so far led to an unambiguous result, either. According to the investigations of *Öpik* and *Pryce* (*67*), the elongated octahedron represents the most favorable coordination, while *Liehr* and *Ballhausen* (*52*) see the most favorable coordination in the compressed octahedron. The analysis of *Pryce, Sinha* and *Tanabe* (*71*) leaves either possibility open, according to the degree of covalency in the complex: a purely electrostatic consideration predicts a distortion with the tetragonal axis compressed. Increasing covalency would favour the elongated coordination. According to the experimental findings, the elongated coordination clearly predominates (Table 1), so that, even in the extreme ionic case, the structure seems to be determined by covalency effects. The only compounds for which compressed octahedra are still discussed are K_2CuF_4, investigated by *Knox* (*48*), and Ba_2CuF_6, investigated by *von Schnering* (*78*). A reinvestigation of the K_2CuF_4 struc-

Table 1. *Structural properties of some $3d^4$ and $3d^9$ transition metal fluorides*

	Structure Type	Me—F [Å]	Ref.
CuF_2	Rutile type, Monoclinic distorted	4F 1.93 2F 2.27	(*15*)
$KCuF_3$	Perovskite type, Tetragonal distorted	2F 1.89 2F 1.96 2F 2.25	(*66*)
Na_2CuF_4	$P2_1/c—C^5_{2h}$	4F 1.91 2F 2.36	(*1*)
K_2CuF_4	K_2NiF_4, distorted	4F > 2F 4F < 2F	(*48*) (*2*)
Ba_2CuF_6	$I\ 2/m—C^3_{2h}$	2F 1.85 4F 2.08	(*78*)
$CsCuAlF_6$	Fd3m	(regular octahedron?)	(*3*)
CrF_2	Rutile type, Monoclinic distorted	2F 1.98 2F 2.01 2F 2.43	(*41*)
$KCrF_3$	as $KCuF_3$	?	(*88*)
Na_2CrF_4	as Na_2CuF_4	?	(*88*)
MnF_3	VF_3-type, Monoclinic distorted	2F 1.79 2F 1.91 2F 2.09	(*36*)
$(NH_4)_2MnF_5$ (Na_2MnF_5)	Pnma-D^{16}_{2h}	2F 1.84 2F 1.85 2F 2.12	(*79*)
K_2NaMnF_6	Elpasolite type, Tetragonal distorted	4F 1.86 2F 2.06	(*49*)

ture by *Babel* (*2*), however, suggests that the elongated structure is more likely, and *von Schnering* points out that this might apply to Ba_2CuF_6 too, in case of finding superstructure reflexes. It is therefore not certain, whether compressed tetragonal distortions in octahedral systems with E_g electronic ground states exist at all.

It is far more difficult to detect a JT distortion in compounds with a T_{1g} or T_{2g} ground state, in which the π-antibonding t_{2g} orbitals are unsymmetrically occupied by electrons. The perturbing potential of the vibrational mode e_g may cause a tetragonal distortion, similar to that of the E_g state, and that of the vibrational mode t_{2g} *a* trigonal distortion along one of the threefold octahedral axes, the resulting symmetry being

D_{3d}. In both cases the electronic ground state splits into a doublet and a singlet. This enables the direction of the distortion to be predicted, as, according to JT, the non-degenerate state must have the lowest energy. But such distortions are seldom revealed by X-ray analysis. There may be various reasons for this. Either the effects of the slight interaction between t_{2g} and ligand orbitals are too small to be detected experimentally, or the structure appears as regular, because the stability gain is approximately equal to the zero-point energy of the vibrational mode or kT. Then the molecule may override the energy barriers between the different potential minima of a distorted coordination and therefore no static deformation will be found. This is the situation encountered in the dynamical JT effect. The only T_{2g} (and T_{1g}) fluorocomplex in which X-ray analysis has so far revealed a definite static distortion is FeF_2 (*11*). In accordance with the JT predictions, the FeF_6 octahedra are compressed (ground state $^5B_{2g}$) with four Fe—F distances of 2.12 Å and two of 1.99 Å (the actual symmetry is not D_{4h} but only D_{2h}, for some of the F—Fe—F angles deviate from 90°). Some infrared absorption data also give experimental evidence of the JT effect. So the splittings of the t_{1u} vibrational levels in $(NH_4)_3TiF_6$ and $Ba_3(CoF_6)_2$ have been assigned to tetragonal MeF_6 clusters (*57, 75*). But no definite conclusions can be drawn about the static symmetry of the ground state from the optical data, as the optical transition is determined by the momentaneous coordination, which need not be identical with the average static coordination. Therefore we cannot say definitely whether the JT effect in the case of octahedral T_{1g} and T_{2g} configurations produces a static distortion, until more detailed structural data of such compounds are available.

III. *d*-Orbital Energy Differences in a Tetragonal Ligand Field

The actual symmetry of the distorted MeF_6 units in most cases is lower than D_{4h}. But to describe the electronic spectra it will be sufficient to consider only the tetragonal field component, as minor structural differences disappear within the width of the absorption bands. The same holds for the splitting of terms through spin-orbit coupling which therefore can be neglected.

The following considerations start from isolated MeF_6 units with the Me^{n+} ion placed at the origin. The F^- ions are situated on the Cartesian axes with the coordinates $x, -x, y, -y, z, -z$, where $|x| = |y| \neq |z|$. The discussion will be restricted to the high-spin configurations d^1, d^4, d^6, and d^9, because their spin-allowed $d-d$ transitions in *a* D_{4h} ligand field are determined directly by the energy differences of the one-electron orbitals.

A. The Purely Ionic Model

1. Point Charge Ligands

The classical model of the ligand field theory neglects any overlap between central ion and ligand orbitals and represents the excess charge of a ligand as a charge point being localized in the ligand nucleus. The d-orbital energies consequently are affected only by the electrostatic potential of the ligands. As long as the hypothesis of non-existing overlap is maintained, it is immaterial whether the ligands are in reality larger than the central ion. The reason is that if the potential of a charged sphere with arbitrary radial (but spherical) charge distribution is calculated at any point outside the sphere, the result will be the same as if the whole charge were localized in the center.

In a tetragonal potential $v_{D_{4h}}$ of six point charges qe_0 with the positions mentioned above, the relative energies of the orbitals d_{xz} and d_{yz} (e_g), d_{xy} (b_{2g}), d_{z^2} (a_{1g}), and $d_{x^2-y^2}$ (b_{1g}) will be

$$
\begin{aligned}
\langle e_g|v_{D_{4h}}|e_g\rangle &= qe_0^2\left[-\frac{2}{3}\overline{r^4}\frac{1}{x^5}-\frac{2}{7}\overline{r^2}\left(\frac{1}{x^3}-\frac{1}{z^3}\right)+\frac{8}{21}\overline{r^4}\left(\frac{1}{x^5}-\frac{1}{z^5}\right)\right]\\
\langle b_{2g}|v_{D_{4h}}|b_{2g}\rangle &= qe_0^2\left[-\frac{2}{3}\overline{r^4}\frac{1}{x^5}+\frac{4}{7}\overline{r^2}\left(\frac{1}{x^3}-\frac{1}{z^3}\right)-\frac{2}{21}\overline{r^4}\left(\frac{1}{x^5}-\frac{1}{z^5}\right)\right]\\
\langle a_{1g}|v_{D_{4h}}|a_{1g}\rangle &= qe_0^2\left[\overline{r^4}\frac{1}{x^5}-\frac{4}{7}\overline{r^2}\left(\frac{1}{x^3}-\frac{1}{z^3}\right)+\frac{12}{21}\overline{r^4}\left(\frac{1}{x^5}-\frac{1}{z^5}\right)\right]\\
\langle b_{1g}|v_{D_{4h}}|b_{1g}\rangle &= qe_0^2\left[\overline{r^4}\frac{1}{x^5}+\frac{4}{7}\overline{r^2}\left(\frac{1}{x^3}-\frac{1}{z^3}\right)-\frac{2}{21}\overline{r^4}\left(\frac{1}{x^5}-\frac{1}{z^5}\right)\right]
\end{aligned}
\quad (1)
$$

The term which raises all energies by the same amount is not considered in Eq. (1). $\overline{r^n}$ is defined as $\int_0^\infty (R_{3d})^2\, r^{n+2}\, dr$ with the radial function R_{3d} of the $3d$ electrons. Because the energy difference between the b_{1g} and b_{2g} orbitals does not depend on the position of the z ligands, its value will be the same as in a regular octahedron with six ligands at the distance x

$$10\,\mathrm{Dq} = \frac{5\, qe_0^2\, \overline{r^4}}{3x^5} \quad (2)$$

To fix all three orbital energy differences, usually two additional tetragonal parameters are defined (*59*)

$$\mathrm{Ds} = \frac{2}{7}\, qe_0^2\, \overline{r^2}\left(\frac{1}{x^3}-\frac{1}{z^3}\right), \qquad \mathrm{Dt} = \frac{2}{21}\, qe_0^2\, \overline{r^4}\left(\frac{1}{x^5}-\frac{1}{z^5}\right) \quad (3)$$

They may be positive or negative, depending on whether the octahedron is elongated or compressed.

The model seems to be most promising for the fluorides. It has also been applied in a semi-empirical manner to the other halides (*20*) and to compounds containing oxygen ligands (*14, 39, 70*), but in all these cases attention is drawn to the deficiencies of the model, which become obvious as soon as one attempts a quantitative evaluation of the ligand field parameters of Eqs. (2) and (3). Table 2 gives the calculated and experimental Dq parameters of fluorides whose Me-F distances are known. Slater-type or Hartree-Fock atomic orbitals, respectively, were used as R_{3d}, the latter in the analytical approximation of *Richardson et al.* (*73*). The 10 Dq parameters calculated with Slater orbitals are comparable in order of magnitude with the experimental results. They are, however, too large at the beginning of the period and too small at the end. Calculations with Hartree-Fock orbitals, though theoretically to be preferred, result in too small energy differences throughout, which is the reason why R_{3d} can be treated in the point charge model as an empirical parameter only. It must be taken from one of the absorption bands in

Table 2. *10 Dq parameters of some 3d fluorides. Point charge calculations were carried out using Hartree-Fock (H.F.) or Slater functions*

	Me—F [Å]	Ref.	10 Dq [cm^{-1}] calculated H. F.	Slater	10 Dq [cm^{-1}] Experim.	Ref.
$[CrF_2]$[a])	2.00	(*41*)	3870	16200	11000	(*18,63*)
MnF_2	2.12	(*11*)	2310	8580	7800	(*84*)
$[FeF_2]$	2.12	(*11*)	1840	5760	6900	(*34*)
CoF_2	2.04	(*11*)	1760	4760	7200	(*54*)
NiF_2	2.00	(*11*)	1580	3690	7400	(*56*)
$[CuF_2]$	1.93	(*15*)	1550	3180	8900	(*55,64*)
TiF_3	1.97	(*82*)	4350	27500	16000 K_2NaTiF_6	(*13*)
VF_3	1.94	(*40*)	3640	21300	15900 $(NH_4)_3VF_6$	(*10*)
CrF_3	1.90	(*41*)			14600	(*28,87*)
K_2NaCrF_6	1.93	(*50*)	2970	15300	16500	(*27*)
$[K_2NaMnF_6]$	1.85	(*49*)	2910	12700	17400 K_3MnF_6	(*33,63*)
K_2NaFeF_6	1.91	(*50*)	2020	7680	14000 FeF_6^{3-}	(*43*)
CoF_3	1.89	(*37*)	1730	5700	11400 K_3CoF_6	(*17*)

a) Compounds in square brackets do not form regular octahedra. The experimental and calculated 10 Dq parameters are based on the four Me—F distances in the *x, y* plane.

order to calculate the other band positions. Such a procedure has to a certain extent been justified by the investigations of *Drickamer et al.* (*23, 58, 83*) in which the x^{-5} dependence of 10 Dq, required by the electrostatic model, sometimes has been found by experiment, although only within a very small range of Δx.

2. Spherical Ligand Charges

Because of the exponential radial decrease of the ligand and the central ion orbitals, the hypothesis of non-existent overlap cannot be maintained in a more accurate calculation. Even if the overlap is very small, it will change the relative d-orbital energies to such an extent that the point charge model is totally destroyed. This was shown the first time by *Kleiner* (*47*) who calculated 10 Dq in $Cr(H_2O)_6^{3+}$ from electrostatic principles but included the delocalization of the charge on the oxygen ligands by using Slater orbitals for the oxygen electrons. The value of 10 Dq had the wrong sign, because the positive nuclear charge attracted the e_g electrons more strongly than the ligand electrons repelled them.

If the same calculations are carried out for the fluorides, the results are no better (*65, 85, 86*). In Fig. 1 the energy differences between the

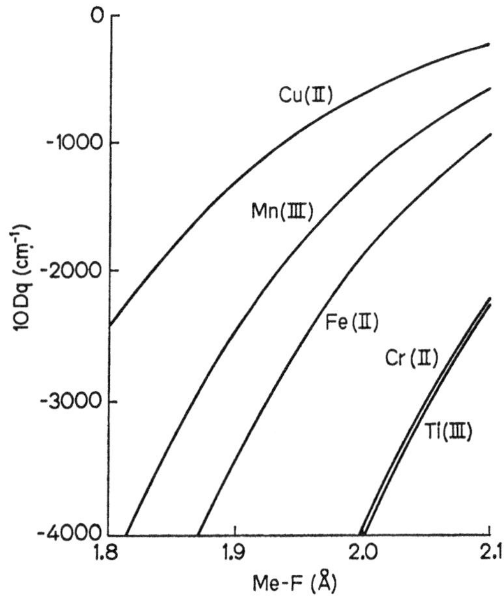

Fig. 1. Electrostatic 10 Dq parameters of some octahedral MeF_6-complexes, calculated by considering the spatial extension of the ligand electrons. Hartree-Fock functions have been used for both the *3d* and F^- orbitals

Coulomb interaction integrals $\langle e_g|v_{O_h}|e_g\rangle - \langle t_{2g}|v_{O_h}|t_{2g}\rangle = 10$ Dq are plotted against the Me—F internuclear distances. Hartree-Fock radial functions in *Richardson's* approximation (*73*) have been used for the 3*d* electrons, and v_{O_h} was built up from the ninefold positively charged fluorine nucleus and from *Sugano* and *Shulman's* (85) analytical Hartree-Fock wave functions of all ten F^- electrons. All octahedral 10 Dq values become negative at the Me—F equilibrium distances, the slopes of the Dq-lines being opposite to the predictions of the point-charge model and opposite to the experimental results (*23, 58, 83*). There is therefore no longer any doubt that the electrostatic model is inadequate from the theoretical point of view and that the positive results of point charge calculations are either accidental or originate from a formal but not physical analogy with a more realistic description of the interaction between the 3*d* and ligand electrons.

B. The Model of Weak Covalent Bonding

It is a consequence of the overlap between 3*d* and ligand electrons that their exchange interactions have primarily to be considered, in addition to their Coulomb interactions, which seem to be of minor importance for the 3*d*-orbital splittings. Therefore the usual molecular techniques have to be applied to solve the energy problem. In the LCAO—MO approximation the wave functions of the cluster $[MeL_x]$ are described as normalized linear combinations $\psi = N(a\varphi_M + b\chi_L)$ of the metal orbitals φ_M and symmetry-adapted linear combinations χ_L of the ligand orbitals φ_L. The molecular orbitals involved in octahedral ligand field transitions are the predominantly d-like π-antibonding t_{2g} orbitals and σ-antibonding e_g orbitals (δ-bonds are not considered). Approximate energies of these molecular orbitals are obtained by solving the corresponding secular equation $|H_{ij} - EG_{ij}| = 0$. Here H_{ij} are the interaction integrals, G_{ij} are the group overlap integrals (if off-diagonal), i.e. the total overlap of a metal orbital and a linear combination of ligand orbitals ($\equiv G_{ML}$). The simplest method of solving the equation is to estimate the diagonal elements H_{ii} as valence state ionization energies of the metal ion (H_M) and the ligand (H_L). According to the somewhat intuitive suggestion of *Wolfsberg* and *Helmholz* (*90*), the off-diagonal elements H_{ij} can be approximated as

$$H_{ij} = k\, G_{ML}\, (H_M + H_L)/2 \qquad (4)$$

with the constant $k \approx 2$.

Starting from this relation, *Jørgensen et al.* (*45*) proposed a model in which the antibonding energy effect ΔE_a upon the d-like molecular orbitals is supposed to be proportional to the square of the group overlap integral G_{ML}. This can be shown by solving the secular equation with the help of Eq. (4) and $G_{ML} \ll 1$, $H_L \ll H_M$ (*44*):

$$\Delta E_a \approx [(k/2-1)\,H_M + k/2H_L]^2\, G_{ML}^2/(H_M - H_L) = G_{ML}^2 \cdot \varepsilon. \quad (5)$$

Eq. (5) therefore is valid for predominantly ionic complexes only. The metal ion orbitals usually are approximated as products of an angular and a radial function, and the G_{ML} are expressed in terms of diatomic overlap integrals $S(\varphi_M, \varphi_L)$, e.g. $S(3d_{z^2}, \sigma_z)$ for σ-overlap (σ_z is a σ-orbital of a ligand located on the z-axis) and $S(3d_{xy}, \pi_x)$ for π-overlap.

In an octahedral complex the G_{ML} are then

$$\begin{aligned} G(e_g, \chi_\sigma) &= \sqrt{3}\, S(e_g, \sigma) \\ G(t_{2g}, \chi_\pi) &= 2\, S(t_{2g}, \pi) \end{aligned} \quad (6)$$

neglecting ligand-ligand overlap. If the latter effect is included in the normalization constant of the molecular orbitals, slightly more complicated relations are obtained (*8*). The octahedral ligand field parameter is now obtained as

$$10\,Dq = 3u_\sigma - 4u_\pi \quad (7)$$

with

$$u = S_{ML}^2 \cdot \varepsilon \quad (8)$$

Because ε is assumed to be constant in this simple MO-model, a quantity proportional to 10 Dq can easily be calculated from diatomic overlap integrals. Fig. 2 gives the results for a hypothetical $[CuF_6]^{4-}$ octahedron. The Me-F dependence of 10 Dq now has the correct sign, but the x^{-5} law, which sometimes has been verified by experiment (*23, 58, 83*), is realized only within a very small range of Me-F distances.

In complexes of lower symmetry usually more than one linear combination of ligand orbitals will have non-vanishing overlap integrals with a d-orbital of given symmetry, and the quantity $G^2(\varphi_M, \chi_L)$ in Eq. (5) should be substituted by $\sum_i G^2(\varphi_M, \chi_{Li})$. For a tetragonal complex with identical ligands at distances $|x| = |y| \neq |z|$ the molecular orbital combinations are

Symmetry	Metal orbitals	Ligand orbitals
a_{1g}	d_{z^2}	$-1/2\,(\sigma_x+\sigma_y+\sigma_{-x}+\sigma_{-y})$, $1/\sqrt{2}\,(\sigma_z+\sigma_{-z})$
b_{1g}	$d_{x^2-y^2}$	$1/2\,(\sigma_x-\sigma_y+\sigma_{-x}-\sigma_{-y})$
e_g	d_{xz}	$1/\sqrt{2}\,(\pi_x+\pi_{-x})$, $1/\sqrt{2}\,(\pi_z+\pi_{-z})$
	d_{yz}	$1/\sqrt{2}\,(\pi_y+\pi_{-y})$, $1/\sqrt{2}\,(\pi_z+\pi_{-z})$
b_{2g}	d_{xy}	$1/2\,(\pi_x+\pi_y+\pi_{-x}+\pi_{-y})$

and so the antibonding energy expressions are:

$$\begin{aligned} E_{a_{1g}} &= u_{\sigma x}+2u_{\sigma z} \\ E_{b_{1g}} &= 3u_{\sigma x} \\ E_{e_g} &= 2u_{\pi x}+2u_{\pi z} \\ E_{b_{2g}} &= 4u_{\pi x} \end{aligned} \tag{9}$$

The splitting of the octahedral e_g orbitals is caused by differences of σ-antibonding energies exclusively and that of the t_{2g} orbitals by differences of π-antibonding energies only.

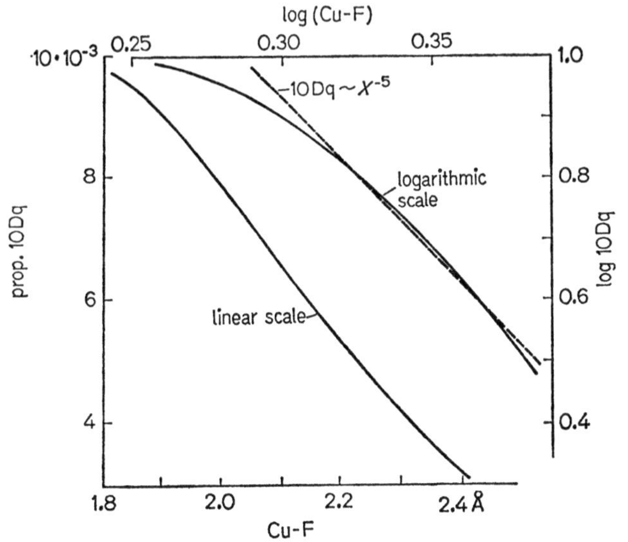

Fig. 2. The calculated group overlap integrals $G^2(e_g,\chi\sigma)-G^2(t_{2g},\chi_\pi)\sim 10\,Dq$ of a $[CuF_6]^{4-}$ octahedron. Only 2p-orbitals have been used in the ligand σ set

Clearly, the chief use of Eq. (9) is to give a qualitative picture of the term splittings. To obtain energy differences, the procedures outlined by *Wolfsberg* and *Helmholz* (*90*) and *Mulliken* (*61*) or non-empirical calculation methods should be applied. The Wolfsberg-Helmholz method was reviewed, e.g. by *Ballhausen* and *Gray* (*7*). It has become a simple standard method in the MO-treatment of transition metal compounds and has been applied, for instance, to octahedral fluorocomplexes (*12, 13*) and also to complexes of lower symmetry containing different simple ligands (*6, 32, 62*). Although the calculated term splittings agree fairly well with the experimental ones, it should be borne in mind that the results depend very much upon the choice of linear combinations of the ligand orbitals (*9*).

The more rigorous non-empirical MO-calculations are based on the work of *Sugano* and *Shulman* (*85*) and have been carried on by *Watson* and *Freeman* (*89*), *Sugano* and *Tanabe* (*86*), and *Offenhartz* (*65*). All calculations are restricted to octahedral fluorocomplexes and the agreement between theory and experiment is not too good. In short, much remains to be done before all the problems of ligand field splitting are understood theoretically.

IV. Discussion of the Absorption Spectra

Practically all the absorption data reported are taken from diffuse reflectance spectra, as single-crystal spectra of the compounds in question have hardly been available until recently. The investigations therefore are restricted to energies and relative intensities of the absorption bands, and no comment can be given about oscillator strengths and polarizations of the transition dipole moments. If, however, the spectra of many compounds are compared, including data for other transition metal halides or oxides, a fairly complete survey of ligand field interactions can be given.

A. The Electronic Configurations $3d^4$ and $3d^9$

1. Static Distortions

The tetragonal distortion of the MeF_6 unit in the crystal is so distinct in most examples (see Table 1) that we must start from a D_{4h} term diagram to describe the electronic transitions. As the elongated coordination obviously predominates, the B_{1g} term will be the ground state, and, according to Eqs. (1)—(3) or Eq. (9), respectively, the following transition energies are to be expected:

	electrostatic	covalent	
$B_{1g} \longrightarrow A_{1g}$	4 Ds + 5 Dt	$2u_{\sigma x} - 2u_{\sigma z}$	
$\longrightarrow B_{2g}$	10 Dq	$3u_{\sigma x} - 4u_{\pi x}$	(10)
$\longrightarrow E_g$	10 Dq + 3 Ds − 5 Dt	$3u_{\sigma x} - 2u_{\pi x} - 2u_{\pi z}$	

In a d^9 electronic system all transitions are determined by these relations, in a high-spin d^4 system, however, only the quintet-quintet transitions. The latter may be distinguished from the quintet-triplet transitions by their band positions, as in chromium(II) fluorides, or at least by their band intensities, as in manganese(III) fluorides. The intensity criterion must be handled with some care, however, as the maximum extinction coefficients of intercombination bands may well reach the order of magnitude of spin-allowed bands. Thus the absorption maxima at higher energies, which are of interest here, cannot be definitely localized; least in MnF_3 (*18*) but neither in the other manganese(III) compounds (see Fig. 6).

All three absorption bands which are expected according to Eq. (10) can be detected in most of the examples reproduced in Figs. 3—6. But they are never completely separated from each other and sometimes one band is totally hidden by the other two, so that it does not even appear as a shoulder.

To assign the bands and to calculate the ligand field splittings certain simplifications are introduced.

a) The Eqs. (10) are assumed to be correct.

b) For a given cation, the ligand field parameters are determined only by the positions of the fluoride ions of the first coordination sphere.

c) The complex is assumed to be ionic in that sense that the radial functions of the ions can be used to calculate the overlap integrals.

If the distortions in the complex are weak, the energy order of the excited states will be $A_{1g} < B_{2g} < E_g$. Increasing elongation of the z distances, however, will raise the A_{1g} term above B_{2g} or even above E_g. In the extreme case of a square-planar complex, the relative term energies E_i are determined by the following criteria:

			electrostatic	covalent
1.	$E_{A_{1g}} > E_{B_{2g}}$	if	$Ds > \frac{5}{28}$ 10 Dq	$4u_\pi > u_\sigma$
2.	$E_{A_{1g}} > E_{E_g}$	if	$Ds > \frac{3}{7}$ 10 Dq	$2u_\pi > u_\sigma$
3.	$E_{B_{2g}} > E_{E_g}$	if	$Ds < \frac{2}{21}$ 10 Dq	not possible

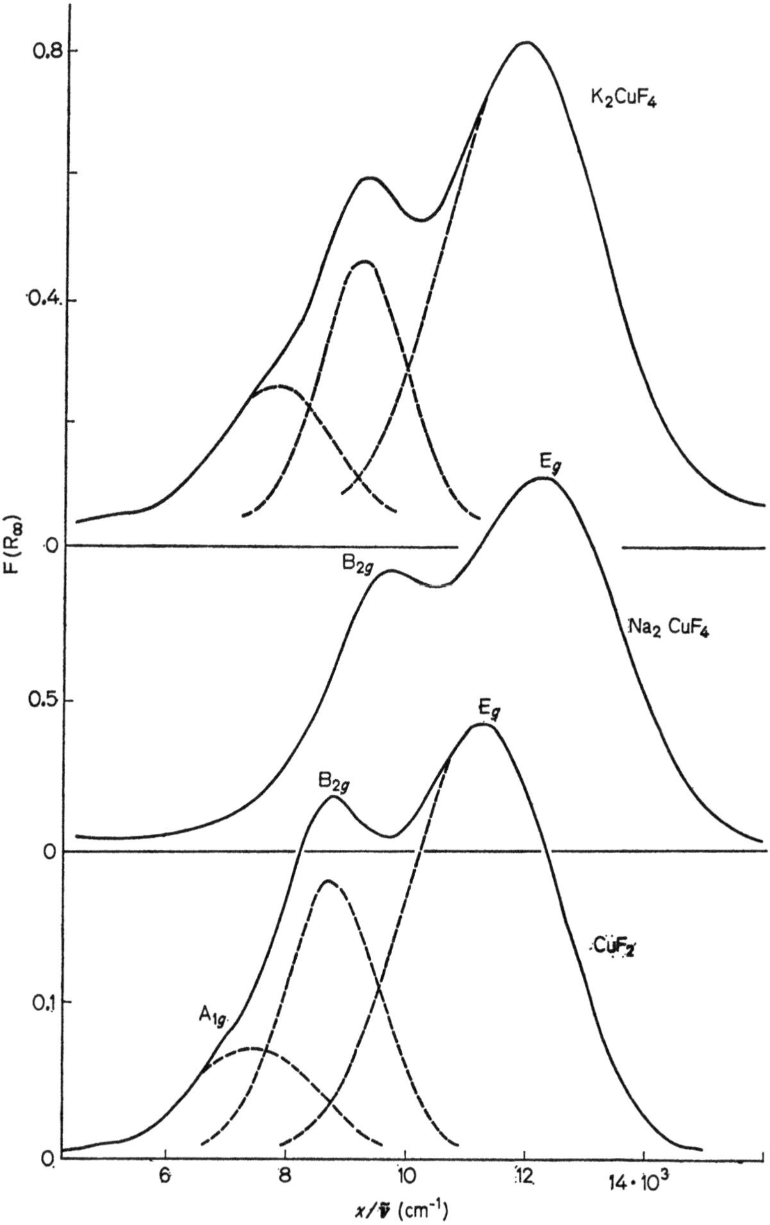

Fig. 3. Reflectance spectra of some Cu(II) fluorides. Elongated octahedral CuF_6-symmetry.
The correct abscissa symbols of Figs. 3-6, 8 should be: $\tilde{\nu}$

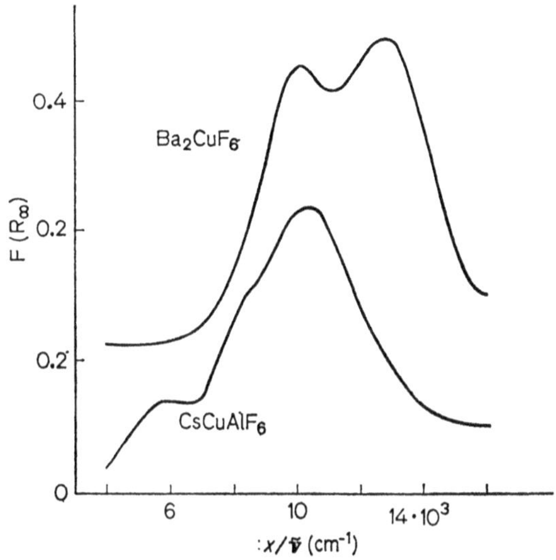

Fig. 4. Reflectance spectra of Cu(II) fluorides. Compressed (Ba_2CuF_6) or regular ($CsCuAlF_6$) octahedral CuF_6-symmetry. $CsCuAlF_6$ is identical with $CsCuGaF_6$ (both according to Babel, unpublished work)

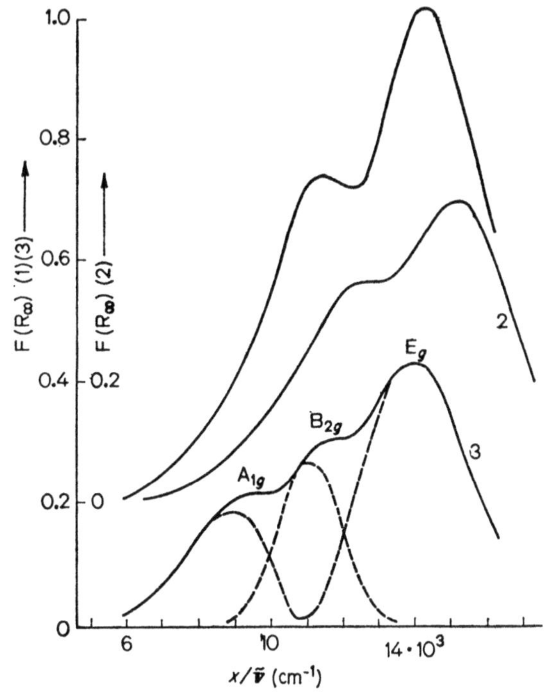

Fig. 5. Reflectance spectra of CrF_2 (1), Na_2CrF_4 (2) and $KCrF_3$ (3). Only the range of the quintet-quintet transitions is reproduced

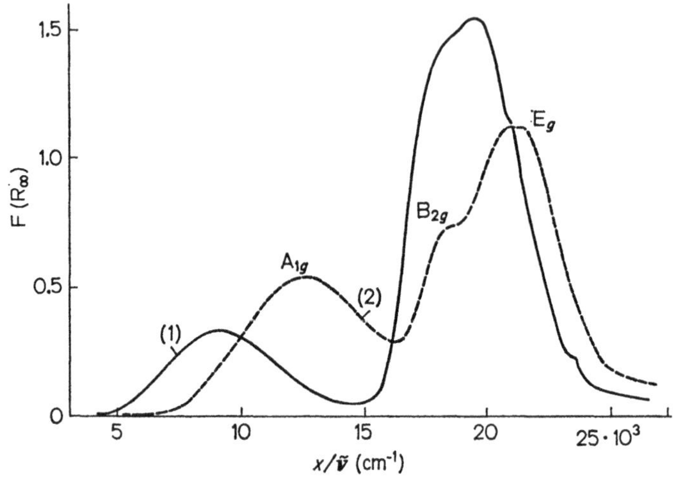

Fig. 6. Reflectance spectra of K_3MnF_6 (1) and Na_2MnF_5 (2)

The third case has only a formal meaning, because an inversion of the octahedral T_{2g} components is impossible in the weak covalent model and would occur in the electrostatic model only at very small ligand distances, which cannot be realized. Because the tetragonal field component is always smaller in six-coordinated complexes than in square-planar complexes, the second case $E_{A_{1g}} > E_{E_g}$ is improbable too. Therefore we have to find only the order of the A_{1g} and B_{2g} terms. Some hints are supplied by crystal structure data: an elongation of the octahedra means primarily a variation of the z distances whilst the x distances remain about constant. Therefore in a very rough approximation, the transition $B_{1g} \rightarrow B_{2g}$ is not affected by the distortions in both models, so that the band whose energy remains about constant can be assigned to this transition. This seems to be the middle band for all compounds in which three bands are observed. It is situated at about 8500—9500 cm^{-1} in the copper(II) fluorides, at about 11000—12000 cm^{-1} in the chromium(II) fluorides, and at about 17500—18000 cm^{-1} in the manganese(III) fluorides. The low energy band then corresponds to the transition $B_{1g} \rightarrow A_{1g}$. Its position depends most strongly upon the degree of distortion which can be seen if CuF_2 is compared with Na_2CuF_4 (Fig. 3), and K_3MnF_6 with Na_2MnF_5 (Fig. 6). The long wave shoulder in Na_2CuF_4 obviously disappears completely under the B_{2g} band.

If the position of the B_{2g} band is known, one can try to calculate the effect of the tetragonal field component on the positions of the A_{1g} and E_g bands, using the B_{2g} band as a reference energy. The results obtained for the $[CuF_6]^{4-}$ complex are given in Fig. 7. Point charge calculations started from hydrogen-like $3d$ radial functions with the effective nuclear charge $z = 6.37$, determined from the energy of the $B_{1g} \rightarrow B_{2g}$ maximum using the value of 1.93 Å for the short Cu-F distances. The overlap integrals in the model of weak covalent bonding have been calculated using the wave functions of *Richardson* (*73*) and *Sugano* (*85*), but considering only the $2p$ fluoride functions in the σ linear combinations of the ligand orbitals. Ligand-ligand overlap has been included in the normalization constant. Both models lead to comparable splittings of terms, and this astonishing fact explains why point charge calculations sometimes have been so successful.

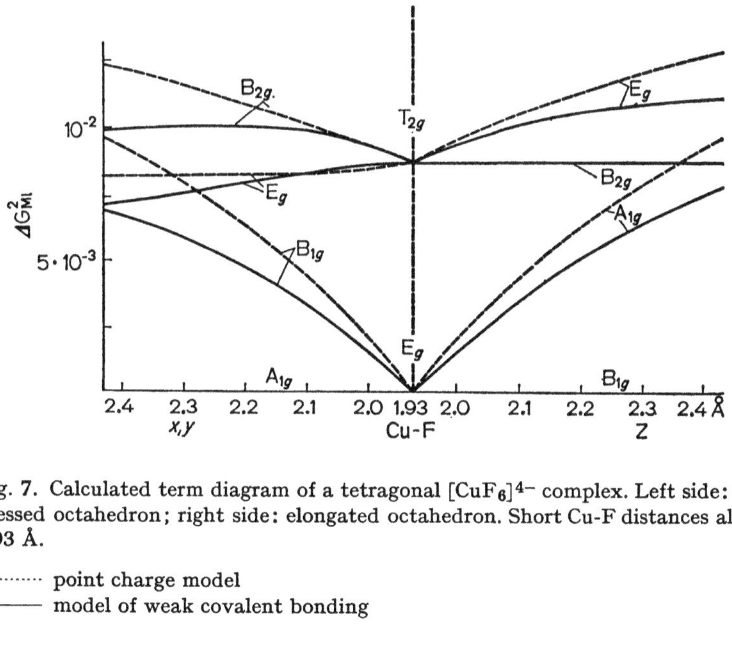

Fig. 7. Calculated term diagram of a tetragonal $[CuF_6]^{4-}$ complex. Left side: compressed octahedron; right side: elongated octahedron. Short Cu-F distances always 1.93 Å.

.... point charge model
——— model of weak covalent bonding

Some experimental and calculated transition energies of compounds with known Me-F distances are listed in Table 3. The order of magnitude of the tetragonal field effect is reproduced quite well but quantitative agreement is not obtained and is not to be expected, because the Me-F distances are accurate only within about $\pm 0{,}02$ Å and because the absorption maxima do not represent exactly the energy difference between two electronic levels.

Table 3. *Experimental and calculated ligand field transitions of some $3d^4$- and $3d^9$-[MeF_6] complexes*

	Absorption maxima [cm^{-1}]	Ref.	Experimental parameters [cm^{-1}]	Energy ratios a) experimental b) point charge model c) covalent bonding
CuF_2	$^2A_{1g}$ (7500)[a])		Dq 880	a) 0.85/1/1.29
(similar	$^2B_{2g}$ 8800	*(55, 64, 77)*	Ds 1435	b) 0.86/1/1.37 Z = 6.37[b])
to $KCuF_3$)	^2E_g 11350		Dt 350	c) 0.70/1/1.24
Na_2CuF_4	$^2A_{1g}$?		Dq 955	a) ? /1/1.28
	$^2B_{2g}$ 9550	*(64, 77)*	Ds —	b) 1.02/1/1.43 Z = 6.33
	^2E_g 12200		Dt —	c) 0.86/1/1.29
K_2CuF_4	$^2A_{1g}$? (7800)		Dq 920	
	$^2B_{2g}$? 9200	*(64, 77)*	Ds 1500	—
	^2E_g? 11900		Dt 360	—
Ba_2CuF_6	$^2B_{1g}$?			a) ? /1/1.27
	^2E_g 10050		?	b) 0.67/1/1.27 Z = 6.37
	$^2B_{2g}$ 12800			c) 0.53/1/1.28
$CsCuAlF_6$	$^2A_{1g}$ 5800		Dq 830	
	$^2B_{2g}$ (8300)		Ds 1110	—
	^2E_g 10250		Dt 275	—
CrF_2	$^5A_{1g}$?		Dq 1095	a) ? /1/1.31
	$^5B_{2g}$ 10950	*(18, 63)*	Ds —	b) 0.89/1/1.36 Z = 5.76
	^5E_g 14300		Dt —	c) 0.95/1/1.51
K_3MnF_6	$^5A_{1g}$ 9000		Dq 1740	a) 0.52/1/1.13
(K_2NaMnF_6)	$^5B_{2g}$ (17400)	*(63)*	Ds 1600	b) 0.47/1/1.15 Z = 5.62
	^5E_g 19600		Dt 520	c) 0.50/1/1.28
	(^5E_g 21000)	*(33)*		
Na_2MnF_5	$^5A_{1g}$ 12500		Dq 1800	a) 0.69/1/1.17
($(NH_4)_2$	$^5B_{2g}$ (18000)	*(22, 63)*	Ds 2210	b) 0.59/1/1.19 Z = 5.63
MnF_5)	^5E_g 21000		Dt 730	c) 0.67/1/1.35

a) Shoulders in brackets.
b) Effective charge, calculated from $B_{1g} \rightarrow B_{2g}$.

The spectra of K_2CuF_4 and Ba_2CuF_6, if analyzed in the same manner, may give some insight into the problem as to whether these compounds do crystallize in a compressed octahedral coordination or not. In the case of K_2CuF_4, the band positions are almost the same as in CuF_2 or $KCuF_3$. Therefore one would expect a similar CuF_6 coordination, and in fact the energy calculations using point charges (*64*) as well as the overlap model disagree with experiment if the structural data of the compressed octahedron are used. *Babel* (*2*) in his X-ray analysis has indeed found the superstructure reflexes which indicate an elongated CuF_6 coordination in K_2CuF_4, so that the ligand field calculations could be brought to a good end. A closer reinvestigation of the K_2CuF_4 structure would be desirable.

In the case of Ba_2CuF_6, the spectrum gives the incomplete information of only two bands (Fig. 4). If the coordination is assumed to be compressed, they should be $A_{1g} \rightarrow E_g$ and $A_{1g} \rightarrow B_{2g}$, because the transition $A_{1g} \rightarrow B_{1g}$ is the weakest one in all examples. The calculated relative transition energies agree in both models with the observed maxima (Table 3), and if the experimental values of CuF_2 (effective charge = 6.37 or $\varepsilon = 1.02 \cdot 10^6$ cm^{-1} (Eq. 5)) are used, agreement is obtained also on an absolute scale (electrostatic: 6770, 10060, 12760 cm^{-1}; covalent: 4900, 9300, 11900 cm^{-1}). Definite confirmation of the compressed coordination, however, is not given by the absorption spectrum, because the B_{1g} band does not occur in the region of the calculated energy. On the other hand, the spectrum of Ba_2CuF_6 is comparable to that of Na_2CuF_4, a compound with an extremely elongated coordination. A model of the elongated coordination in Ba_2CuF_6 can be constructed if the long Cu-F distances of 2.08 Å actually represent the sum long + short = 2 × 2.08 Å (superstructure). The Cu-F distances then would be 4 × 1.85 Å and 2 × 2.31 Å, if exact D_{4h} symmetry is assumed. Such distortions correspond to those within Na_2CuF_4 and could also explain the Ba_2CuF_6 spectrum, even the shift to higher energies, because the Cu-F distances in the *x,y* plane are now extremely short.

2. Dynamic Effects

The static aspect of the JT theorem, which has been treated in a number of papers (*4*, *51*, *52*, *53*), results in different energies for both the E_g components; these are expressed in the simplest manner by harmonic potential energy surfaces

$$E_{1,2} = {}^1/_2 \cdot k\, r^2 \pm {}^1/_2 \cdot \alpha r \tag{11}$$

The quantities r and k are the normal coordinates and force constants associated with the e_g vibrational mode, α are multiples of radial integrals which are tabulated by *Ballhausen* and *Ancmon* (*5*). If the two energy surfaces are well separated, the molecule will remain on the lower potential surface with a maximum energy gain of $-\alpha^2/8k$ at $r = \alpha/2k$ and with an associated transition energy $E_1 \rightarrow E_2$ of $\alpha^2/2k$. Eq. (11) does not include the energy variation with respect to the cyclic coordinate and therefore the elongated and compressed octahedra will be mixed together. If, however, second-order terms are not disregarded in the energy equation, there will exist in addition three rotational energy barriers which may force the complex into a distorted coordination. The levels of these potential maxima relative to the thermal energy determine whether the complex structure will be statically distorted or not.

Microsymmetry and absorption spectrum of the pure phases of copper (II), chromium(II) or manganese(III) fluorides normally correspond to one another. Therefore the problem may be treated in terms of the static JT effect. If, however, mixed crystals are examined, this is no longer possible. The crystal structures of e.g. $KMg_{1-x}Cu_xF_3$ and $KZn_{1-x}Cu_xF_3$ have been indexed as cubic in the range $0 \leqq x \leqq 0.3$ and $0 \leqq x \leqq 0.5$, but the absorption spectra show the three ligand field bands of a tetragonal complex (*77*). The same holds for $Zn_{1-x}Cu_xF_2$ (*77*) and also for some stoichiometric compounds of the type $AMe^{II}Me^{III}F_6$ (A = alkali) (*3*), such as $CsCuAlF_6$ or $CsCuGaF_6$, the spectra of which are given in Fig. 4. Similar phenomena have been observed in oxygen complexes, e.g. in $Cu_xMg_{0,7-x}Zn_{0,3}O$ (*72*), and also in pure compounds such as manganese(III) acetylacetonate in which the MnO_6-symmetry is practically octahedral (*60*), while the absorption spectrum shows three spin-allowed bands with positions very similar to those of statically distorted complexes (*19, 24*).

According to these spectroscopic results, the coordination of the transition metal ions may be regarded as regular only in the average. An explanation is offered by the dynamic JT effect, even in the crystalline state. No quantitative formulation is possible if only the isolated complex is considered, because obviously the crystal structure is also influenced by the closed shell cations (in the case of manganese(III) acetylacetonate perhaps by the bidentate ligands). Therefore the forces of the environment have to be added to the distorting JT forces. The potential minima of Eq. (11) are displaced by this interaction to smaller r values, furthermore they are flattened, which makes an exchange among the three minima of the extremely elongated (or compressed) coordinations more probable. Some evidence of such behaviour is given by the spectra of the above-mentioned mixed copper(II) fluorides and oxides. As long as the X-ray structure is "regular", the splittings of the octahedral states are relatively small and we get spectra like that of $CsCuAlF_6$ (Fig. 4) (*72, 74*). If, with increasing copper(II) concentration, the X-ray structure of the mixed crystals shows static distortion, the splittings become larger, i.e. the A_{1g} and E_g bands are shifted to higher energies, and the spectral properties are now like those of the pure copper(II) fluorides in Fig. 3.

B. The Electronic Configurations $3d^1$ and $3d^6$

The situation described in the last chapter as exceptional: regular octahedral microsymmetry but with spectral properties indicating lower symmetry, has to be regarded as normal in the case of the electronic configurations $3d^1$ and $3d^6$. Hardly any absorption spectra of compounds with regular octahedral coordination show the octahedral transition $T_{2g} \rightarrow E_g$ exactly as one band. It is at least asymmetrically broadened or, what is more common in the solid state, split into two components. This behaviour is conventionally explained by a tetragonal dynamic JT distortion in the E_g excited state (a trigonal distortion, which would be possible in the ground state, does not split the E_g state).

Only a few examples are known in which the ground state is statically distorted according to X-ray or infrared analysis (see Chap. II). These distortions, which are relatively small compared to those in the complexes with $3d^4$ or $3d^9$ central ions, need not necessarily be due to JT forces, because they may occur also in compounds with non-degenerate ground states, e.g. in Na_3AlF_6 and related compounds. The origin of a distortion is irrelevant if only the resulting spectral properties are examined. The calculation procedures, as outlined in the preceding chapters, can hardly be applied to the scarce experimental data. Firstly, the Me-F distances of the distorted compounds are known only in the case of FeF_2 and, secondly, the lowest of the three absorption bands with energies

	electrostatic	covalent	
$B_{2g} \longrightarrow E_g$	$-3\ Ds + 5\ Dt$	$2u_{\pi z} - 2u_{\pi x}$	
$\longrightarrow B_{1g}$	$10\ Dq$	$3u_{\sigma x} - 4u_{\pi x}$	(12)
$\longrightarrow A_{1g}$	$10\ Dq - 4\ Ds - 5\ Dt$	$2u_{\sigma z} + u_{\sigma x} - 4u_{\pi x}$	

has not so far detected in the pure fluorides. Therefore the tetragonal ligand field parameters cannot be obtained from spectroscopic data alone, but may be calculated if ESR or magnetic susceptibility results are incorporated.

The ligand field spectrum of FeF_2 clearly consists of two bands (*34*), see Fig. 8 and Table 4. According to the compressed octahedral FeF_6 coordination, which is assumed to be tetragonal for our purposes, the low-energetic band is assigned to $^5B_{1g}$, the other to $^5A_{1g}$. Using the electrostatic point charge model and a hydrogen-like $3d$ radial function, we can get $Z_{eff} = 6.01$ from the maximum of the $^5B_{1g}$ band. The calculated transition energies are then 1400 cm^{-1} for $^5B_{2g} \rightarrow {}^5E_g$ and 10480 cm^{-1}

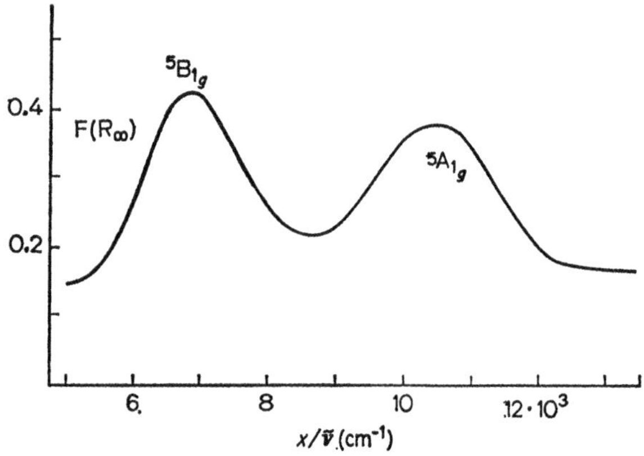

Fig. 8. Reflectance spectrum of FeF_2

for ${}^5B_{2g} \rightarrow {}^5A_{1g}$. The latter energy corresponds exactly to the experimental absorption maximum at 10500 cm^{-1}. If the model of weak covalent bonding is applied by analogy with the procedure which led to the results of Table 3, the energy ratios $E_g/B_{1g}/A_{1g}$ will be ?/1/1.52 by experiment and 0.194/1/1.33 by calculation.

The compounds $(NH_4)_3TiF_6$ and $Ba_3(CoF_6)_2$, whose tetragonal microsymmetry is apparent from infrared spectral data (*57, 75*), both show a considerable energy difference of 4200 cm^{-1} (*75, 76*) and 6500 cm^{-1} (*17*) between the B_{1g} and A_{1g} band. The A_{1g} band in $Ba_3(CoF_6)_2$, however, is very broad with the absorption maximum at 16700 cm^{-1} and a shoulder at about 14500 cm^{-1}, so that this special problem does not seem to be completely solved. But in any case the energy separation is more pronounced in the statically distorted compounds than in the compounds with regular X-ray microsymmetry (Table 4).

This is obvious from a comparison of the spectral data of FeF_2 with those of some other Fe^{2+} fluorides, which contain regular $[FeF_6]$ octahedra (*91*).

The differences in band splitting may be interpreted in a similar way as in the copper(II) fluorides. Dynamic effects are manifest in the spectra as smaller distortions compared to the static case. A smaller distortion of a compressed octahedron means a decrease in the long and an increase in the short distances (Fe-F distances in e.g. the cubic $KFeF_3$ are ≈ 2.06 Å compared to 1.99 and 2,12 Å in FeF_2). Therefore the energy of the B_{1g} band is raised, whereas the A_{1g} band is much less affected, because here the increasing cubic field strength is compensated by the decrease in the tetragonal component (Eq. 12; table 4).

Table 4. *Absorption maxima* [cm^{-1}] *of some Ti(III), Co(III) and Fe(II) fluorides (reflectance spectra)*

	$B_{2g} \rightarrow B_{1g}$	$B_{2g} \rightarrow A_{1g}$	Separation	Ref.
$(NH_4)_3TiF_6$	15110	19010	3900	(*13*)
	15030	19240	4210	(*75,76*)
Na_2KTiF_6	16000	18900	2900	(*13*)
NaK_2TiF_6	16100	18900	2800	(*13*)
$Ba_3(CoF_6)_2$	10200	16700	6500	(*17*)
K_3CoF_6	11400	14500	3100	(*17*)
Na_2KCoF_6	11800	14400	2600	(*17*)
FeF_2	6900	10500	3600	(*34,91*, fig. 8)
$Fe_{0.1}Zn_{0.9}F_2$	7700	10300	2600	(*91*)
$KFeF_3$	7600	10000	2400	(*91*)
$KFe_{0.1}Zn_{0.9}F_3$	7700	9900	2200	(*91*)

One of the most important conclusions which can be drawn from the MO-interpretation of the absorption spectra is that π-bonding in the fluorocomplexes is about $^1/_4$ to $^1/_3$ as effective as σ-bonding. This can be shown for d^4 and d^9 complexes if the octahedral T_{2g} splittings are compared with the octahedral E_g splittings. In d^1 and d^6 complexes the excited component of the T_{2g} state is presumably situated in the infrared and therefore, because of its broadness, it might be difficult to find it. Its energy can be obtained, however, if the temperature dependence of the magnetic susceptibility is analyzed (*29*), as has been done by *Schläfer* (*76*) for some titanium(III) complexes. According to this investigation the T_{2g} state in $(NH_4)_3TiF_6$ is split by 910 cm^{-1}. This value is unusually high compared with other Ti(III) compounds, but is in agreement with the large energy separation of the excited E_g state in the optical spectrum.

That is really π-bonding which is responsible for the splitting of the T_{2g} term may be shown if one fluorine ligand in a MeF_6-complex with a highly charged d^1 central ion is substituted by oxygen. The vibration spectra in those compounds indicate a high π-bonding character of the Me-O bond (*21*). The ligand field spectra are expected to correspond to tetragonally compressed octahedra (formally the non-cubic component has only half the value), but now the $B_2 \rightarrow E$ transition may be shifted to such high energies that it becomes observable in the absorption spectrum. Good examples are the oxovanadium(IV) complexes from which among others the spectrum of $[VOF_5]^{3-}$ has been examined (*68, 69, 81*).

Three bands are observed at about 9500, 15500, and 24000 cm^{-1}, whose assignment, however, does not seem to be absolutely clear. The ligand field transitions to E, B_1, A_1 are assumed to be represented either by all three bands (*6, 38, 46*) or, as seems more plausible, only by the first and second (*25, 68, 80, 81*), so that the third must be a charge transfer transition. Apart from these difficulties, the model of weak covalent bonding, although it can be used only for a crude estimation in this case, may give some idea about the relative magnitude of σ- and π-bonding. The energy of the E band with $u_{\pi ox.} - u_{\pi fl.} = 9500$ cm^{-1} is not much lower than the energy of the B_1 band with $3u_{\sigma fl.} - 4u_{\pi fl.} \approx 12000$ to 15500 cm^{-1}, and the difference $A_1 - B_1 = u_{\sigma ox.} - u_{\sigma fl.}$ is about equal to the energy of the E band in the first assignment, or is roughly zero in the second one. In any case $u_{\pi ox.}$ has at least the magnitude of $u_{\sigma fl.}$ and in the second assignment also the magnitude of $u_{\sigma ox.}$

The bonding and spectral properties of the fluorides have so far been correlated only for complexes which can be treated as one-electron systems. In the many-electron case, the spectra of fluoride compounds give little evidence of deviation from octahedral symmetry, although some excited states could be split by at least dynamical distortions. Such effects, however, are observed when cubic complexes with a higher degree of covalency are examined. The threefold-degenerate excited states of e.g. tetrahedral $[CoHal_4]^{2-}$ (*16*) and even more of $[FeHal_4]^-$ (*30, 31, 35*) are split to such an extent that, apart from spin-orbit coupling, the dynamical deviations from tetrahedral symmetry should be taken into account. Little attention has been given to such a concept which could explain unexpected bands in the vicinity of degenerate states in many cases.

References

1. *Babel, D.:* Z. Anorg. Allgem. Chem. *336*, 200 (1965).
2. — Struct. Bonding, Vol. *3*, 1. Berlin–Heidelberg–New York: Springer 1967.
3. — *Pausewang, G., Viebahn, W.:* Z. Naturforsch. *22b*, 1219 (1967).
4. *Ballhausen, C. J.:* Introduction to Ligand Field Theory, p. 193 ff. New York–San Francisco–Toronto–London: McGraw Hill Comp. 1962.
5. — *Ancmon, E. M.:* Kgl. Danske Videnskab. Selskab, Mat.-Fys. Medd. *31*, (9) (1958).
6. — *Gray, H. B.:* Inorg. Chem. *1*, 111 (1962).
7. — — Molecular Orbital Theory. New York–Amsterdam: W. A. Benjamin 1965.
8. — — Ref. 7, p. 110.
9. — — Ref. 7, p. 131.
10. — *Winter, F.:* Acta Chem. Scand. *13*, 1729 (1959).
11. *Baur, W. H.:* Acta Cryst. *11*, *488* (1958).
12. *Bedon, H. D., Hatfield, W. E., Horner, S. M.:* Inorg. Chem. *4*, 743 (1965).
13. — *Horner, S. M., Tyree, S. Y.:* Inorg. Chem. *3*, 647 (1964).

14. *Benford, R. L., Calvin, M., Benford, G.:* J. Chem. Phys. *26*, 1165 (1957).
15. *Billy, H., Haendler, H. M.:* J. Am. Chem. Soc. *79*, 1049 (1957).
16. *Cotton, F. A., Goodgame, D. M. L., Goodgame, M.:* J. Am. Chem. Soc. *83*, 4690 (1961).
17. — *Meyers, D.:* J. Am. Chem. Soc. *82*, 5023 (1960).
18. *Clark, R. J. H.:* J. Chem. Soc. *1964*, 417.
19. *Davis, T. S., Fackler, J. P., Weeks, M. J.:* Inorg. Chem. *7*, 1994 (1968).
20. *Day, P.:* Proc. Chem. Soc. *1964*, 18.
21. *Dehnicke, K., Pausewang, G., Rüdorff, W.:* Z. Anorg. Allgem. Chem. *366*, 64 (1969).
22. *Dingle, R.:* Inorg. Chem. *4*, 1287 (1965).
23. *Drickamer, H. G.:* J. Chem. Phys. *47*, 1880 (1967).
24. *Fackler, J. P.*, Davis, *T. S., Chawla, I. D.:* Inorg. Chem. *4*, 130 (1965).
25. *Feltz, A.:* Z. Anorg. Allgem. Chem. *354*, 225 (1967).
26. *Ferguson, J., Guggenheim, H. J., Tanabe, Y.:* J. Phys. Soc. Japan *21*, 692 (1966).
27. — *Knox, K., Wood, D. L.:* J. Chem. Phys. *35*, 2236 (1961).
28. — — — J. Chem. Phys. *37*, 193 (1962).
29. *Figgis, B. N.:* Trans. Faraday Soc. *57*, 198 (1961).
30. *Gill, N. S.:* J. Chem. Soc. *1961*, 3512.
31. *Ginsberg, A. P., Robin, M. R.:* Inorg. Chem. *2*, 817 (1963).
32. *Gray, H. B., Ballhausen, C. J.:* J. Am. Chem. Soc. *85*, 260 (1963).
33. *Hatfield, W. E., Parker, W. E.:* Inorg. Nucl. Chem. Letters, *1*, 7 (1965).
34. — *Piper, T. S.:* Inorg. Chem. *3*, 1295 (1964).
35. *Hathaway, B. J., Holah, D. G.:* J. Chem. Soc. *1964*, 2408.
36. *Hepworth, M. A., Jack, K. H.:* Acta Cryst. *10*, 345 (1957).
37. — — *Peacock, R. D., Westland, G. J.:* Acta Cryst. *10*, 63 (1957).
38. *Horner, S. M., Tyree, S. Y., Venezky, D. L.:* Inorg. Chem. *1*, 844 (1962).
39. *Ilse, F. E., Hartmann, H.:* Z. Physik. Chem. *197*, 239 (1951).
40. *Jack, K. H., Gutman, V.:* Acta Cryst. *4*, 246 (1951).
41. — *Maitland, R.:* Proc. Chem. Soc. (London) *1957*, 232.
42. *Jahn, H. A., Teller, E.:* Proc. Roy. Soc. (London) *A 161*, 220 (1937); *A 164*, 117 (1938).
43. *Jørgensen, C. K.:* Absorption Spectra and Chemical Bonding in Complexes, London–New York: Table 27, 28. Pergamon Press 1962.
44. — Orbitals in Atoms und Molecules, p. 93. New York–London: Academic Press 1962.
45. — *Pappalardo, R., Schmidtke, H.-H.:* J. Chem. Phys. *39*, 1422 (1963).
46. *Kilty, P. A., Nicholis, D.:* J. Chem. Soc. A *1966*, 1175.
47. *Kleiner, W. H.:* J. Chem. Phys. *20*, 1784 (1952).
48. *Knox*, K.: J. Chem. Phys. *30*, 991 (1959).
49. — Acta Cryst. *26*, A 45 (1963).
50. — *Mitchell, D. W.:* J. Inorg. Nucl. Chem. *21*, 253 (1961).
51. *Liehr, A. D.:* Progr. Inorg. Chem. *3*, 281 (1962).
52. — *Ballhausen, C. J.:* Ann. Phys. *3*, 304 (1958).
53. *Lohr, L.:* Inorg. Chem. *6*, 1890 (1967).
54. *Ludi, A., Feitknecht, W.:* Helv. Chim. Acta *46*, 2226 (1963).
55. — — Helv. Chim. Acta *46*, 2238 (1963).
56. *Lulé, A.:* Dissertation, Bonn 1968.
57. *Meyers, M. D., Cotton, F. A.:* J. Am. Chem. Soc. *82*, 5027 (1960).
58. *Minomura, S., Drickamer, H. G.:* J. Chem. Phys. *35*, 903 (1961).
59. *Moffit, W., Ballhausen, C. J.:* Ann. Rev. Phys. Chem. *7*, 107 (1956).
60. *Morosin*, B., *Brathovde, J. R.:* Acta Cryst. *17*, 705 (1964).

61. *Mulliken, R. S.:* J. Chem. Phys. *23*, 1841 (1955).
62. *Nakamoto, K., Fujita, J., Kobayaski, M., Tsuchida, R.:* J. Chem. Phys. *27*, 439 (1957).
63. *Oelkrug, D.:* Ber. Bunsenges. Physik. Chem. *70*, 736 (1966).
64. — Z. Physik. Chem. N. F. *56*, 325 (1967).
65. *O'Donnell Offenhartz, P.:* J. Chem. Phys. *47*, 2951 (1967).
66. *Okazaki, A., Suemune, Y.:* J. Phys. Soc. Japan *16*, 176 (1961).
67. *Öpik, U., Pryce, M. H. L.:* Proc. Roy. Soc. (London) A *238*, 425 (1957).
68. *Ortolano, T. R., Selbin, J., McGlynn, S. P.:* J. Chem. Phys. *41*, 262 (1964).
69. *Pausewang, G., Rüdorff, W.:* Z. Anorg. Allgem. Chem. *364*, 69 (1969).
70. *Piper, T. S., Carlin, R. L.:* J. Chem. Phys. *33*, 1208 (1960).
71. *Pryce, M. H. L., Sinha, K. P., Tanabe, Y.:* Mol. Phys. *9*, 33 (1965).
72. *Reinen, D.:* Z. Naturforsch. *23a*, 521 (1968).
73. *Richardson, J. W., Nieuwpoort, W. C., Powell, R. R., Edgall, W. F.:* J. Chem. Phys. *36*, 1057 (1962).
74. *Scatturin, V., Corliss, L., Elliott, N., Hastings, J.:* Acta Cryst. *14*, 19 (1961).
75. *Schläfer, H. L., Fritz, H. P.:* Spectrochim. Acta *23A*, 1409 (1967).
76. — *Lenz, W., Staab, J.:* Z. Physik. Chem. N. F. *62*, 290 (1968).
77. *Schmitz-Du Mont, O., Grimm, D.:* Z. Anorg. Allgem. Chem. *355*, 280 (1967).
78. *von Schnering, H. G.:* Z. Anorg. Allgem. Chem. *353*, 13 (1967).
79. *Sears, D. R.:* Ph. D. Thesis, Cornell University 1958.
80. *Selbin, J., Morpurgo, C.:* J. Inorg. Nucl. Chem. *27*, 673 (1965).
81. — *Ortolano, T. R., Smith, F. J.:* Inorg. Chem. *2*, 1315 (1963).
82. *Siegel, S.:* Acta Cryst. *9*, 684 (1956).
83. *Stephens, D. R., Drickamer, H. G.:* J. Chem. Phys. *35*, 429 (1961).
84. *Stout, H. W.:* J. Chem. Phys. *31*, 709 (1959).
85. *Sugano, S., Shulman, R. G.:* Phys. Rev. *130*, 517 (1963).
86. — *Tanabe, Y.:* J. Phys. Soc. Japan *20*, 1155 (1965).
87. *Viebahn, W., Rüdorff, W., Hänsler, R.:* Chimia *23*, 503 (1969).
88. *Vollmer, G.:* Dissertation, Tübingen 1966.
89. *Watson*, R. E., *Freeman, A. J.:* Phys. Rev. *134*, 1526 (1964).
90. *Wolfsberg, M., Helmholz, L.:* J. Chem. Phys. *20*, 837 (1952).
91. *Brokopf, H., Reinen, D., Schmitz-Du Mont, O.:* Z. Physik. Chem. N. F. *68*, 228 (1969).

Received April 17, 1970

Structure and Bonding in Inorganic Derivatives of β-Diketones

Dr. D. W. Thompson

Department of Chemistry, College of William and Mary, Williamsburg, Virginia 23185, USA

Table of Contents

Introduction

Since the preparation of acetylacetone and similar β-dicarbonyl compounds in the latter half of the nineteenth century, organic chemists have had considerable interest in their properties and reactions. The best known phenomenon of β-dicarbonyls is their participation in tautomeric equilibria where an apparently pure compound is actually a mixture of two substances having keto and enol structures. Early studies with β-dicarbonyls revealed that these compounds exhibited chemical properties indicative of both the dicarbonyl and unsaturated hydroxy ketone structures. In an unsuccessful attempt to explain this anomalous behavior *Laar* (*1*) in 1885 coined the word "*tautomerism*" which has survived to the present. By the turn of the century it was generally held that the dichotomous properties of β-dicarbonyls resulted from the "pure" compound existing as a mixture of both keto and enol tautomers. Since

these early days tautomeric equilibria have been the subject of continued interest. Major investigations of this phenomenon were reported as recently as 1964 (*2*).

Although the organic chemist's interest in acetylacetone and like molecules is widely appreciated, inorganic chemists also have long been fascinated by the variety of compounds that form from the reaction of metallic and metalloidal elements with β-diketones. As early as 1887 the preparation of several metal derivatives of acetylacetone was described (*3*). After this initial work interest in inorganic derivatives of acetylacetone increased steadily and by the turn of the century alkali metal, iron, copper, nickel, cobalt, platinum and other metal complexes of acetylacetone had been prepared. By 1914 *Morgan* and *Moss* (*4*), in what appears to be the first review of acetylacetonate complexes, cited some 39 known complexes. From that time to the immediate present inorganic chemists have had a vigorous and sustained interest in metallic and metalloidal derivatives and complexes of acetylacetone and other similar β-diketones. The extent of this interest can be appreciated by reference to a recent review by *Fackler* (*5*) in which he states "*complexes with β-ketoenols have been reported for all of the nonradioactive metallic or metalloidal elements in the periodic table*", and "*the number of β-diketones known to form metal complexes certainly exceeds one hundred, and more potential ligands are prepared each year.*"

This article is designed to convey salient features of the inorganic and coordination chemistry of acetylacetone in order to provide perspective and appreciation of the continuing interest in inorganic derivatives of β-diketones and to illustrate the rich variety of coordinative interactions that can occur between a metal and a ligand.

Bonding and Structure of Chelated Acetylacetonates

To an inorganic chemist one of the most fascinating features of the chemistry of acetylacetone and other congeneric diketones is the variety of ways in which these β-diketones and especially their enolate anions can bond to metallic or metalloidal atoms to give varied molecular structures. Attention here will be focused where possible on acetylacetonates which most often can be thought of as prototypes for other β-ketoenolate complexes.

Acetylacetone is a weak acid ($pK_a \sim 8.8$) (*6*), and the 3-proton can be lost easily to give the enolate anion. This enolate anion has a five-atom π network extending over the two oxygen and three non-terminal carbon atoms. Six electrons occupy the resulting π-type molecular orbitals. The enolate anion thus has a delocalized symmetric (C_{2v}) structure (see later discussion) and is most often represented as illustrated in Fig. 1A. Numerous derivatives of this anion can be formed giving rise to a variety of metal-to-ligand bonding structures. Also, acetylacetone as the neutral molecule is a difunctional Lewis base and can coordinate to acceptor (Lewis acid) molecules.

By far the most frequently occurring acetylacetonate derivatives are those in which the enolate anion is coordinated to a central metal atom through both oxygen atoms. A classic example of this is the well-known *tris(acetylacetonato)iron(III) complex* (Fig. 1B) whose structure has been determined by single crystal X-ray structure analysis (7). Crystallographic data show that the two chelate ring C—C bond distances are

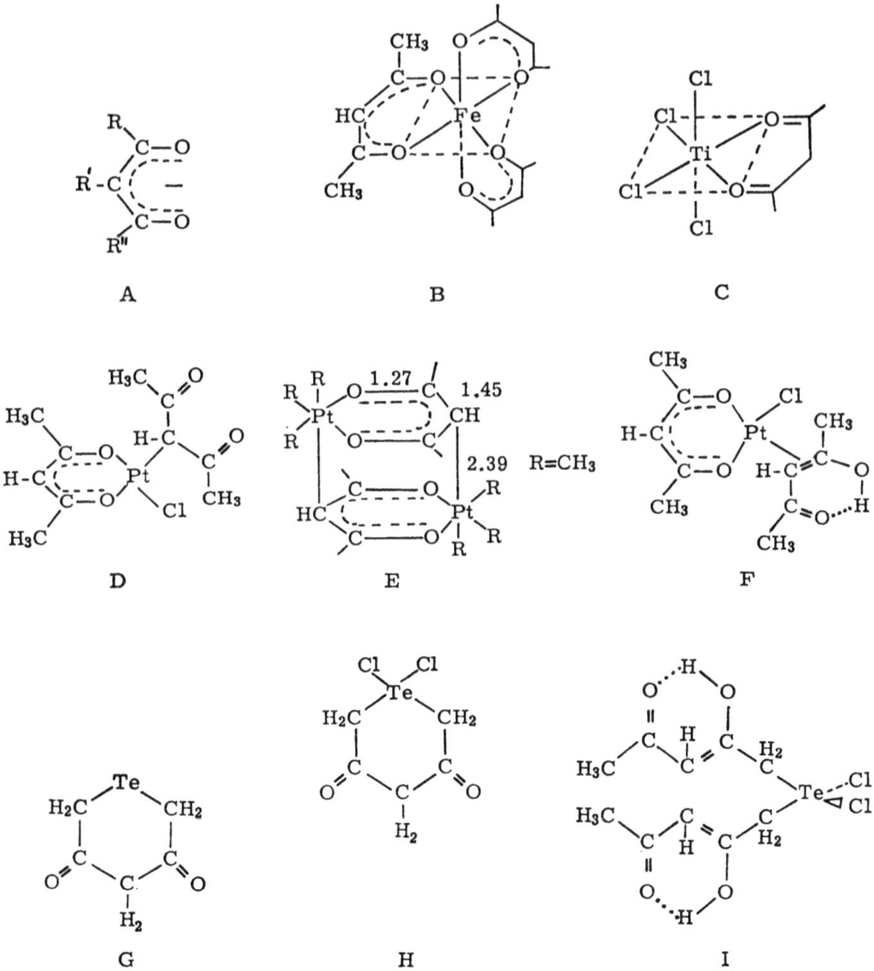

Fig. 1 A

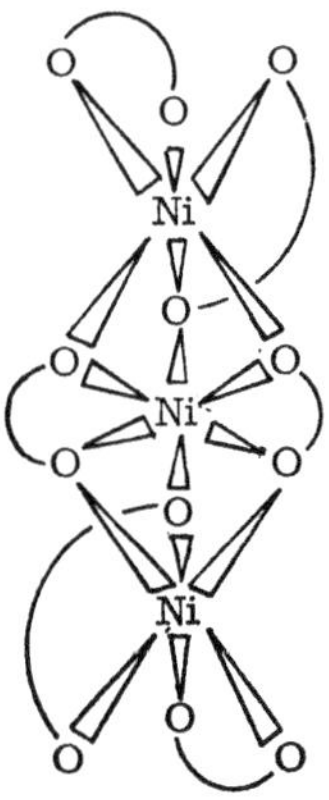

J

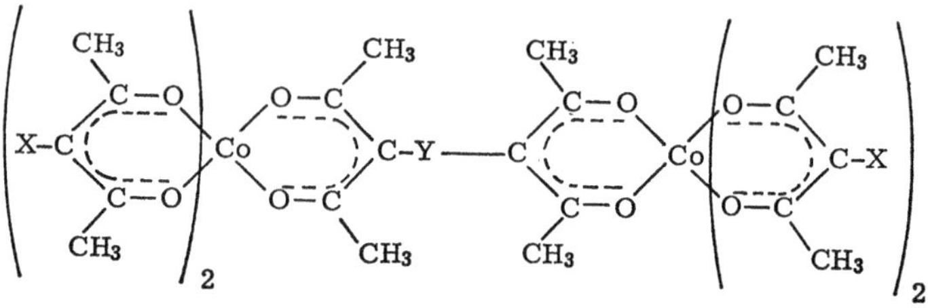

K

$Y = -CH_2SCH_2-$, $X = -Cl$
$X = -CH_2-$, $X = -NO_2$

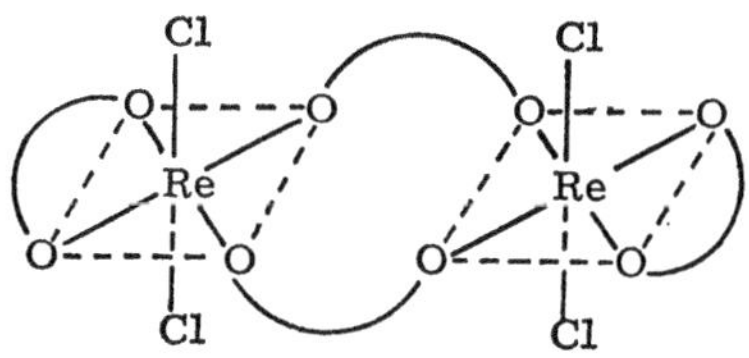

L

Fig. 1B

equal as are the two C—O bond distances and verify the C_{2v} symmetry of the acetylacetonate ligand. Further, crystallographic studies (*8*) for numerous other oxygen-chelated acetylacetonate complexes suggest that all coordinated enolate anions have C_{2v} symmetry. The ubiquity of oxygen-chelated acetylacetonate complexes can be appreciated from the fact that such complexes have been reported for all the main group transition elements (except technetium) and lanthanide elements (except promethium) as well as numerous main group and actinide elements. Various aspects of this subject have received extensive treatment in several review articles (*9—13*).

A second type of oxygen-chelated complex that can be formed with acetylacetone is the simple Lewis acid-base adduct. In these complexes acetylacetone does not lose its acidic proton to form an enolate anion, but rather as the neutral molecule in the keto tautomer donates electrons from the oxygens of each carbonyl to an acceptor or acidic species. Examples of this type of complex are the six-coordinate adducts formed between typically strong Lewis acids as tin tetrachloride or titanium tetrachloride and acetylacetone (*14*). An illustrative example is shown in Fig. 1C. Compounds of this type have also been found with the nickel(II) ion and cobalt(II) bromide (*15, 16*). These donor-acceptor complexes are often reactive with respect to loss of the acidic ring proton to give the chelated acetylacetonate complex. For example, the titanium tetrachloride-acetylacetone adduct when refluxed in inert organic solvents loses hydrogen chloride to give the enolate complex, trichloro(acetylacetonato)titanium(IV). Although the number of donor-acceptor complexes that have been isolated is small, nevertheless, they probably form as intermediates in the formation of many of the chelated enolate derivatives.

Exemplificative of a third type of oxygen-bonded acetylacetonate derivative is the compound formed from the reaction trimethylchlorosilane and acetylacetone (*17*). The product from this reaction is *2-trimethylsiloxy-2-pentene-4-one*. It is to be noticed that this silicon compound contains a "dangling" ligand even though several chelated acetylacetonates of silicon are known. *Dangling complexation* of potential

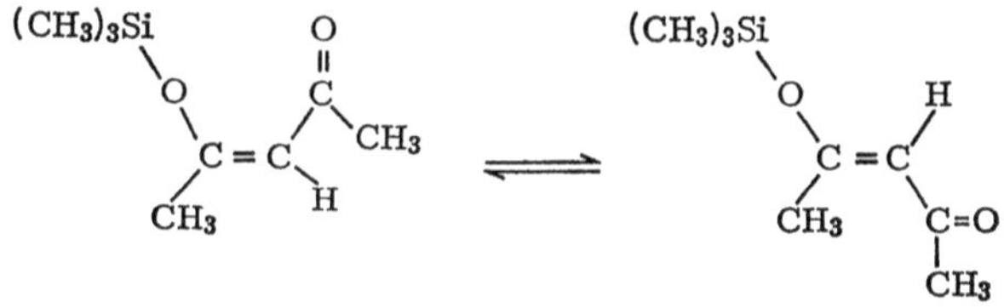

cis trans

2-trimethylsiloxy-2-pentene-4-one

chelating ligands is rare in coordination chemistry. Dialkyldichlorosilanes undergo similar reactions although trichloromethylsilane reacts to give the oxygen-chelated enolate complex. Recently, *Howe, Collins*, and *Pinnavaia* (*18*) reported that 2-trimethylsiloxy-2-pentene-4-one exists in a *cis-trans* equilibrium as shown above. From nmr data the *cis* isomer was found to undergo rapid intramolecular rearrangement at room temperature (one possible mechanism is illustrated) which averaged the nonequivalent methyl environments of the acetylacetonate ligand. On the other hand, the *trans* isomer remains stereochemically rigid as

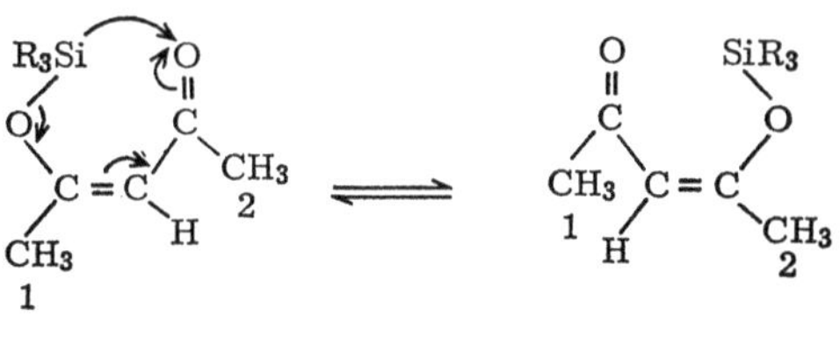

high as 120 °C. The difference in behavior was ascribed to the ease of formation of a five-coordinate (oxygen-chelated enolate) intermediate in the *cis* isomer as compared to the *trans* form.

Far less numerous than oxygen-chelated acetylacetonate derivatives are those in which the metal atom is bonded directly to the unique (*3*) carbon atom rather than to the enolate oxygens. Although metal complexes of this type have all been characterized recently (since 1962), the first carbon-bonded acetylacetonate complex, potassium chlorobis(acetylacetonato)platinate(II),

$$K[Pt(acac)_2Cl],$$

was reported by *A. Werner* (*19*) in 1901. The

$$[Pt(acac)_2Cl] \text{ anion}$$

has been thoroughly characterized (*20—22*), by nmr, ir, and single crystal X-ray analysis and is square planar having one 3-carbon-bonded and one oxygen-chelated acetylacetonate ligand in coordination sphere. The complex is illustrated in Fig. 1D.

Several additional carbon-bonded complexes have been reported and a short review has appeared recently (*23*). Metals most likely to form carbon-bonded complexes are second and third row transition elements at the right hand side of the periodic table with an increased tendency toward formation in going from the second to the third row. Most noticeable among these is platinum. Although *Werner* prepared the first 3-carbon-bonded acetylacetonate complex, he mistakenly assumed that it contained one oxygen-chelated ligand and one oxygen-bonded "danglar". It is interesting to note that in all the 3-carbon-bonded complexes of β-diketonate anions known, the diketone is in the keto form, although both *cis* and *trans* carbonyl configurations have been found. A possible exception to the above is a manganese complex containing a carbon-bonded hexafluoro-acetylacetonate ligand which has been reported (*24*) to exist in the enol form; however, the structure of this complex has been questioned.

A second type of 3-carbon-bonded acetylacetonate complex is illustrated by the *trimethyl(acetylacetonato)platinum(IV) dimer*. Although the X-ray crystal structure analysis has not been carried out for this specific complex, both X-ray and neutron diffraction studies have been done for the analogous trimethyl(nonane-4,6-dionato)platinum(IV) complex (*25, 26*). This and other similarities suggest that the acetylacetonate complex has the same structure. The structure of the nonane-4,6-dionate complex is illustrated in Fig. 1E along with some selected bond lengths. Compounds of this kind have been referred to as "bridge-bonding" complexes by *Gibson* (*23*).

In these bridge-bonded β-ketoenolate complexes the enolate ligand is best considered as a tridentate ligand with the third donor atom, in addition to the usual two carbonyl oxygen atoms, being the central carbon atom of the chelate ring. The precise nature of bonding in these bridge-bonded complexes is unsettled at present. *Gibson* has suggested that the bonding is "*probably best represented as intermediate between the keto and enol forms of the diketone*", i.e. the bonding is intermediate between the two limiting forms.

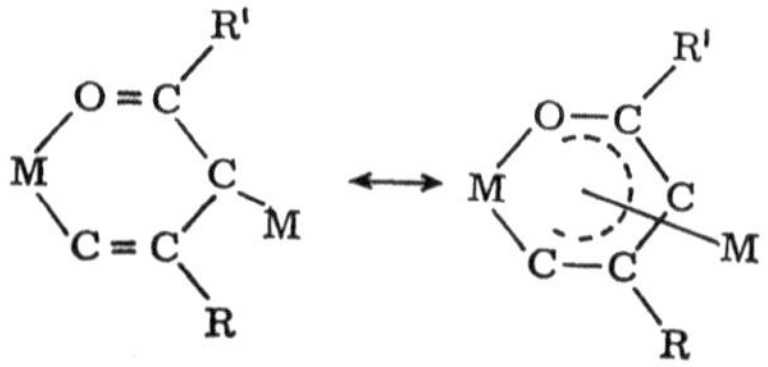

Alternatively, since the platinum atom in trimethyl(nonane-2,4-dione)platinum(IV) is directly above the carbon atom, one could conceive of this as a *pseudo-π-allylic complex* over the three ring carbon atoms. A final description of bonding in these bridge-bonded complexes can be given in terms of valence-bond theory which suggests that there is a predominence of a third resonance contributing form (II) which enhances the donor ability of the unique ring carbon atom (*27*).

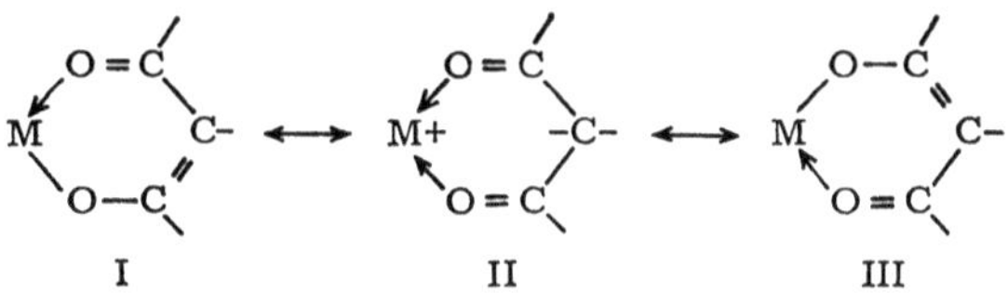

Fig. 2

Additional complexes which appear to have bridge-bonded structures are

acetylacetonato(tetrafluoroethylene) (acetonitrile)rhodium(I),

(C_2F_4) $(CH_3CN)Rh(acac)_2$, (*28*),

trisilver dinitrate tris(acetylacetonato)nickelate(II),

$AgNi(acac)_3 \cdot 2\,AgNO_3 \cdot H_2O$, (*29*),

and bis(acetylacetonato)copper(II), $Cu(acac)_2$ (*30*).

The structures of the latter two complexes have been determined from X-ray diffraction data.

Recently, *Allen, Lewis, Long,* and *Oldham* (*31a*) have isolated the first *π-bonded metal complex* of a neutral acetylacetone molecule. Potassium chlorobis(acetylacetonato)platinate(II),

$$K[Pt(acac)_2Cl],$$

reacts with strong acids in aqueous solution to give a complex which analyzes as $HPt(acac)_2Cl$. The structure shown in Fig. 1F was originally postulated. However, after further ir and nmr studies, *Gibson, Lewis,* and *Oldham* (*31b*) proposed that complex contained a delocalized five-atom *π*-bonded enol tautomer as illustrated below.

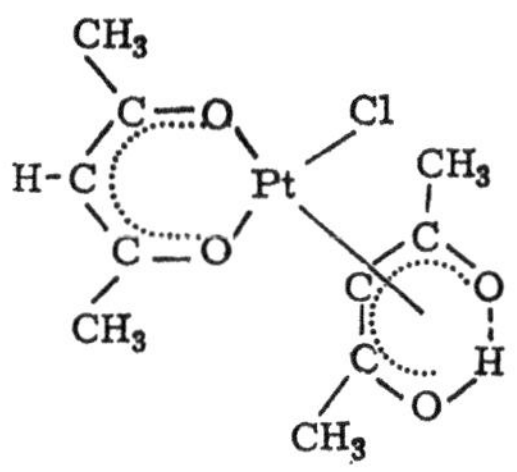

A later detailed infrared study (*32*), of $HPt(acac)_2Cl$ by *Nakamoto* and *Behnke* has been interpreted as most consistent with the structure originally suggested in reference (*31a*) (Fig. 1F). Final word on this interesting complex will have to await an X-ray crystal structure determination.

A final and unusual type of carbon-bonded acetylacetonate is that in which the *metal atom forms a derivative at a terminal or 1-carbon.* The only compounds in this case are three tellurium derivatives shown in Fig. 1G, H, I. These compounds were first prepared by *Morgan* and *Drew* (*33*) in the 1920's, and on the basis of chemical evidence they correctly postulated the structure of 1G, but not of 1H or 1I. Recent spectroscopic studies by *Dewar et al.* (*34*) have confirmed the structure of 1G and determined the structures illustrated in Fig. 1H and 1I. It is interesting to note here that sulfur and selenium form bonds only to the 3-carbon atom. At present it is difficult to rationalize the loss of hydrogen for a terminal methyl group which is many orders of magnitude less reactive than the 3-protons.

The third and final class of acetylacetonate derivatives to be discussed are those containing bridging enolate ligands giving rise to *oligomeric or polymeric complexes*. Within this class four types of polymeric complexes may be distinguished. These are:

1. oxygen-chelated oxygen-bridged complexes as shown in Fig. 1 J;
2. oxygen-chelated complexes of tetraketones as illustrated in Fig. 1 K;
3. oxygen-bonded two-center β-diketonato complexes — see Fig. 1 L; and
4. bridge-bonded complexes as trimethyl(acetylacetonato)platinum(IV) which has already been discussed above in the 3-carbon-bonded category.

A prime example of an oxygen-chelated oxygen-bridged complex is bis(acetylacetonato)nickel(II), $Ni(acac)_2$. This compound has been known for many years; however, only recently has it been precisely characterized. Early investigators (*35*) found that $Ni(acac)_2$ was paramagnetic in the solid state. It was later found to be paramagnetic in benzene, chloroform, and methanol as well (*36*). Prior to the late 1950's these data were most commonly interpreted as suggesting a tetrahedral structure, since a planar geometry would most likely be diamagnetic.

In the late 1950's evidence began to appear that called into question the presumed tetrahedral structure of $Ni(acac)_2$. About 1959 *Cotton* and his associates (*37*) found or noted in the literature that the magnetic moments of $Ni(acac)_2$ and its dihydrate adduct were virtually identical; that the visible spectra of $Ni(acac)_2$ and $Ni(acac)_2 \cdot 2\,H_2O$ were similar (*38*); and that the spectral and magnetic data did not correlate with data for known tetrahedral nickel complexes. Thus, they concluded (*37*) that "*it is clear that the latter* (i.e., tetrahedrally coordinated nickel) *does not occur in Ni(acac)₂ under any known conditions*." About this same time *Bullen*, *Mason*, and *Pauling* (*39*) reported the X-ray crystal structure of $Ni(acac)_2$. A trimeric structure was found and is illustrated in Fig. 1 J. Thus $Ni(acac)_2$ is octahedral and paramagnetic which explains its spectral and magnetic similarities to the dihydrate known to be octahedrally coordinated by isomorphism with $Co(acac)_2 \cdot 2\,H_2O$ (*40*) whose octahedral structure was determined by single crystal X-ray analysis (*41*). The trimeric structure was also found to exist in benzene solution from cryoscopic measurements (*42*).

In 1961, *Cotton* and *Fackler* (*43*) further found that $Ni(acac)_2$ exists in a monomer-trimer equilibrium in diphenylmethane over a 80—200 °C range. At low temperatures the green octahedrally coordinated trimeric species predominates, but at 200 °C the monomeric, square planar nickel complex is stable giving a red solution. Other β-diketonate-nickel(II) complexes found to participate in a monomeric-trimer equilibrium from spectral and magnetic studies were

bis(2,6-dimethyl-3,5-heptanedionato)nickel(II), and
bis(3-phenylacetylacetonato)nickel(II).

Bis(2,2,6,6-tetramethyl-3,5-heptanedionato)nickel(II) was found to be diamagnetic under all conditions presumably due to steric interactions of the bulky tertiary butyl groups.

Other complexes which polymerize via chelated oxygen-bridges are $Zn(acac)_2$ (*44*), and $Co(acac)_2$ (*45*), which are trimeric and tetrameric in the solid state respectively.

Beyond the self-association aspect of bis(acetylacetonates) of the type $M(acac)_2$, there has been interest in these complexes as Lewis acids, and a review has recently appeared concerning this subject (*46*). Complexes of the type $M(acac)_2$ react with basic ligands such as pyridine and water to become six-coordinate adducts. Indeed, the self-association of $M(acac)_2$ molecules as $Ni(acac)_2$ can be thought of in terms of intermolecular Lewis type donor-acceptor interactions.

Another class of *polymeric β-diketonate derivatives* involves the joining of two β-diketonate ligands through the terminal methyl groups or through the center carbon to give a tetradentate ligand which can span two metal centers. *Collman* and *Barker* (*47*) have reported the synthesis of the two molecules illustrated in Fig. 1 K. There are also reports in the patent literature of attempts to prepare coordination polymers with oligomeric β-diketonate ligands. For example, *Wilkins* and *Wittbecker* (*48*) reported in 1953 that tetraketones form linear polymers of the types

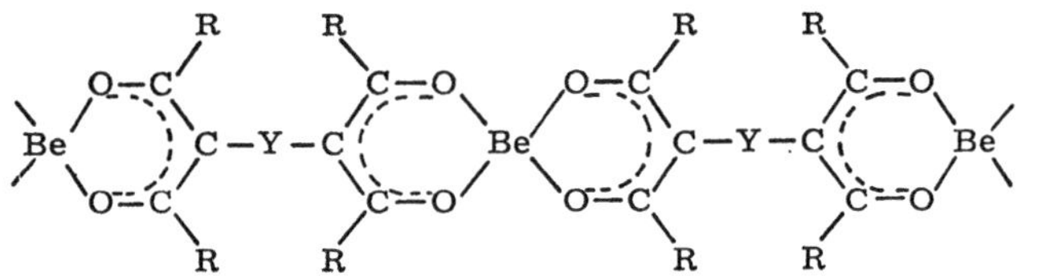

and

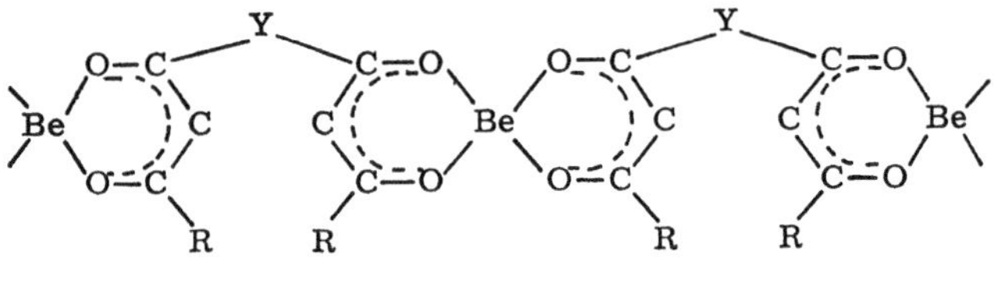

where X and Y represent a variety of R groups. To date it appears that this kind of polymer is not of any practical importance and neither does it seem to be of academic interest. More extensive studies in this area may prove otherwise.

A third type of polymerization may occur through β-diketonate ligands which span two metal centers *via* oxygen coordination. However, only one compound has been suggested as an example of this type of bonding.

In 1965 *Wilkinson et al.* (*49*) reported the preparation of a rhenium complex, $Re_2Cl_4(acac)_4$, which they "*considered to have acetylacetonato-groups bridging the metal atoms*" — (see Fig. 1L). However, little evidence was cited to support this structure and a final decision must await further studies.

A last class of polymeric complexes is that of "*bridge-bonded" ligands* which was discussed earlier under the section on carbon-bonded acetylacetonates.

As a summary we have just discussed various modes of bonding and structural patterns found in β-ketoenolate compounds. Somewhat arbitrarily this subject was divided into three parts:

1. oxygen-bonded β-ketoenolate derivatives;
2. carbon-bonded enolates; and
3. β-ketoenolate interaction giving rise to oligomeric or polymeric complexes.

Since interest in the inorganic derivative chemistry of β-diketones shows no sign of diminishing at present, it is not unlikely that additional unusual bonding and structural phenomena will be uncovered.

The Electronic Structure of Chelated Acetylacetonates

The electronic structure of oxygen-chelated acetylacetonate complexes has been the subject of considerable attention. Especially interesting have been the questions of aromaticity or benzenoid resonance and ring-currents arising *via* the participation of appropriate metal d orbitals in $d\pi$–$p\pi$ bonding with the ligand. Both chemical and physical studies, which are discussed below, have provided evidence concerning the electronic structure of the chelate ring.

Morgan and *Moss* (*4*) in 1914 appear to have presented the first theoretical discussion of bonding in metal acetylacetonates. These authors wrote:

"*These substances* [β-diketonate metal derivatives] *may be represented by formulae [I] and [III], from which it will be seen that structural isomerism*

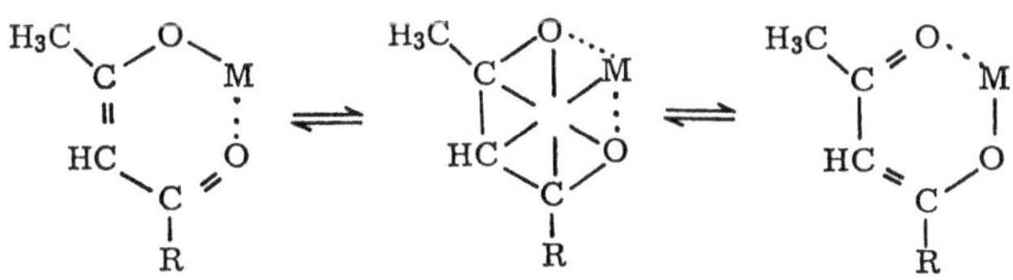

is possible if the arrangement is regarded as static in either case, but inasmuch as at present there is no experimental evidence that isomerism of this kind exists among these compounds, it is much more likely that these two formulae are analogous to the two phases of Kekule's benzene formula. The configuration of metallic acetylacetones and their analogues may oscillate between these two extreme positions, or a rearrangement may lead to an intermediate centric distribution of the chemical affinities corresponding with the Armstrong-Baeyer[1]) *centric formula for benzene and its homologues."*

In 1945 *Calvin* and *Wilson* (*50*) first suggested the importance of "benzenoid resonance" in metal chelates of β-ketoenolates. In their classic paper, "Stability of Chelate Compounds", *Calvin* and *Wilson* attempted to "*determine the influence of certain structural factors upon the stability of the chelate complexes of divalent copper in which four atoms bound to the metal are all oxygen.*" After determining stability constants for the copper chelate complexes, the authors conclude for β-diketonate ligands that "*in addition to their character as bases, a benzenoid resonance effect involving copper plays a very important role in determining the stability of these compounds.*" The canonical forms which *Calvin* and *Wilson* considered important are shown below.

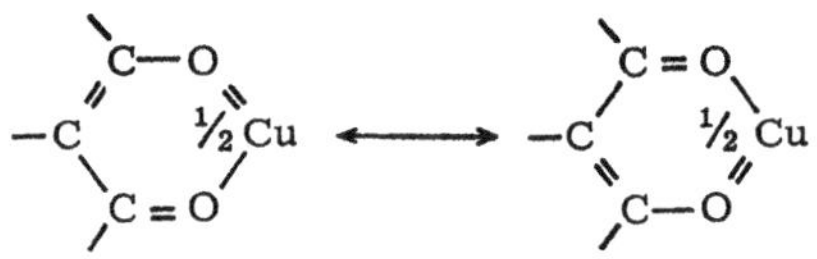

The question of benzenoid resonance in acetylacetonate chelate rings lay dormant after *Calvin* and *Wilson's* paper appeared. In 1956, however, an X-ray crystallographic analysis of tris(acetylacetonato)iron(III) was reported (*7*). This study showed that the acetylacetonate ligand has C_{2v} symmetry with the pairs of M—O, C—O, and C—C bond distances being equal within experimental error. This unambiguously excluded the existence of localized double bonds in the enolate anion. Further, the six-membered ring formed with the iron is essentially planar and the ring C—C bond distances are *ca.* 1.375 Å which is quite close to the 1.39 Å C—C bond distances in benzene. A more detailed report from *Lingafelter* and *Braun* (*8*) showed that for thirteen acetylacetone complexes the ligand has essentially C_{2v} symmetry, that the six-membered ring is planar, and that the average ring C—C bond distance is 1.39 Å. Thus

[1]) See *Fieser, L. F.*, and *Fieser, M.*: Organic Chemistry, 3rd ed., p. 529 for an explanation of Armstrong-Baeyer theory. New York: Reinhold Publishing Corp. 1956.

there is no question that the acetylacetonate anion can be properly considered as a resonance hybrid of the canonical forms illustrated in Fig. 2 or in terms of molecular orbital theory as having a delocalized π electronic system over at least the five ring ligand atoms. Although crystallographic data are all consistent with the idea of "benzenoid resonance", they are not conclusive since the C_{2v} symmetry of the acetylacetonate anion may be explained by resonance stabilization in the ligand alone.

In spite of the benzenoid resonance suggestion by *Calvin* and *Wilson* in 1945 and much activity with regard to the chemistry of acetylacetonate complexes, *Holm* and *Cotton* (*51*) in 1958 were correct in stating that "*there is still relatively little concerning their* (acetylacetonate complexes) *electronic structures which may be stated with certainty*". From an extensive nmr and uv spectral study of acetylacetonate complexes it was concluded that "*the nmr spectra of diamagnetic complexes provide no support for* (but do not exclude) *the postulate of benzenoid resonance in the chelate ring*..." Specifically, it was found that the chemical shifts of the chelate ring protons were quite similar to shifts observed for olefinic protons. No downfield shifts were observed in the region for aromatic or benzene-type protons, and further, no correlation between the ring-proton chemical shifts and π-bonding ability of the central metal was apparent. However it was pointed out that little is known about aromatic heterocycles containing more than one hetereoatom. On the other hand *Holm* and *Cotton* did state that one of "*several...parameters*" influencing that uv spectra was "*the ability of the metal to participate in $d\pi$-$p\pi$ overlap with the π-system of the chelate ring*".

Since 1958 substantial progress has been made in understanding the complicated electronic spectral properties of acetylacetonate complexes. The acetylacetonate anion and alkali metal ("ionic") acetylacetonates exhibit a single, intense ($\varepsilon \sim 10^4$) and nearly symmetric band near 34.500 cm^{-1}. Molecular orbital calculations of both the Hückel- and self-consistent field — LCAO types (*52, 53*) lead to the assignment of this band as a π-π^* transition. (A second high-energy π-π^* transition (π_3—π_5 or π_2—π_4) is predicted but apparently occurs outside the normal ultraviolet region.)

Of greater interest and complexity are the spectra of the more "*covalent*" *acetylacetonate complexes*. Especially interesting are the complexes of transition metal ions having partially filled d orbitals which can interact with appropriate ligand orbitals as well as give rise to various charge-transfer transitions. Several semi-empirical theoretical analyses coupled with experimental observations have provided a reasonable framework within which to understand the spectral properties of metal acetylacetonates.

Barnum (*54*) has carried out spectral measurements and Hückel LCAO-MO calculations for acetylacetonate complexes with trivalent transition metal ions. The results for $Co(acac)_3$ are shown graphically.

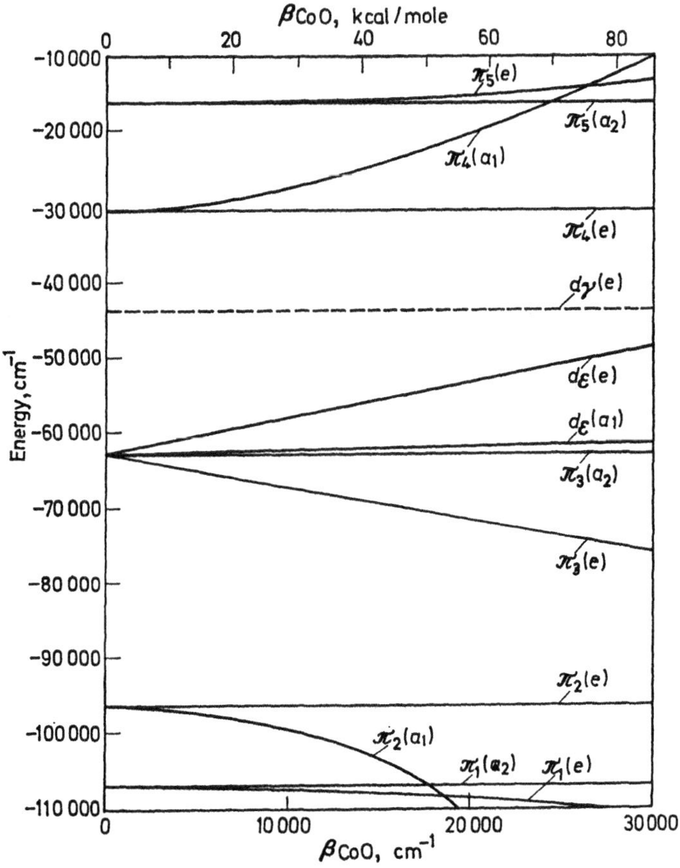

Fig. 3. Effect of metal-oxygen π-bonding on the energy levels of $Co(acac)_3$

As metal-ligand π-interaction increases it can be seen that the γ_5 subshell and all of the π-orbitals are split into their various components. Observed spectral bands were then assigned in terms of three types of transitions:

$$1.\ \pi_3 \longrightarrow \pi_4;\ 2.\ \gamma_5 \longrightarrow \pi_4;\ \text{and}\ d \longrightarrow d.$$

From the values of β_{M-O} used to obtain reasonable agreement between theory and experiment rough estimates of the metal-oxygen π-bond energies were made. These π-band energies ranged from near zero to about 40 kcal/mole going from titanium through cobalt. However, an estimate of 40 kcal/mole for the π-bond energy appears too high in light of recent thermochemical studies (*55—57*) which have determined the total metal-oxygen bond energies for the chromium, manganese, and iron *tris*-acetylacetonates to about 55, 44, 45 kcal/mole, respectively.

Barnum's assignment of the charge transfer bands as metal to ligand transitions rather than vice versa was based on the fact that the bands shift to higher energy on going from $Ti(acac)_3$ to $Co(acac)_3$. Since the trivalent metal ions become increasing difficult to oxidize ($Ti \rightarrow Co$) the transition must involve "oxidation" of the central ion, i.e. $d_{\gamma 5} \rightarrow \pi_4$ charge transfer.

Although ligand to metal electron transfer transitions are theoretically possible the results of the calculations and band shifts through this series of complexes were taken to exclude all but the $d_{\gamma 5} \rightarrow \pi_4$ charge-transfer transitions. It was suggested that the $\pi_3 \rightarrow d_{\gamma 5}$ transitions were forbidden.

Jørgensen (*58*), while recognizing validity in Barnum's treatment, has expressed reservation over the fact that the MO energies are rather insensitive to the number of electrons in the partially filled d shell. For transition metal halides, whose spectral properties have been rather successfully interpreted, the electron transfer bands show a strong dependence on the number of electrons in the partially filled shell (*59*). Assuming that both "oxidizing" and "reducing" charge transfer transitions are possible, that significant energy changes do occur on increasing the number of d electrons, that that the 23.200 cm^{-1} band in $Fe(acac)_3$ and the 17.600 cm^{-1} band in $Ti(acac)_3$ are due to $\pi \rightarrow \gamma_5$ and γ_5—π^* transitions, respectively, Jørgensen was able to calculate band positions for the tris-acetylacetonates of the first row transition elements through cobalt which are in good agreement with experiment (*58*). Thus the complicated aspects of the absorption spectra in the near ultraviolet can be reasonably explained by four types of charge transfer transitions

$$\pi \longrightarrow \gamma_5,\ \pi \longrightarrow \gamma_3,\ \gamma_5 \longrightarrow \pi^*,\ \gamma_3 \longrightarrow \pi^*$$

in addition to internal ligand transitions without recourse to multiple π and π^* levels. Furthermore, *Jørgensen's* results indicate that the assignment of certain bands as d-d transition may have to be revised.

If the optical electronegatively (*58, 60*) of iron in $Fe(acac)_3$ is taken to be 2.3, then the optical electronegativity of the highest filled π-orbitals of acetylacetonate anion is 2.7. This value is in agreement with that suggested from $Cu(acac)_2$ if one assumes that the shoulder near 26,000 cm^{-1} in Cu(acac) is a charge transfer band $\pi \rightarrow d_{xy}$. However, *Cotton* and coworkers (*61*) have assigned this band as a $d_{xy} \rightarrow \pi^*$ transition. The value 2.7 would give the acetylacetonate anion a reducing character somewhere between bromide and iodide.

In conclusion, from experimental and theoretical studies of the electronic absorption spectra of transition metal acetylacetonate complexes it seems reasonable to assume that *there is significant participation of symmetry appropriate metal d orbitals with the π-system of the ligand.*

In 1959 *Forman, Murrell,* and *Orgel* (*62*) forcefully demonstrated that metal d orbitals do interact with the π system of the chelate ring. For the paramagnetic tris(acetylacetonato)vanadium(III) complex these workers found that the ring protons exhibited large contact shifts indicative of significant unpaired π-electron density at the unique carbon in the chelate ring. In a more extensive study of paramagnetic acetylacetonates *Eaton* (*63*) (1965) also found large contact shifts which unambiguously indicated dπ-pπ metal-ligand interaction.

In 1962 *Collman, Marshall,* and *Young* (*64*) reported the nmr spectra for 3-substituted tris(acetylacetonate) complexes of the type illustrated, $X \neq Y = H$, Cl, NO_2, etc. They found that complexes of this type gave

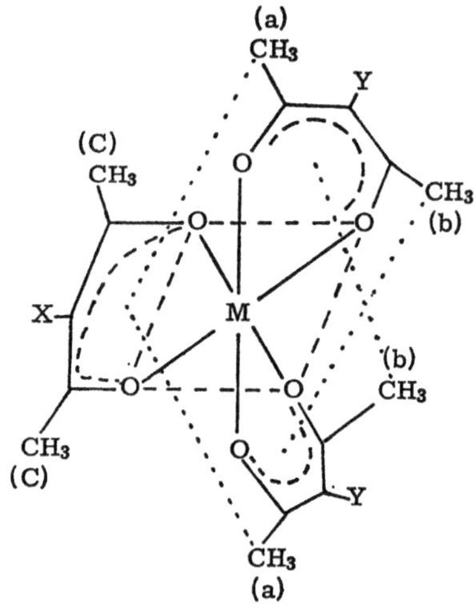

three methyl resonance peaks of equal intensity. This was explained as being due to long-range magnetic anistropy arising from ring current differences between X and Y substituted rings. Specifically, it can be seen that methyl groups (a) are above the X-substituted chelate rings and methyl groups (b) are above the Y-substituted chelate ring; the (c) methyl groups are, of course, unique being attached to the X-substituted ring. Thus, three methyl resonances are observed with a small splitting between the Y-methyl groups (1—2 Hz at 60 MHz). The interpretation by *Collman et al.* of the Y-methyl splittings has been challenged by *Linck* and *Sievers* (*65*) who point out that "*complexes of this symmetry* (as illustrated above) *have three non-equivalent methyl groups, and thus the* (observation of three peaks) *are expected-symmetry alone demands the non-equivalence, whether caused by a trans effect, by aromatic character of the chelate ring, or by another mechanism*".

Collman and associates (*66*) also have carried out extensive investigations of the chemistry of acetylacetonate anion when complexes to a metal. They find that the ring behaves similarly to aromatic compounds. Specifically, they find that:

"*The central hydrogen atom on these chelate rings can be replaced by several groups under apparently electrophilic conditions... Halogenation, thiocyanogenation, arylsulfenylation, chlorosulfenylation, nitration, Friedal-Crafts acylation, formylation, chloromethylation, and dimethylamino-methylation have been realized.*"

These studies unambiguously show that chemically speaking *the chelate ring is aromatic*. Whether this aromatic-like behavior involves the metal in conjugation with the ligand cannot be stated with certainty.

In 1963 *Hester* (*67*) observed that the ring proton chemical shift for tris(acetylacetonato)silicon(IV) chloride hydrochloride was noticeably downfield from shifts for numerous other tris(acetylacetonato) complexes, —6.35 ppm versus ca. —5.5 ppm. This large downfield shift was interpreted by *Hester* as evidence for *dπ-pπ* interaction and "*considerable benzenoid resonance*" in the chelate ring. Hester's interpretation has been questioned by several workers (*56, 52b, 5*) principally because of the unknown effect of the positive charge on the complex. In fact, *Fay* and *Serpone* (*68b*) have recently demonstrated a positive charge on the complex does give a downfield shift to the ring proton. As late as 1966 *Linck* and *Seivers* (*56*) were able to state that "*although there is no doubt that electrons from the metal are delocalized into the acetylacetonate ring* (esr studies), *the question of the magnitude of ring currents is not answered*". These authors go on to point out there may be significant *dπ-pπ* interactions without accompanying ring-currents, that is, the presence of several heteroatoms in the chelate ring may drastically reduce the probability of ring-currents even though significant *dπ-pπ* interaction may occur.

Without doubt the most enlightening experiments on the ring-current question have been reported by *Fay* and *Serpone* (*68b*). These authors prepared several cationic bis(acetylacetonato)boron(III) complexes all of which exhibited large downfield shifts somewhat lower than those observed for cationic tris(acetylacetonato)silicon(IV) complexes. Since boron has no energetically appropriate orbital available for π-bonding with the chelated acetylacetonate ligand, no ring-current is possible. Furthermore, the downfield shifts in several cationic complexes could be theoretically accounted for in terms of electric field effects at the ring proton. Thus the suggestion that *"considerable benzenoid resonance"* is responsible for the downfield ring proton chemical shift in tris(acetylacetonato)silicon(IV) complexes is at best tenuous. Fay and Serpone concluded from their work that *"point charge model calculations* (for electric field effects on resonance positions) *indicate that the shifts can be satisfactorily accounted for in terms of the electric field caused by the charge on the ion, and there is no need to invoke ring currents"*. On the other hand, these same authors have presented evidence to indicate the $d\pi$-$p\pi$ occurs to a significant extent in group IV transition metal-acetylacetonate complexes (*68*).

A final experiment of interest to the relation between metal acetylacetonates and benzene-type molecules has recently been reported by *Singh* and *Sahai* (*69*). These authors have found that

$Be(acac)_2$, $Al(acac)_3$, $Sc(acac)_3$, $Zr(acac)_4$, and $Th(acac)_4$

form charge-transfer complexes with iodine similar to benzene and its derivatives. Although no heats of complexation were determined, equilibrium constants are somewhat similar to those found for the interaction of organic aromatics with iodine. It is uncertain, however, what part the metal plays in this charge-transfer interaction. It may be that the dominant interaction occurs with the ligand whose electronic structure is perturbed slightly by the metal. It would be interesting to be able to know the strength of the interaction between the acetylacetonate anion and iodine. Further work in this area should be of interest.

Although the final word may not yet be in on the question of benzenoid resonance or aromaticity, ring-currents, and $d\pi$-$p\pi$ bonding, it is fair to conclude that present data are consistent with the following:

1. ring-currents are either non-existent or do not influence properties to any presently measurable extent;
2. $d\pi$-$p\pi$ interactions do occur and do observably influence molecular properties; and
3. the chelate ring is best considered as aromatic in nature with regard to chemical reactivity of the coordinated ligand.

Conclusion

Although interests in inorganic derivatives of β-diketones are several and varied, the above discussions were confined to two major topics:

1. geometrical structures of β-diketone derivatives and 2. the electronic structure of chelated β-diketonate complexes. Furthermore, since this paper dealt solely with matters of academic interest, it might be appropriate to cite several possible practical applications of β-ketoenolate complexes. Possible applications exist in the areas of liquid organometallic lasers, unusual phosphors, catalysts for polymerization, and the separation of rare earth elements.

Acknowledgements: The author wishes to express thanks to Professors *C. Djordjevic*, *M. D. Schiavelli*, and *E. Katz* for helpful discussions.

References

1. *Laar, C.:* Ber. *18*, 648 (1885).
2. a) *Burdett, J. L., Rodgers, M. T.:* J. Am. Chem. Soc. *86*, 2105 (1964);
   b) — — Can. J. Chem. *43*, 1516 (1965).
3. *Combes, A.:* Compt. Rend. *105*, 871 (1887).
4. *Morgan, G. T., Moss, H. W.:* J. Chem. Soc. *105*, 189 (1914).
5. *Fackler, J. P., Jr.:* Progr. Inorg. Chem. *7*, 361 (1967).
6. *Hashimoto, F., Tanaka, J., Nagakura, S.:* J. Mol. Spectry. *10*, 410 (1963).
7. *Roof, R. B.:* Acta Cryst. *9*, 78 (1956).
8. *Lingafelter, E. C., Braun, R. L.:* J. Am. Chem. Soc. *88*, 2951 (1966).
9. *Moshier, R. W., Sievers, R. E.:* Gas Chromatography of Metal Chelates, Appendix I. Pergamon 1965.
10. a) *Collman, J. P.:* In: Reactions of Coordinated Ligands, p. 78. Washington, D. C.: American Chemical Society 1963;
    b) — Angew. Chem. Intern. Ed. Engl. *4*, 132 (1965).
11. *Selbin, J.:* Chem. Rev. *65*, 153 (1965).
12. *Lingafelter, E. C.:* Coord. Chem. Rev. *1*, 151 (1966).
13. *Bonati, F.:* Organometal. Chem. Rev. *1*, 379 (1966).
14. *Allred, A. L., Thompson, D. W.:* Inorg. Chem. *7*, 1196 (1968).
15. *Van Leeuwen, P. W. N. M.:* Rec. Trav. Chim. *87*, 396 (1968).
16. *Kakamura, Y., Kawaguchi, S.:* Chem. Commun. 716 (1968).
17. *West, R.:* J. Am. Chem. Soc. *80*, 3246 (1958).
18. *Howe, J. J., Collins, W. T., Pinnavaia, T. J.:* J. Am. Soc. Chem. *91*, 5378 (1969).
19. *Werner, A.:* Ber. *34*, 2584 (1901).
20. a) *Figgis, B. N., Lewis, J., Long, R. F., Mason, K., Nyholm, R. S., Pauling, P. J., Robertson, G. B.:* Nature *195*, 1278 (1962);
    b) *Mason, R., Robertson, G. B., Pauling, P. J.:* J. Chem. Soc. *485* (1969).
21. *Lewis, J., Long, R. F., Oldham, C.:* J. Chem. Soc. 6750 (1965).
22. *Nakamoto, K., Behnke, G. T.:* Inorg. Chem. *6*, 440 (1967).
23. *Gibson, D.:* Coord. Chem. Rev. *4*, 225 (1969).
24. *Hartman, F. A., Kilmer, M., Wjocieki, A.:* Inorg. Chem. *6*, 34 (1967).
25. *Swallow, M. A. G., Truter, M. R.:* Proc. Roy. Soc. (London) A, *254*, 205 (1960).
26. a) *Hargreaves, R. W., Truter, M. R.:* Chem. Commun. *473* (1968).
    b) — J. Chem. Soc. A 2282 (1968).
27. *Wilkinson, G., Cotton, F. A.:* Advanced Inorganic Chemistry, 2nd ed., p. 1036. New York: McCraw-Hill 1966.

28. *Parshall, G. W., Jones, F. N.:* J. Am. Chem. Soc. *87*, 535 (1965).
29. *Watson, W. H., Lin, C.:* Inorg. Chem. *5*, 1074 (1966).
30. *Dahl, L.*, communication to *Robertson, I., Truter, M. R.:* J. Chem. Soc. A 309 (1967). This latter reference report the crystal structure fo bis(3-methylacetylacetonato)copper(II).
31. a) *Allen, G., Lewis, J., Long, R. F., Oldham, C.:* Nature *202*, 589 (1964);
b) *Gibson, D., Lewis, J., Oldham, C.:* J. Chem. Soc. A 72 (1967).
32. *Behnke, G. T., Nakomoto, J.:* Inorg. Chem. *7*, 2030 (1968).
33. a) *Morgan, G. T., Drew, H. D. K.:* J. Chem. Soc. *117*, 1456 (1920).
b) — — *121*, 922 (1922).
34. *Dewar, D. H., Fergusson, J. E., Hentschel, P. R., Wilkins, C. J., Williams, P P.:* J. Chem. Soc. 688 (1964).
35. *Asmussen, R. W.:* Thesis Copenhagen, 1944, Reported in reference 37.
36. *Holm, R. H.:* Ph. D. Thesis, Massachusetts Institute of Technology, 1959. Cited in reference 37.
37. *Fackler, J. P., Cotton, F. A.:* J. Am. Chem. Soc. *83*, 2818 (1961).
38. a) *Jørgensen, C. K.:* Acta Chem. Scand. *9*, 1362 (1955). *Jørgensen* observed that the visible spectra for $Ni(acac)_2 \cdot 2\ C_2H_5OH$ and $Ni(acac)_2$ were virtually identical and was the first to correctly suggest that $Ni(acac)_2$ contained octahedral nickel.
b) *Maki, A.:* J. Chem. Phys. *29*, 162 (1958).
39. a) *Bullen, G. J.:* Nature *177*, 537 (1956) — only very preliminary results reported;
b) — *Mason, R., Pauling, P.:* Nature *189*, 291 (1961);
c) — — — Inorg. Chem. *4*, 456 (1965) — complete structural data.
40. *Holm, R. H., Cotton, F. A.:* J. Phys. Chem. *65*, 321 (1961).
41. *Bullen, G. J.:* Acta Cryst. *12*, 103 (1959).
42. *Graddon, D. P., Watton:* Nature *190*, 906 (1961).
43. *Fackler, J. P., Cotton, F. A.:* J. Am. Chem. Soc. *83*, 3775 (1961).
44. *Bennett, M. J., Cotton, F. A., Eiss, R., Elder, R. C.:* Nature *213*, 174 (1967).
45. *Cotton, F. A., Elder, R. C.:* J. Am. Chem. Soc. *86*, 2294 (1964).
46. *Graddon, D. P.:* Coord. Chem. Rev. *4*, 1 (1969).
47. a) *Collman, J. P., Barker, R. H.:* reported in Angew. Chem. Intern. Ed. Engl. *4*, 132 (1965);
b) *Barker, R. H., Collman, J. P., Marshall, R. L.:* J. Org. Chem. *29*, 3216 (1964).
48. *Wilkins, J. P., Wittbecker, E. L.:* U.S. Patent 2,659,711 (Nov. 17, 1953).
49. *Grove, D. E., Johnson, N. P., Lock, C. J. L., Wilkinson, G.:* J. Chem. Soc. 490 (1965).
50. *Calvin, M., Wilson, K. W.:* J. Am. Chem. Soc. *67*, 2003 (1945).
51. *Holm, R. H. Cotton, F. A.:* J. Am. Chem. Soc. *80*, 5658 (1958).
52. *Belford, R. L., Martell, A. E., Calvon, M.:* J. Inorg. Nucl. Chem. *2*, 11 (1956).
53. *Forster, L. S.:* J. Am. Chem. Soc. *86*, 3001 (1964).
54. a) *Barnum, D. W.:* J. Inorg. Nucl. Chem. *21*, 221 (1961);
b) — J. Inorg. Nucl. Chem. *22*, 183 (1961).
55. *Hill, J. O., Irving, R. J.:* J. Chem. Soc. A 1413 (1967).
56. — — J. Chem. Soc. 3116 (1968).
57. — — J. Chem. Soc. 1052 (1968).
58. *Jørgensen, C. K.:* Acta Chem. Scand. *16*, 2406 (1962).
59. — Absorption Spectra and Chemical Bonding in Complexes, p. 146ff. and references therein. Reading, Mass.: Addison-Wesley 1962.
60. — Orbitals in Atoms and Molecules, p. 94ff. New York: Academic Press 1962.

61. a) *Cotton, F. A., Harris, C. B., Wise, J. J.:* Inorg. Chem. *6*, 909 (1967);
b) — *Wise, J. J.:* Inorg. Chem. *6*, 917 (1967) and references therein.
62. *Forman, A., Murrell, J. N., Orgel, L. E.:* J. Chem. Phys. *31*, 1129 (1959).
63. *Eaton, D. R.:* J. Am. Chem. Soc. *87*, 3097 (1965).
64. *Collman, J. O., Marshall, R. L., Young, W. L.:* Chem. Ind. (London) 1380 (1962).
65. *Linck, R. G., Sievers, R. E.:* Inorg. Chem. *5*, 806 (1966).
66. Reference **47** a and references therein.
67. *Hester, R. E.:* Chem. Ind. (London) 1397 (1963).
68. a) *Sepone, N., Fay, R. C :* Inorg. Chem. *8*, 2379 (1969);
b) *Fay, R. C., Sepone, N.:* J. Am. Chem. Soc. *90*, 5701 (1968).
69. *Singh, P. R., Sahai, R.:* Australian J. Chem. *23*, 269 (1970).

Received July 23, 1970

The Electronic Spectra of the Hexafluoro Complexes of the First Transition Series

Dr. G. C. Allen* and Dr. K. D. Warren

Department of Chemistry, University College, Cardiff, Wales, Great Britain

Table of Contents

---

* Present address: Central Electricity Generating Board, Berkeley Nuclear Laboratories, Berkeley, Gloucestershire, England.

1. Introduction

(i) Historical Background and Scope of Coverage

The beginnings of interest in the electronic spectra of the hexafluoro complexes of the first transition series can conveniently be dated to the publication of a short communication by *Jørgensen* (*1*), in 1958, dealing with the reflectance spectrum of *potassium hexafluoromanganate(IV)*, K_2MnF_6, in which he pointed out certain advantageous features of the spectra of hexafluoro anions, and made predictions of the spectra of a number of complexes which, at that time, had not been investigated. In the ensuing ten years the electronic spectra of six of the anions of the MF_6^{3-} series (M = Ti, V, Cr, Mn, Fe, and Co) were reported, but the MF_6^{2-} compounds remained unstudied, apart from MnF_6^{2-}. The neglect of the latter series, and of the NiF_6^{3-} and CuF_6^{3-} ions, was due not so much to lack of enthusiasm as to the extremely reactive nature of the species involved, since in these cases the compounds were generally unstable in air and in the presence of moisture, and proved either to be insoluble in, or decomposed by, any of the usual solvents. This therefore necessitated the examination of such compounds in sealed cells, in vacuo or under a suitable inert gas, and the only technique obviously suitable for this purpose, *diffuse reflectance spectroscopy*, was for some time regarded with suspicion and considered incapable of furnishing even qualitative information regarding intensities. Nevertheless, this procedure, when properly handled, can afford very valuable data for compounds not otherwise accessible, and, along with other recently developed methods, (s. Sect. 1 (iv)) has allowed the measurement of the electronic spectra of the outstanding MF_6^{3-} ions and of the MF_6^{2-} ions of V, Cr, Co, and Ni.

The investigations of the last two years have thus completed the spectroscopic study of the hexafluoro anions of the trivalent M(III) species, from Ti to Cu (d^1 to d^8), together with those of the tetravalent M(IV) cations from V to Ni (d^1 to d^6), with the exception of the elusive FeF_6^{2-} ion, which is as yet unknown. One particular advantage which accrues from the study of hexafluoro complexes is embodied in the fact that many high and difficultly accessible oxidation states of the metals of the first transition series are stabilised, almost uniquely, by the fluoride ligand, and it is therefore possible to build up a collection of data for the d^n configurations representing hitherto unavailable valencies. Actually, the vast majority of hexafluoro complexes of the first transition series are MF_6^{2-} or MF_6^{3-} ions, and although some MF_6^{4-} species are known (*2*) (Fe to Zn), most M(II) fluoro complexes which have been studied are of types such as $KNiF_3$, and do not contain discrete MF_6 units, even though the micro symmetry about the metal may be more or less O_h. We have though restricted our specific treatment to those hexafluoro complexes of the higher oxidation states, M(III) and M(IV), which contain such a distinct MF_6 entity.

In Sections 2 and 3 respectively we have reviewed the spectroscopic data for the MF_6^{3-} and MF_6^{2-} (M(III) and M(IV)) anion complexes. In most cases we have summarised briefly the structural data relating to a particular ion, together with the information derived from magnetic studies which frequently provides strong indications for the spectroscopic assignments. Where appropriate we have also dealt with infra-red and Raman spectroscopic evidence.

Our interpretation of the spectra of the MF_6^{3-} and MF_6^{2-} ions follows the familiar lines of *ligand field theory*, with various ramifications as described by *Jørgensen* (*3*), for the *d–d* transitions, and the data are fitted in terms of the crystal field splitting parameter, D*q*, and the Racah inter-electronic repulsion parameters, B and C (s. Sect. 1 (ii)). The effects of spin-orbit coupling are in general quite small in the 3*d* series and are only considered in a few cases.

In Section 4 we have examined phenomena related to the *nephelauxetic effect* — the reduction of the B and C parameters below the free-ion value — and in particular its utility as a measure of covalency and its relationship to the chemical stability of particular oxidation states, whilst in Section 5 we have dealt with the charge-transfer spectra on the basis of the usual qualitative molecular orbital theory. Values of the *optical electronegativities*, $\chi_{opt.}$, derived from the Laporte-allowed bands are listed, and correlations with nephelauxetic effects estimated from the *d–d* transitions are examined. Sections 6 and 7 survey respectively the question of *vibrational fine structure*, which appears to be limited to the M(IV) complexes, and some general considerations about the operation of the *Jahn-Teller effect* in hexafluoro compounds.

(ii) Basic Theory

The basic theory of the electronic spectra and the related properties of transition metal complexes is by now so familiar as to require little further description but we briefly summarise here the two essentially complementary methods of treatment commonly employed. These are

(a) the crystal field theory, and its various more complicated modifications, collectively known as ligand field theory, and

(b) the molecular orbital theory.

The former is preoccupied exclusively with the *d–d* transitions characteristic of such compounds whilst the latter attempts to rationalise the complete range of spectroscopic transitions. The reader is referred to the texts of *Ballhausen* (4) and of *Figgis* (5) for further details.

a) Ligand Field Theory. In the original crystal field theory of *Bethe* (6) the complex is regarded as an entirely ionic aggregate consisting of a positively charged cation surrounded by an octahedral environment of negatively charged ligands. The effect of this octahedral potential, V_{oct}, on the energy levels of the free cation is treated by first-order perturbation theory, from which it may be shown that the five fold degeneracy of the d orbitals is lifted to give a lower three-fold degenerate, t_{2g}, set, and an upper doubly degenerate, e_g, set, which are separated by an energy of 10 Dq, where D and q are parameters arising from the simple electrostatic theory. Although the energy separation between the t_{2g} and e_g sets calculated on this electrostatic model is usually about an order of magnitude too small, due to the complete neglect of covalency effects, the treatment is perfectly satisfactory from a symmetry standpoint and the parameter Dq, obtained from experimental observations, can therefore still be used in a phenomenological sense.

In describing the effect of V_{oct} on the energy levels of d^n configurations two alternative formalisms may be adopted. In the first, the weak field treatment, the perturbing potential is regarded as exerting only small influences on the free-ion energy levels, expressed in terms of the usual L S classification (see *Condon* and *Shortley* (7), or *Ballhausen* (4)), whilst in the latter, the strong field treatment, the octahedral potential is considered to be sufficiently great for the energy levels to be characterised as well defined $t_{2g}^p e_g^q$ configurations. In both cases it is usual, for the first transition series, to retain two terms in the perturbing Hamiltonian, — the octahedral potential, V_{oct}, and the electronic repulsion term,

$$\sum_{i<j} e^2/r_{ij},$$

which is commonly evaluated in terms of the Racah parameters, B, and C (equal to $F_2 - 5\,F_4$, and $35\,F_4$ respectively).

In *the weak field approach* appropriate symmetry adapted combinations of the free-ion (L, M_L, S, M_S) determinental functions are constructed and the matrix elements of V_{oct} determined. Except where a given d^n configuration leads to more than one term state of the same designation the weak field matrices are diagonal in the B and C parameters, and in all cases the resulting energy levels may be obtained by solution of the corresponding secular determinants. In *the strong field treatment* similar symmetry adapted combinations are formed using determinental functions based on the t_{2g} (d_{xy}, d_{xz}, d_{yz}) and e_g (d_{z^2}, $d_{x^2-y^2}$) orbitals, and in this case the matrices are diagonal in the Dq parameter. Since the two sets of basis functions are related by a unitary transform, the weak and strong field methods must yield identical results when both are solved exactly.

In the above fashion the complete matrices for d^2 to d^8 systems (d^1 and d^9 are trivial) have been obtained by *Orgel* (*8*), from a weak field standpoint and by *Tanabe* and *Sugano* (*9*), using the strong field approach, the results for d^6 to d^8 being derived from the d^4 to d^2 cases, using the hole equivalence principle. *Tanabe* and *Sugano* also published diagrams of the results of their calculations, plotting E/B against Dq/B, and using C/B ratios based on the free-ion values, and these plots are now almost universally used in the interpretation of d–d spectra. The reader is therefore referred both to the original article and to a number of text books in which these diagrams are reproduced. (See for example *Figgis* (*5*), *Griffith* (*10*), *Dunn et al.* (*11*), *Kettle* (*12*), and *Cotton* (*13*).)

In general the effects of spin-orbit coupling are not sufficiently great to necessitate its inclusion in the perturbation Hamiltonian in the first transition series. The appropriate spin-orbit matrices are though sometimes useful in understanding the relative intensities of spin-allowed and spin-forbidden transitions, and have been given in both a weak and a strong field basis for d^2 systems by *Liehr* and *Ballhausen* (*14*), whilst the strong field results have been presented by *Eisenstein* (*15*) for d^3 and by *Schroeder* (*16*) for d^4 and d^5 configurations.

The various levels of sophistication at which ligand field theory has been applied to the description of d^n complexes have been summarised by *Jørgensen* (*3, 17*), who also indicated their limits of validity. The most primitive, and obsolete, approximation was to consider only the diagonal elements corresponding to the usual first order perturbation of the d-orbital angular functions. This, the weak or strong field *approximation* was equivalent to the assumption of well defined LS states or well defined $t_{2g}^p e_g^q$ sub-shell configurations. The removal of this restriction, and the derivation of the energy levels from the complete weak or strong field matrices, constituted a more refined treatment, but initially it was supposed that the atomic term distances were fixed and the theory had therefore only one disposable parameter, Dq, and thus corresponded to an Orgel type diagram.

Later it became clear that as well as the first order perturbation acting on the angular function there existed also a further perturbation which had the effect of changing the central field and allowing the expansion of the radial function and a concomitant decrease in the parameters of inter-electronic repulsion (B and C). It was assumed that the angular function was still hydrogenic, i.e. $l=2$, but the theory now contained two parameters, Dq as before, and, taking a fixed C/B ratio, a variable B by which the relative reduction of the electron repulsion integrals, as compared with the free ion values, could be expressed. This level of approximation is thus represented by the familiar Tanabe-Sugano diagram, where the independent variable is Dq/B.

As noted earlier, the failure of the purely electrostatic crystal field theory adequately to predict the magnitude of 10 Dq was soon recognised as being due to the neglect of covalency. It is clear that the relative extent of the involvement of the metal t_{2g} and e_g orbitals in bond formation will profoundly influence the extent of their energetic separation, since, as shown later, the e_g set may interact in a σ-sense with the ligand s and p orbitals, whereas the t_{2g} set interact only in a π-sense with the ligand p orbitals. Such a covalent bonding process would be expected to bring about a reduction in the effective central field of the metal and to allow the spreading out of the metal radial function, and on this basis the observation of substantially reduced B values in many metal complexes can be rationalised.

Jørgensen therefore introduced a parameter, β, equal to the ratio of the observed B value in a given complex to that of the corresponding free ion, known as the nephelauxetic (cloud expanding) ratio, which thus afforded a measure of the extent of the expansion of the radial function of the metal, and also gave some indication of the tendencies towards covalency in the given complex. The behaviour of a given ligand as regards the magnitude of both Dq and β may be discussed according to the spectrochemical and nephelauxetic series respectively (s. Sect. 1 (iii)), but in general one normally expects σ-type interactions to be rather more important than π-bonding contributions, except for ligands such as CN^-, which have their own π-systems. Consequently the e_g set of orbitals should be more involved in covalency than the t_{2g} set with the result that the e_g orbitals should undergo a greater expansion of the radial function, and should show a greater reduction in the value of the parameter B. In such a level of approximation therefore one still assumes a hydrogenic angular function with $l=2$, but allows B and C to vary according to the sub-shell configuration concerned. In the octahedral point group the representations e_g and t_{2g} are frequently abbreviated as γ_3 and γ_5 respectively, and interactions between two e_g electrons would thus involve a parameter B_{33}, between two t_{2g} electrons the parameter B_{55}, and between an e_g and a t_{2g} electron the parameter B_{35}. It thus follows from the above arguments that for the corresponding nephelauxetic ratios we should expect

$$\beta_{55} > \beta_{35} > \beta_{33}.$$

This last modification does not however represent the limit of refinement which can be made to the ligand field theory since it is possible to remove the assumption of hydrogenic angular dependence, and to allow also for complete configuration interaction between the appropriate levels. At this point though we have effectively arrived at a complete molecular orbital interpretation in which the ligands are not regarded merely as a source of not too precisely defined perturbations but are fully involved in the bonding scheme, which is postulated to embrace not only the $3d$ metal orbitals, but also the $4s$ and $4p$ levels, together with both s and p orbitals of the ligands.

b) Molecular Orbital Theory. In this approach we shall consider for simplicity a typical hexafluoro complex anion, the hexafluorochromate (III) species, CrF_6^{3-}, and both σ- and π-type bonding will be included. The basis set is taken to include the $3d$, $4s$, and $4p$ orbitals of Cr and the $2p$ orbitals of F. The lower lying $2s$ orbitals of fluorine will also be involved to a lesser extent but since these orbitals will behave in essentially the same way as the σ-bonding $2p$ orbitals they have been omitted for greater clarity.

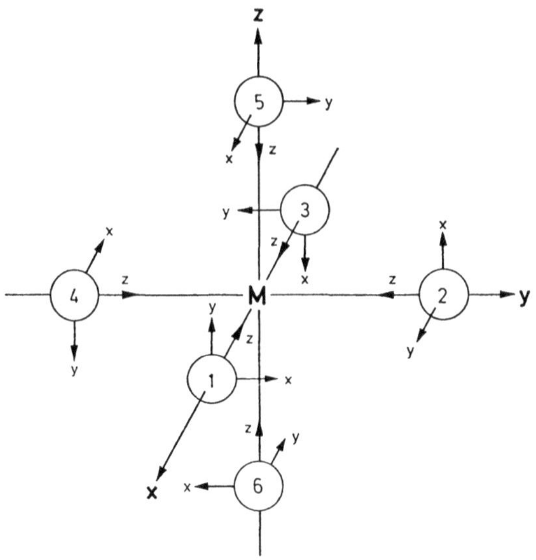

Fig. 1. Coordinate system for an octahedral complex

In a system of O_h point symmetry such as the CrF_6^{3-} ion, the metal orbitals used will constitute four irreducible representations. As before the $3d$ level gives rise to an e_g and a t_{2g} set, whilst the $4s$ and the $4p$ levels transform as a_{1g} and t_{1u} respectively. For each of the six fluorine ligand atoms three $2p$ orbitals are available for bond formation, and these eighteen orbitals may be formed into appropriate symmetry adapted combinations which will be capable of either σ or π-type interaction with metal orbitals belonging to the same irreducible representation. Using the nomenclature of Fig. 1 the normalised combinations listed in Table 1 (i) are found, neglecting the slight effect of ligand-ligand overlap. It is seen that the σ-bonding ligand combinations transform as a_{1g}, e_g, and t_{1u}, whilst the π-bonding symmetry orbitals are t_{1g}, t_{2g}, t_{1u}, and t_{2u} and that the metal d-orbitals are therefore involved in such a way that the e_g set can participate only in σ-bonding whilst the t_{2g} set is restricted to π-bonding. Of the remaining ligand orbitals the a_{1g} combination may interact with the metal 4s level, whilst for the t_{1u} representation both σ and π-bonding ligand combinations are available for bonding.

Table 1 (i).

Representation	Metal Orbital	Ligand Orbitals	
		σ	π
a_{1g}	$4s$	$(6)^{-\frac{1}{2}}(z_1+z_2+z_3+z_4+z_5+z_6)$	—
e_g	$3d_{x^2-y^2}$	$\frac{1}{2}(z_1-z_2+z_3-z_4)$	—
	$3d_{z^2}$	$\frac{1}{2}(3)^{-\frac{1}{2}}(2z_5+2z_6-z_1-z_2-z_3-z_4)$	—
t_{1u}	$4p_x$	$(2)^{-\frac{1}{2}}(z_1-z_3)$	$\frac{1}{2}(y_2+x_5-x_4-y_6)$
	$4p_y$	$(2)^{-\frac{1}{2}}(z_2-z_4)$	$\frac{1}{2}(x_1+y_5-y_3-x_6)$
	$4p_z$	$(2)^{-\frac{1}{2}}(z_5-z_6)$	$\frac{1}{2}(y_1+x_2-x_3-y_4)$
t_{2g}	$3d_{xz}$	—	$\frac{1}{2}(y_1+x_5+x_3+y_6)$
	$3d_{yz}$	—	$\frac{1}{2}(x_2+y_5+y_4+x_6)$
	$3d_{xy}$	—	$\frac{1}{2}(x_1+y_2+y_3+x_4)$
t_{1g}	—	—	$\frac{1}{2}(y_1-x_5+x_3-y_6)$
	—	—	$\frac{1}{2}(x_2-y_5+y_4-x_6)$
	—	—	$\frac{1}{2}(x_1-y_2+y_3-x_4)$
t_{2u}	—	—	$\frac{1}{2}(y_2-x_5-x_4+y_6)$
	—	—	$\frac{1}{2}(x_1-y_5-y_3+x_6)$
	—	—	$\frac{1}{2}(y_1-x_2-x_3+y_4)$

An approximate molecular orbital diagram for *the* CrF_6^{3-} *system* is shown in Fig. 2 and from such a diagram the relationship between the ligand field and molecular orbital approaches may be illustrated. (See, for example, *Orgel* (*78*)). In the former treatment the system is regarded as a d^3 Cr^{3+} ion in an octahedral fluoride environment and a determinental function corresponding to the three t_{2g} orbitals each being singly occupied with parallel spins constitutes the $^4A_{2g}$ ground state. On the other hand the molecular orbital picture shows that all levels below the second t_{2g} orbital are fully occupied and again the ground state corresponds to each of this t_{2g} set being singly occupied with parallel spins. This therefore exemplifies the essential equivalence of the two schemes, but in a molecular orbital treatment the allowed charge-transfer bands may also be readily interpreted: in the case of CrF_6^{3-} it is clear that transitions originating from the t_{1u} and $t_{2u}(\pi)$ and $t_{1u}(\sigma)$ levels and terminating at the e_g and t_{2g} levels all represent Laporte allowed $u \rightarrow g$ transitions. In general, for complexes containing up to five electrons in the t_{2g} level, charge-transfer transitions from both π and σ levels to either t_{2g} or e_g are possible.

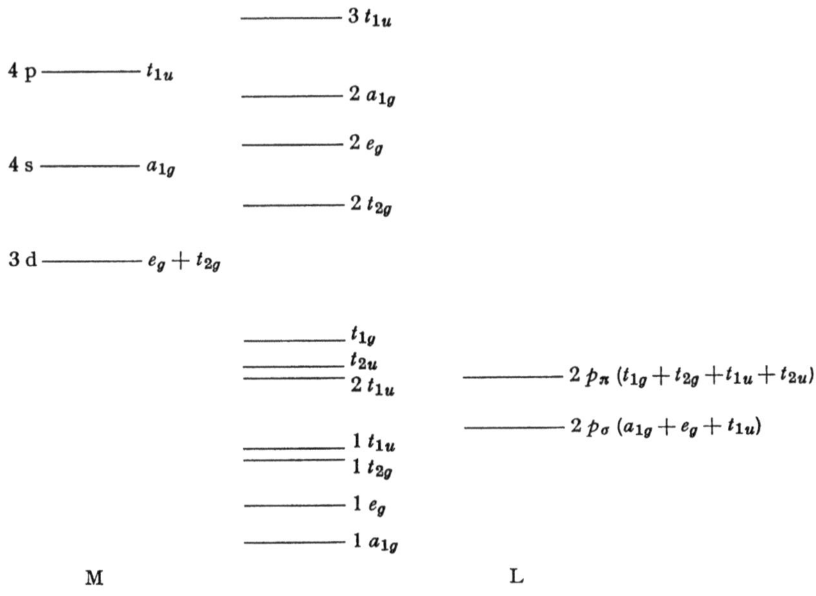

Fig. 2. Molecular orbital energy level scheme for CrF_6^{3-}

On the basis of this molecular orbital treatment it is a fairly simple matter to make approximate predictions as to the composition of the various levels. Following *Figgis* (5) we may classify the metal or ligand contribution to a given molecular orbital as large (80%), moderate (50%), or small (20%), and the usual situation for octahedral complexes is shown in Table 1 (ii). Thus the highest lying orbitals, 2 t_{2g} and 2 e_g are dominantly metal in character, and the occupation of these levels therefore effectively corresponds to the d^n-configurations treated in a phenomenological fashion by the ligand field theory. For the t_{1u} representation both σ- and π-bonding ligand combinations are available, but the latter contribution to the resulting molecular orbitals is normally substantially the smaller for the $1\,t_{1u}$ and $3\,t_{1u}$ levels, and predominant for $2\,t_{1u}$.

Table 1 (ii). *Metal and ligand contributions to the molecular orbitals of an octahedral complex*

Orbital	Metal	Ligand
$1\,a_{1g}(\sigma)$	small-moderate	large-moderate
$1\,e_g(\sigma)$	small-moderate	large-moderate
$1\,t_{1u}(\sigma$ and $\pi)$	small-moderate	large-moderate
$1\,t_{2g}(\pi)$ [a]	very small	very large
$2\,t_{1u}(\sigma$ and $\pi)$	very small	very large
$1\,t_{2u}$	none	complete
$1\,t_{1g}$	none	complete
$2\,t_{2g}(\pi)$ [a]	very large	very small
$2\,e_g^*(\sigma)$	large-moderate	small-moderate
$2\,a_{1g}^*(\sigma)$	large-moderate	small-moderate
$3\,t_{1u}^*(\sigma$ and $\pi)$	large-moderate	small-moderate

[a]) Assuming filled ligand orbitals.

As a result of various one-electron type calculations (see below) it has frequently been assumed that the difference in orbital energy between the $2\,t_{2g}^*$ and $2\,e_g^*$ levels corresponds exactly to 10 Dq. However, as has been pointed out by *Offenhartz* (*19*), 10 Dq is properly defined as this difference in the absence of inter-electronic repulsions among these same two molecular orbitals, and he thus defined 10 Dq as being given by the expression

$$10\,\mathrm{D}q = \varepsilon_t - \varepsilon_e - \sum_d (\mathrm{J}_{td'} - \mathrm{K}_{td'}) + \sum_d (\mathrm{J}_{ed'} - \mathrm{K}_{ed'})$$

where the sums extend over singly occupied orbitals of maximum d character. *Offenhartz* further showed, for the closed shell t_{2g}^6 case (and by an heuristic extension for open shell systems also), that the conventional molecular orbital method for the calculation of excitation energies yielded an identical result for the formulation of 10 Dq as that given by the above relationship.

However, whilst the qualitative interpretation of the molecular orbital scheme is quite straightforward, it is a matter of considerable difficulty to obtain useful quantitative results. A great many calculations have been carried out by variants of the *Wolfsberg-Helmholz* (*20*) approach (see for example *Ballhausen* and *Gray* (*21*) and *Basch et al.* (*22*)) in which the secular determinants

$$|H_{ij} - S_{ij}E| = 0$$

are solved for each irreducible representation by an iterative charge and configuration consistent procedure, the required H terms being derived from valence state ionisation potential data. Such a method is though necessarily a one-electron procedure and the transition energies are obtained simply as the differences of orbital energies: thus 10 Dq is here merely the orbital energy difference between the t_{2g} and e_g levels, and repulsion terms are ignored.

On the whole calculations by the Wolfsberg-Helmholz method have about the same degree of reliability as the related one-electron Hückel method has for conjugated organic systems, and it would therefore be desirable for calculations to be based on some method which takes explicit account of electron repulsion terms, such as, for example, the well known *Roothaan* (*23*) procedure. The difficulties which beset such an attempt have been surveyed in detail by *Dahl* and *Ballhausen* (*24*) who showed that the principal problem lies in the evaluation of the vast number of multi-centre electron repulsion integrals which occur for transition metal complex systems, and in the choice of a systematic collection of simplifying assumption in such a way as to preserve the rotational invariance of the basis set, and to avoid the pitfall of allowing an electron to interact with itself. These authors tended to favour a modification of the CNDO method of *Pople et al.* (*25*) for such calculations, but the assumption that all repulsion integrals are taken as *s* type (i.e. that 3*d*, 4*s*, and 4*p* integrals would differ only by virtue of the radial functions) may well limit the general utility of such calculations.

As far as we have been able to ascertain there are in the literature no wholly satisfactory molecular orbital treatments of MF_6^{3-} or MF_6^{2-} ions. The most extensive set of calculations on these complexes was compiled by *Fenske et al.* (*26*) for the MF_6^{3-} series with M = Ti, V, Cr, Mn, Fe, and Co, but their procedure has been severely criticised by *Dahl* and *Ballhausen* (*24*) on the grounds of an incorrect assumption concerning the exchange operator, omission of the ligand 2*s* orbitals from the basis set, and neglect of electron repulsion terms in evaluating 10 Dq. However, the situation is not altogether unpromising and the recent analysis of the problem of molecular orbital calculations for inorganic systems presented by *Brown* and *Roby* (*27*) holds out some hope of advances in this field.

(iii) Characteristics of Fluorine as a Ligand

As originally pointed out by *Jørgensen* (*1*) there are a number of very good reasons why the study of the hexafluoro complexes of M(III) and M(IV) cations should be especially informative from a spectroscopic point of view. In the first place the fluoride ion is one of the least nephelauxetic of all ligands, and shows therefore a correspondingly weaker tendency towards covalency, and the hexafluoro complexes are frequently representative of high and often unusual oxidation states of the metals concerned. Indeed, in some cases the higher valencies appear only to be stabilised for certain to any appreciable extent in such hexafluoro anions. Moreover, fluoride ligands give rise to a relatively small value of 10 Dq for a given metal cation, and this behaviour, coupled with the rather high energies at which charge-transfer bands are found, tends to lead to spectra in which the Laporte-allowed transitions are far enough above the *d*–*d* transitions for the latter to be unobscured. The generally large energies found for the charge-transfer bands of hexafluoro complexes arise from the very considerable electronegativity of the ligand because, as shown by *Jørgensen* (*28*) (s. Sect. 5), there exists a strong connection between the position of the lowest charge-transfer transition and the ease of reduction of, or electron transfer to, the central metal atom in the given oxidation state. As the oxidising character of the metal cation increases the charge-transfer bands will therefore tend to move to lower energies, but in most cases the *d*–*d* bands will be clearly distinguishable from the Laporte-allowed transitions.

The general behaviour of hexafluoro complexes will though be more readily clarified if we consider the position of the fluoride ligand in the two series commonly used to classify the characteristics of ligands — the spectrochemical series, and the nephelauxetic series. As regards the former we may arrange the ligands in order of increasing Dq developed for any constant metal ion, or, alternatively, we may arrange the metal cations in a similar series according to the Dq developed for any constant ligand. For the ligands a representative selection is

$$—I^- < —Br^- < —Cl^- < —SCN^- < —F^- < \text{urea} < —OH^- < \text{oxalate}^{2-} < H_2O$$
$$< —NCS^- < \text{pyridine} < NH_3 < \text{ethylenediamine} < —NO_2^- < —CN^-$$

whilst for the metal ions of the 3*d* series we find

$$Mn^{2+} < Ni^{2+} < Co^{2+} < Fe^{2+} < V^{2+} < Fe^{3+} < Cr^{3+} < V^{3+} < Co^{3+} < Mn^{4+}$$

although this latter series is not so well defined in places as the former. Note that in the latter series there is a strong dependence on the valency of the metal ion, even though this would not be predicted on the simple electrostatic theory, and this is presumed to be due to the greater involvement in covalent bonding shown by the metal orbitals in order to preserve approximate electroneutrality in the complex.

The four halide ligands all lie close to the bottom end of the spectrochemical series and in this way they contrast sharply with ligands such as the nitrite and cyanide ions. This is because if we restrict the valence orbitals of the halogens to the n*s* and n*p* levels there are accessible for molecular orbital formation only *filled* ligand orbitals which are capable of π-type interaction. Since these will generally lie lower than the metal t_{2g} set the net result is to raise the energy of the latter and thereby to decrease 10 D*q*. On the other hand, ligands such as $-NO_2^-$ and $-CN^-$ have vacant π-type anti-bonding orbitals accessible for molecular orbital formation which will this time lie above the metal t_{2g} set. In this case the result of the interaction will be to depress the energy of the metal t_{2g} orbitals, and to increase 10 D*q*, thereby accounting for the relative positions of these types of ligands in the spectrochemical series.

The order of the halide ions in the spectrochemical series is in fact that which would be predicted from the electrostatic crystal field theory, but this is probably coincidental. If one starts from a molecular orbital standpoint, assuming functions appropriate to neutral metal, M, atoms, and neutral, F, ligands, the order can be rationalised by assuming that the most electronegative fluorine, which displays the most ionic behaviour, will interact to the greatest extent by σ-bonding with the ligands, thereby bringing about the greatest raising of the metal $2\,e_g^*$ orbitals and the largest value of 10 D*q*. The relatively small absolute magnitude of the D*q* values shown by the halide ligands is consistent with the raising of the metal $2\,t_{2g}^*$ orbital by the π-interaction with the filled ligand orbitals, and it is probable that the extent of π-bonding is roughly constant for all the halides. The overall inference is therefore that the magnitude of D*q* for the halide ligands is determined largely by σ-bonding involving the $2\,e_g^*$ molecular orbital with a small roughly constant π-interaction, and this conclusion is supported by *Jørgensen* (*3*, *28*), largely as a result of studies of the electron-transfer spectra of hexahalo complexes. This work shows for example that whereas, for a given M^{n+} cation the charge transfer bands of the $MCl_6^{(6-n)-}$ ion lie some 25 *k*K. lower than for the corresponding hexafluoro compound, the bands of the MCl_6, MBr_6, and $MI_6^{(6-n)-}$ species differ only by some 15 *k*K. overall. For further details concerning the rationalisation of the spectrochemical series in terms of electronegativities, polarisabilities, σ- and π-bonding propensities, and the like, the reader is referred to the text-books of *Jørgensen* (*3, 28*).

However, besides arranging the ligands in a series depending upon the D*q* value developed for any constant metal ion it is also possible to construct a similar series in which the ligands are ordered according to a decreasing value of β, the nephelauxetic ratio. From studies of complexes of the $3d$ block such a series is quite well established and a typical selection is

$$-F^- > H_2O > \text{urea} > NH_3 > \text{ethylenediamine} > \text{oxalate}^{2-} > -NCS^- > -Cl^-$$
$$> -CN^- > -Br^- > -I^-$$

but the corresponding series of metal cations, for any constant ligand, is rather more fragmentary, the available data giving

$$Mn^{2+} > V^{2+} > Ni^{2+} > Co^{2+} > Cr^{3+} > Fe^{3+} > Co^{3+} > Mn^{4+}$$

for the $3d$ elements. It is apparent that although there is a not unreasonable parallel between the spectrochemical series and the nephelauxetic series for the metal cations, the two series show no correlation whatever as regards the ligands, but this former feature is understandable if the Dq variation

$$M^{2+} < M^{3+} < M^{4+}$$

does indeed reflect increasing covalent tendencies. The ligand series however involve a number of factors which are not constant throughout the series: for example for ligands like $-CN^-$ increased π-bonding would tend to increase Dq but decrease β whereas for the halide ligands greater π-bonding would decrease both parameters. If then we consider only the closely related halide ligands the relationship between the spectrochemical and nephelauxetic series can be established. Thus the most electronegative fluorine should be most involved in σ-bonding, thereby raising $2\,e_g^*$ and increasing Dq, whilst at the same time constituting the most ionic ligand and exhibiting the largest value of β. There are unfortunately very few MCl_6^{n-} complexes known for the first transition series and the information concerning the position of Cl^- in the nephelauxetic series has largely been obtained indirectly. Nevertheless, such anions as are known do in fact show substantially smaller β values than the corresponding MF_6^{n-} entities (and smaller Dq values) and the data for these parameters are reviewed at greater length in Section 4.

The data reported in this review enable an extensive compilation to be made of nephelauxetic parameters for MF_6^{3-} and MF_6^{2-} complexes of the $3d$ series. Since we are here dealing directly with a series of anions, all octahedral or nearly so, for a constant ligand, F^-, it is possible to establish with some degree of confidence a nephelauxetic series for the M^{3+} and M^{4+} cations of the $3d$ block, and to examine the relationship of these values to those derived for the M^{2+} cations, and to the known chemical characteristics of the various M(III) and M(IV) oxidation states. In interpreting such data we must take note of the fact that it is possible to postulate two types of contribution towards the nephelauxetic effect — central field covalency, and symmetry restricted covalency. The former relates to the screening effect of the ligands on the central field of the metal cation which results in a decrease of the effective positive charge on the metal and a consequent expansion of the radial function, whilst the latter arises from the actual involvement of the metal t_{2g} and e_g orbitals in molecular orbital formation with ligand orbital combinations of the appropriate symmetry, which should also lead to a reduction in the parameters of inter-electronic repulsion. The relative contributions which these two effects are postulated to make will influence considerably the conclusions which may be drawn concerning effective charge densities on the metal and the ligand contributions to the $2\,t_{2g}^*$ and $2\,e_g^*$ orbitals and this aspect is also discussed in more detail in Section 4.

Finally it may be observed that the fluorine ligand also constitutes a favourably simple situation as far as theoretical molecular orbital treatments are concerned. In such calculations a monatomic ligand is always preferable to a more complex entity such as H_2O or NH_3 (which usually require some extraneous assumptions in the evaluation of the relevant integrals), and for fluorine it is a reasonably safe assumption to limit the ligand basis set for the valence orbitals to the 2*s* and 2*p* levels.

(iv) Special Experimental Techniques

Unlike the hexahalo complexes of the 4*d* and 5*d* series, many of which have been extensively investigated in solution, the hexafluoro anions of the first transition series are generally rather *unstable* towards water, this tendency increasing towards the end of the series and with increase in the oxidation number. Because of this the normal solution techniques are virtually inapplicable, although a study of NiF_6^{2-} in anhydrous HF has been reported (*29*), and recourse must be made to methods which can utilise solid samples. Such methods include measurements on dry powders between KBr discs (*30*), which do not appear very successful, and investigations of similarly confined fluorocarbon mulls (*31*), which seem to be much more satisfactory. However, the one generally available method for such materials is diffuse reflectance spectroscopy, which the authors have adopted for their own studies.

For the more stable complexes, e.g. K_3VF_6, it is not necessary for the sample to be studied in a closed system, but for the more reactive species, such as the MF_6^{2-} anions, and those of Ni(III) and Cu(III), the materials have been sealed under vacuum in silica cells. For diffuse reflectance measurements the intensities are expressed in terms of the well known (*32*) Kubelka-Munk function

$$F_R = (1 - R_{diff})^2/2\, R_{diff} = k/s$$

where $R_{diff} = I/I_0$ is the relative diffuse reflectance referred to a non-absorbing standard (e.g. MgO, enclosed, where appropriate, in a similar silica cell), and k is the absorption coefficient defined by $I = I_0\, e^{-ks}$. Here s is the scattering coefficient for the substance in question and is an empirical quantity whose dependence on the wavelength of the incident light and on the geometry of the particles is difficult to predict *a priori*, since the theoretically calculated value of 2, obtained by *Kortüm* (*32*), probably does not relate too closely to the actual experimental conditions most commonly used.

However, it is necessary to make some estimate of the value of s if any quantitative information is to be obtained from reflectance data. This is because k is related to the extinction coefficient by the relationship $k = \varepsilon c$, where c = density/molecular weight, from which we derive $F_R = \varepsilon c / 0.434\, s$. For some substances s may be found experimentally by comparing the transmission spectrum of the compact material with that of the same substance in the powdered state, but this method is only applicable to fairly stable materials, and is not available here. Nevertheless, a value somewhat less than 2 should be the more probable since the sample is usually less densely packed in reflection than in transmission.

Furthermore, although s does not appear to vary widely from one substance to another, its dependence on wavelength is very sensitive to *particle size*. For weakly absorbing compounds, with particle sizes greater than 2—3μ, the scattering coefficient can reasonably be assumed to be independent of wavelength in the visible region, but above about 25 *k*K. it often rises sharply above the value of about 2 and is strongly dependent on particle size. Such behaviour was found by *Kortüm* and *Schottler* (*33*) in measurements on powdered $KMnO_4$ mixed with $KClO_4$ and $BaSO_4$: they reported $s = 1.72 \pm 3\%$ and showed that the Kubelka-Munk equation was of general validity, provided that both the total absorption and the particle size were sufficiently small for one to ignore the regular component of the reflected light.

In their own work therefore the authors have adopted the value of 1.72 for s and have thereby obtained approximate extinction coefficients for transitions up to about 25—30 *k*K. Above this value of course it is likely that s increases quite significantly, which would tend to reduce the apparent intensities of the charge-transfer bands for which no useful quantitative information can therefore be derived. The quantitative information which can be obtained from diffuse reflectance spectra is of course necessarily limited, even in favourable cases, but it should be pointed out that it is F_R which is the linear function of ε, and *not* the measured optical density, and that therefore the practice of presenting spectra as plots of arbitrary absorbance against wavelength hinders the deduction of even this minimal amount of quantitative evidence.

2. Electronic Spectra of Hexafluorometallate(III) Complexes

For the trivalent cations of the first transition series all the MF_6^{3-} anions, from TiF_6^{3-} to CuF_6^{3-}, representing the configurations d^1 to d^8, are known. In this Section our attention is directed primarily towards the d–d transitions which occur in the electronic spectra of these ions, the Laporte-allowed charge-transfer bands being discussed in more detail in Section **5**, although some correlations between molecular orbital calculations, oxidation state stabilities etc., and the positions of the charge-transfer excitations are indicated.

For the ions listed we survey the most recent spectroscopic investigations and present a critical compilation of values of the fitting parameters, Dq, and B. For the d^1 system, TiF_6^{3-}, only the Dq value can, of course, be obtained, and it is similarly difficult to estimate B for MnF_6^{3-} and CoF_6^{3-} since in these cases essentially only one spin-allowed excitation can be identified with confidence. Assignments are also complicated by the operation of significant Jahn-Teller effects for the Mn, Co, and Ni complexes, with smaller influences encountered for the Ti and V compounds.

For the d^4, d^5, d^6 and d^7 (Mn(III), Fe(III), Co(III), and Ni(III)) complexes, the possibility of either high- or low-spin ground states arises, but only in the case of Ni(III) is the latter behaviour found. In all cases though there is evidence that the Dq/B value lies quite close to the high-spin—low-spin crossover point, and we put forward below our reasons for believing that there is no anomaly encompassed by the assignment of only the Ni(III) complex as low-spin.

The values of β, the nephelauxetic ratio, derived from the values of B obtained, are listed in Section 4, together with the assumed free-ion data upon which they are based, and the trends exhibited by this parameter are discussed in full at that point. Although throughout the series the trivalent cation gives rise to a substantially greater Dq value than that generally found for M^{2+} cations, the generally weak nephelauxetic character of the fluoride ion ensures that in most cases B is fairly high, ca. 0.8 times the free-ion value, and only for Ni(III) is the β value sufficiently low for Dq/B to correspond to low-spin behaviour, i.e. towards the end of the series where the low stability of the M(III) state is becoming pronounced.

(i) Hexafluorotitanate(III), TiF_6^{3-}

The potassium salt of this anion, K_3TiF_6, was first obtained by *Ehrlich* and *Pietzka* (*34*) by the reduction of K_2TiF_6 in hydrogen to „K_2TiF_5", followed by hydrolysis of the product and separation of the soluble K_3TiF_6 from the insoluble $KTiF_4$. It was reported as a violet solid, rather unstable to atmospheric oxidation, with a magnetic moment of 1.70 B.M. at 233 °K. The salt was also obtained by *Bright* and *Wurm* (*35*), who in addition prepared K_2NaTiF_6 and Na_3TiF_6, all the salts being obtained by electrolysis of appropriate NaCl–KCl melts containing corresponding amounts of K_2TiF_6. In neither case however were any spectroscopic measurements carried out although the latter investigators did establish that K_2NaTiF_6 had a cubic, K_2NaAlF_6 type, lattice.

Bedon, Horner and *Tyree* (*36*) obtained Na_2KTiF_6 and K_2NaTiF_6 by a similar electrolytic procedure, and also prepared $(NH_4)_3TiF_6$ by the action of excess acidified NH_4F on $TiCl_3$ solution. They measured the electronic spectra of these salts between 5 and 50 *k*K. in KCl and KBr discs, and these results together with the magnetic moments at 293 °K are shown in Table 2 (i). For the d^1 Ti^{3+} ion the single *d–d* band expected ($^2T_{2g} \rightarrow {}^2E_g$) is frequently split into two components, as in the present case, and this is usually regarded as indicative of a tetragonal Jahn-Teller splitting of the excited ^2E_g state since only a small energy separation would be expected to arise from a Jahn-Teller distortion of the $^2T_{2g}$ ground state (see, for example *Van Vleck* (*37*)). *Bedon et al.* (*36*) therefore took the mean of the two observed bands as 10 D*q*, giving respectively 1745, 1740 and 1706 cm.$^{-1}$ for D*q* in the Na_2K, K_2Na and $(NH_4)_3$ salts, postulating that the larger splitting in the ammonium salt was due to deviations from octahedral symmetry inherent in the crystal lattice. No charge-transfer bands were reported to occur below 50 *k*K., and the limiting value of the optical electronegativity to which this leads

Table 2 (i). *Spectroscopic and magnetic data for the hexafluorotitanate(III) anion*

Salt	*d–d* Bands (*k*K).	Magnetic Moment at 293 °K (B.M.)
Na_2KTiF_6	16.00, 18.90	1.86
K_2NaTiF_6	16.10, 18.90	2.06–2.09
$(NH_4)_3TiF_6$	15.11, 19.01	1.70–1.77

is not unreasonable in view of the high electronegativity of F^- and the relative instability of the Ti^{2+} ion. The observation is also consistent with the finding of *Fowles* and *Russ* (*38*) that in the corresponding $TiCl_6^{3-}$ ion the charge-transfer region begins at about 20—25 *k*K. *Bedon et al* (*36*) also showed the presumed position of the charge-transfer transitions to be in accordance with the predictions of Wolfsberg-Helmholz type MO calculations, which indicated π-bonding in the complex to be appreciable, but in view of the limited theoretical foundation for such calculations (see, for example *Dahl* and *Ballhausen* (*24*)), both these conclusions must be regarded as tentative. (Similar reservations also apply to the more advanced treatment of *Fenske et al.* (*26*)).

(ii) Hexafluorovanadate(III), VF_6^{3-}

The greenish complex fluorides of vanadium(III) have been known since about 1890, and salts of the type M_3VF_6 are fairly readily obtained. The potassium compound for example is prepared (*39*) by a straightforward reduction of V_2O_5 in HF with KI. Only the ammonium and the potassium sodium salts, $(NH_4)_3VF_6$, and K_2NaVF_6 have been examined structurally, and these have (*40*) the cubic, α-K_3FeF_6 and K_2NaAlF_6 type lattices respectively. Both the potassium and ammonium salts have though been the subject of diffuse reflectance spectroscopic studies. The latter compound was examined by *Ballhausen* and *Winther* (*41*) who located the $^3T_{1g} \rightarrow {}^3T_{2g}$ and $^3T_{1g} \rightarrow {}^3T_{1g}$ transitions at 14.8 and 23.25 *k*K. respectively, with the spin forbidden $^3T_{1g} \rightarrow {}^1E_g$, $^1T_{2g}$ band at 10.2 *k*K. and these results were confirmed by *Machin* and *Murray* (*42*) with exactly the same bands being found by *Clark* (*43*) for K_3VF_6. Fitting of the two spin-allowed bands yields the parameters $Dq = 1610$ cm.$^{-1}$, and $B_{35} = 650$ cm.$^{-1}$, but the reasons for the position of the spin-forbidden transition have given rise to some controversy.

If the energies of the ^1E_g, $^1T_{2g}$ levels above the ground state are calculated using the parameters derived from the spin-allowed bands, they are both estimated to lie at about 9.2 *k*K., i.e. about 1 *k*K. lower than the observed value. Since this transition is an intra-subshell (t_{2g}^2) excitation it is possible to interpret this result on the basis that B_{55} terms are usually somewhat greater than B_{35} values, because the γ_3 radial functions will be more expanded due to σ-bonding effects than the γ_5 set, which can only participate in π-type interactions. In the strong field approximation the ^1E_g, $^1T_{2g}$ levels lie at $6\,B + 2\,C$ above the ground state which, taking the *Tanabe* and *Sugano* (*9*) value for C/B, yields the result $B_{55} = 700$ cm.$^{-1}$. However, *Ballhausen* and *Winther* (*41*) argued that there were both experimental and theoretical reasons for believing that all cubic V(III) complexes were to a greater or lesser extent distorted trigonally from the regular octahedral symmetry, due to the operation of the Jahn-Teller effect (s. Sect. 7), and they therefore analysed their results on the assumption that the effective symmetry about the V^{3+} ion was lowered to C_{3v} or D_3. In these symmetries the ^1E_g level is not split but remains as ^1E, the $^1T_{2g}$ state yields a $^1A_1$ and a further ^1E level, and the resulting first order interaction between the two ^1E states ensures a substantial separation between them. *Ballhausen* and *Winther* postulated the superposition of a trigonal ligand field component for the VF_6^{3-} ion such that the two ^1E levels would be split by about 1.5 *k*K., thereby explaining the relatively high position of the spin-forbidden band, but they were unable to locate the lower of the two ^1E levels, although this would certainly be expected to be weaker in intensity. Calculations using the trigonal d^2 matrices of *Pryce* and *Runciman* (*44*), or a more convenient strong field form (*45*) based on the t_{2g}^0, $t_{2g}^{\pm}$, $e_g^{\pm}$ set of *Ballhausen* (*4*), show that the trigonal splitting of the spin-allowed bands should be small, compared to that of the ^1E levels, but that the $^3T_{1g}$ ground state should undergo a splitting comparable to that of the ^1E_g, $^1T_{2g}$ states. It is therefore possible, from measurements of the temperature dependence of the magnetic moment, to assess the magnitude of

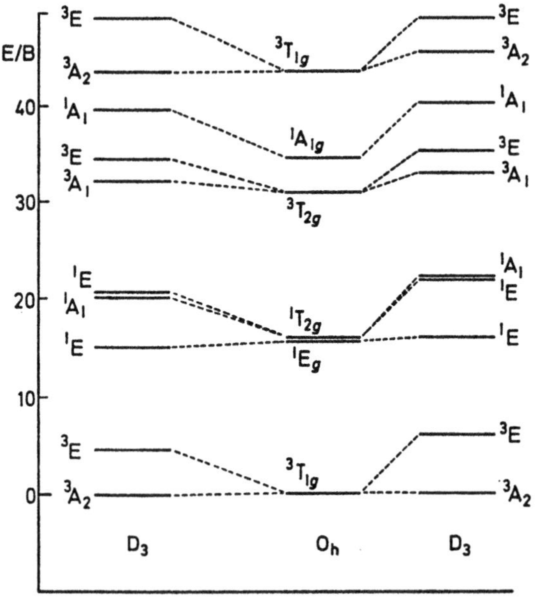

Fig. 3. Energy level diagram for a d^2 system in O_h and D_3 symmetry Left $K' = K$, right $K' = 0$ where $K = D\sigma + 3\,D\tau$ and $K' = D\sigma - (9/4)\,D\tau$

such a ground state splitting and to determine whether this is in agreement with the requirements of the spectroscopic assignment just described.

A full treatment of the magnetic behaviour of $^3T_{1g}$ ground states has been given by *Figgis et al.* (*46*) and shows that for O_h symmetry a magnetic moment close to the spin-only value would be expected at about room temperature, but that this should show a marked temperature dependence. Conversely, a temperature independent moment would correspond to a substantially distorted situation. In the event *Machin* and *Murray* (*42*) found that for $(NH_4)_3VF_6$ the moment (2.79 B.M. at room temperature) decreased quite sharply with temperature, that the $^3A_2$ component of the $^3T_{1g}$ state lay lowest, but that the trigonal splitting could not amount to more than about 150 cm.$^{-1}$. Consequently, the explanation advanced by *Ballhausen* and *Winther* (*41*) no longer appears tenable and the data are therefore assigned as in Table 2 (ii), with the parameters indicated.

Table 2 (ii). *Spectroscopic data for the hexafluorovanadate-(III) anion*

Band Position (*k*K).	Assignment
10.20	$^3T_{1g} \rightarrow {}^1E_g, {}^1T_{2g}$
14.80	$^3T_{1g} \rightarrow {}^3T_{2g}$
23.25	$^3T_{1g} \rightarrow {}^3T_{1g}$

$Dq = 1610$ cm.$^{-1}$, $B_{35} = 648$ cm.$^{-1}$, $B_{55} = 688$ cm.$^{-1}$.

Although their measurements extended from 5 to 50 *k*K. *Machin* and *Murray* (*42*) found no trace of the $^3T_{1g} \rightarrow {}^3A_{2g}$ transition, which would be calculated to appear at about 31 *k*K., but the band should in any case be rather weak since it corresponds to a two-electron excitation. No charge-transfer bands either were reported, but, as for Ti(III), the implication that these transitions lie at higher energies is consistent with the strongly reducing character of V(II), and with the observation (*38*) of such an excitation at 23.25 *k*K. in VBr_6^{3-}. For VCl_6^{3-} *Gruen* and *McBeth* (*47*) found the charge-transfer region to begin at about 22 *k*K. As for TiF_6^{3-} the absence of charge-transfer bands below 50 *k*K. is in accordance with the molecular orbital treatment of *Fenske et al.* (*26*), which also predicts considerable π-bonding contributions, but their procedure, although more sophisticated than the WH method, has also been criticised by *Dahl* and *Ballhausen* (*24*).

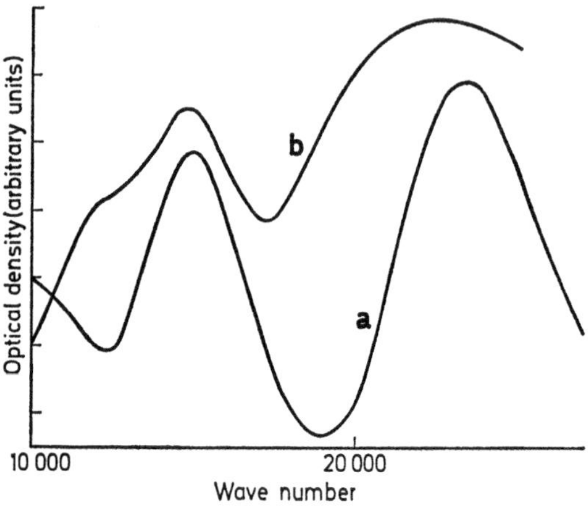

Fig. 4. Electronic spectrum of K_3VF_6. (A) K_3VF_6, (B) VF_3 (*42*)

(iii) Hexafluorochromate(III), CrF_6^{3-}

Anions of this type have been known since the last century (*48*), and the CrF_6^{3-} ion is fairly readily obtained, for example as the potassium salt (*49*), by the fusion of $K_2CrF_5.H_2O$ with KHF_2. It has been fairly extensively investigated structurally (*40*), the K_2Na salt being cubic, K_2NaAlF_6 type, the $(NH_4)_3$ and K_3 salts cubic, α-K_3FeF_6 type, (the latter with a low temperature tetragonal form), and the Na_3 salt monoclinic. For the potassium salt a magnetic moment of 3.8—3.9 B.M. at 298 °K is reported by *Peacock* (*50*), which is consistent with the essentially temperature independent spin-only value expected (*51*) for Cr^{3+}.

Several spectroscopic investigations of the CrF_6^{3-} anion have been carried out. A preliminary investigation of the single crystal spectrum of K_2NaCrF_6 was made by *Ferguson, Knox* and *Wood* (*52*), and in a later paper the more detailed assignments shown in Table 2 (iii) were presented (*53*).

The fitting parameters were derived assuming $C/B = 4$, and using only a single B value. In his textbook *Jørgensen* (*3*) quotes an unpublished study of the CrF_6^{3-} ion by *Schäffer*, the results of which are also shown in Table 2 (iii), from which values for Dq and both B_{35} and B_{55} were derived. *Jørgensen's* assignments differ somewhat from those of *Ferguson et al.* (*52, 53*), but are in good agreement with those of *Schläfer, Gausmann*, and *Zander* (*54*), derived from a diffuse reflectance and fluorescence study of $(NH_4)_3CrF_6$. The fitting parameters of these latter authors differ noticably from those of *Jørgensen* (*3*), but since these were obtained using only the diagonal matrix elements we have preferred to adopt *Jørgensen's* values, in whose derivation proper allowance was made for the off-diagonal terms.

Table 2 (iii). *Spectroscopic data for the hexafluorochromate(III) anion*

Band Position	Assignment[a]	Band Position	Assignment[b]	Band Position	Assignment[c]
15.0	$^4A_{2g} \rightarrow {}^2T_{1g}$	14.9	$^4A_{2g} \rightarrow {}^4T_{2g}$	15.06	$^4A_{2g} \rightarrow {}^4T_{2g}$
16.3	$^4A_{2g} \rightarrow {}^4T_{2g}$	15.7	$^4A_{2g} \rightarrow {}^2E_g$	15.67	$^4A_{2g} \rightarrow {}^2E_g$
22.0	$^4A_{2g} \rightarrow {}^2T_{2g}$	16.4	$^4A_{2g} \rightarrow {}^2T_{1g}$	22.78	$^4A_{2g} \rightarrow {}^4T_{1g}$
22.6	$^4A_{2g} \rightarrow {}^4T_{1g}$	22.0	$^4A_{2g} \rightarrow {}^2T_{2g}$	35.10	$^4A_{2g} \rightarrow {}^4T_{1g}$
36.0	$^4A_{2g} \rightarrow {}^4T_{1g}$	22.7	$^4A_{2g} \rightarrow {}^4T_{1g}$	$Dq = 1506$ cm.$^{-1}$,	
$Dq = 1610$ cm.$^{-1}$,		34.4	$^4A_{2g} \rightarrow {}^4T_{1g}$	$B_{35} = 643$ cm.$^{-1}$,	
$B = 760$ cm.$^{-1}$		$Dq = 1520$ cm.$^{-1}$,		$B_{55} = 746$ cm.$^{-1}$	
		$B_{35} = 820$ cm.$^{-1}$,			
		$B_{55} = 900$ cm.$^{-1}$			

a) Ref. (*52*). b) Ref. (*3*). c) Ref. (*54*).

Jørgensen's placing of the ^2E_g level *above* the $^4T_{2g}$ is moreover confirmed by the luminescence studies (*54*). Thus, *Schläfer, Gausmann* and *Witzke* (*55*) have put forward rules for the observation of fluorescence and phosphorescence, and have shown that this is closely connected with the magnitude of 10 Dq. This follows since when the ligand (e.g. F^-) lies sufficiently low in the spectrochemical series the $^4T_{2g}$ state (the lowest excited quartet level) may lie below the ^2E_g level, leading, correctly, to the prediction that only fluorescence should be observed. *Schläfer, Gausmann* and *Zander* (*54*) investigated a number of complexes partially substituted by ligands with larger Dq values and in these cases phosphorescence *was* detected, showing that the doublet level now lay *below* the lowest excited quartet state.

No spectroscopic measurements extending fully into the charge-transfer region have been reported, but the spectrum of *Ferguson et al.* (*52, 53*) shows that the higher intensity bands do not begin much below 50 kK., and the optical electronegativity value thus implied is once more consistent with the known reducing character of the divalent ion (Cr^{2+}). The result is again not in conflict with the M.O. predictions of *Fenske et al.* (*26*) (c.f. VF_6^{3-} and TiF_6^{3-}).

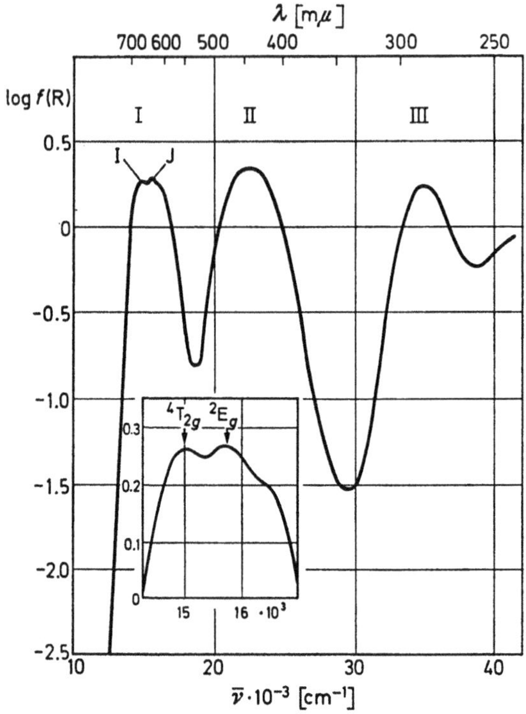

Fig. 5. Electronic spectrum of $(NH_4)_3CrF_6$ (*54*)

(iv) Hexafluoromanganate(III), MnF_6^{3-}

Potassium hexafluoromanganate(III) has been prepared by *Peacock* (*49*) by a method similar to that used for K_3CrF_6. Thus, fusion of the fairly readily obtained $K_2MnF_5.H_2O$ with KHF_2, as before, yields K_3MnF_6 as a purple-blue solid. With a d^4 electronic configuration the Mn^{3+} ion presents the possibility of either high-spin or low-spin behaviour, but the observed magnetic moment (*49*), clearly indicates the former to be the case: the experimental value of 4.95 B.M. at room temperature is in excellent agreement with the almost temperature independent spin-only moment required for a ^5E_g $(t_{2g}^3\, e_g)$ ground state. In fact, low-spin behaviour is rare for Mn(III) and only ligands lying very high in the spectrochemical series, such as CN^-, are known to give rise to a low-spin $^3T_{1g}$ ground level. However, a ^5E_g configuration, containing a single electron in the doubly degenerate e_g level, would be expected to show a substantial tetragonal Jahn-Teller distortion, (s. Sect. 7) and structural measurements support this belief. Thus, K_3MnF_6 has a tetragonal,

β-$(NH_4)_3ScF_6$ type lattice in which the doubling of the cell constants has been attributed to the Jahn-Teller effect, whilst K_2NaMnF_6 provides more definite evidence of this, having a tetragonally distorted, cubic K_2NaAlF_6 type lattice, in which the MnF_6 octahedra are elongated along the four-fold axis to give two long Mn-F bonds of 2.06 Å, and four short bonds of 1.86 Å.

Nevertheless, the interpretation of the electronic spectrum of the MnF_6^{3-} anion has occasioned very considerable discussion over the last decade or so. The earliest investigation was carried out by *Furlani* and *Ciana* (*56*) who ascribed the single band at 21.7 *k*K., in the spectrum of Mn^{3+} in HF, to the $^5E_g \rightarrow {}^5T_{2g}$ transition of the MnF_6^{3-} ion. This however gives $Dq = 2170$ cm.$^{-1}$, which is substantially higher than any of the other values reported for M(III) hexafluoro complexes (s. Sect. 4 (ii)), and also fails to take any account of possible Jahn-Teller effects. On the other hand it was suggested by *Dingle* (*57*) in 1962 that a Jahn-Teller splitting of the ^5E_g ground state could well account for the rather low (ca. 10 *k*K. or less) energy bands observed for a large number of formally octahedral Mn(III) complexes, and the same author (*58*) adduced further evidence for this view from a single crystal study of $(NH_4)_2MnF_5$. In this compound the symmetry of the coordination sphere of F^- ions about the Mn^{3+} cation is D_{4h}, and *Dingle* argued that, since its spectrum was very similar in shape to that found by him for MnF_6^{3-} in solution, it was reasonable to suppose that the MnF_6^{3-} ion was also tetragonally distorted. It is though always uncertain in solution whether the absorbing species is really MnF_6^{3-}, and more reliance can therefore be placed on reflectance data, such as those of *Hatfield* and *Parker* (*59*). These authors found a moderate intensity band at 18.0 *k*K., a weaker band at 9.0 *k*K., a very weak band at 29.0 *k*K., and strong charge-transfer transitions at about 40 *k*K. The 18 *k*K. peak was reported to be rather broad with a half-height width of 3.1 *k*K., and it was suggested that both this feature and the observation of a peak at 9 *k*K. might be due to Jahn-Teller splitting, a view also taken by *Oelkrug* (*60*) as a result of his study of a number of Cr(II) and Mn(III) complexes. *Oelkrug* carried out reflectance measurements on K_3MnF_6 between 2 and 35 *k*K. and his spectrum, expressed in F_R units, is essentially similar to that of *Hatfield* and *Parker*. The main feature of the spectrum is a very broad absorption with a peak at 17.4 *k*K. and a well marked shoulder at 19.6 *k*K., whilst a somewhat weaker and very broad band appears at lower energies with a maximum at 9.0 *k*K. Comparison with Cr(II) complexes in which the metal ion is octahedrally coordinated to F^- indicated an extremely similar shape for all the spectra and the data were analysed on the basis of a tetragonal distortion of the ^5E_g ground state, elongation, rather than compression, along the four-fold axis being assumed. The splitting of the ^5E_g and $^5T_{2g}$ levels resulting from the consequent descent in symmetry from O_h to D_{4h} is shown diagramatically in Fig. 6, and it is easily shown that, since the electron repulsion terms involved are the same for all the four levels, the relative energies of the resulting states are given by the expressions:

$$E\,({}^5B_{1g}) = -6\,Dq - 2\,Ds + Dt \qquad E\,({}^5A_{1g}) = -6\,Dq + 2\,Ds + 6\,Dt$$

$$E\,({}^5B_{2g}) = 4\,Dq - 2\,Ds + Dt \qquad E\,({}^5E_g) = 4\,Dq + Ds - 4\,Dt$$

where the ${}^5B_{1g}$ level lies lowest. If the bands are assigned as in Table 2 (iv) one readily obtains the parameters $Dq = 1740$ cm.$^{-1}$, $Ds = 1600$ cm.$^{-1}$, and $Dt = 520$ cm.$^{-1}$, where Ds and Dt are the usual tetragonal crystal field parameters as defined by *Ballhausen* (*4*). The spectrum is reproduced in Fig. 7.

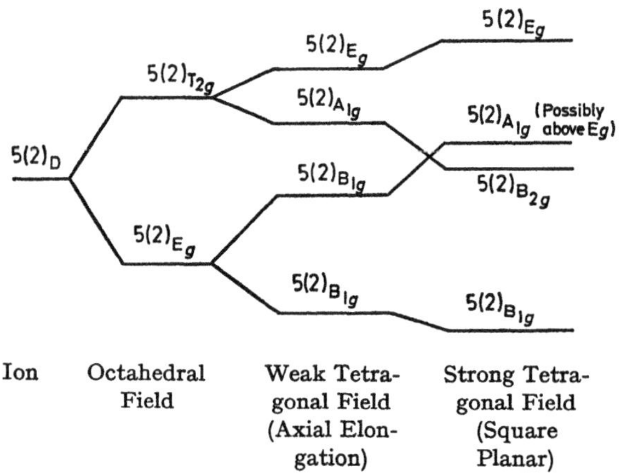

Fig. 6. Energy level diagram for a high-spin d^4 system in O_h and D_{4h} symmetry (*61*)

Table 2 (iv). *Spectroscopic data for potassium hexafluoromanganate(III)*

Band Position	Assignment
9.0	${}^5B_{1g} \rightarrow {}^5A_{1g}$
17.4	${}^5B_{1g} \rightarrow {}^5B_{2g}$
19.6	${}^5B_{1g} \rightarrow {}^5E_g$

A similar interpretation of the spectrum was also adopted by *Davis, Fackler* and *Weeks* (*61*), who carried out a systematic survey of a large number of Mn(III), and some Cr(II) and Cu(II), complexes with a view towards establishing definitely the origin of the low energy transitions observed in almost all Mn(III) compounds. Before reaching the conclusion that the band resulted from a ground state Jahn-Teller splitting, they considered three other possible explanations: (i) that it was due to a splitting of the upper $^5T_{2g}$ level, (ii) that it was due to a spin-forbidden excitation, and (iii) that it represented a charge-transfer transition. The first possibility was rejected on the grounds that the Jahn-Teller splitting of the $^5T_{2g}$ state should be much smaller than that of the ^5E_g level, since the former involves a t_{2g} orbital degeneracy (s. Sect. 6). Furthermore, even if this were the origin of the 9 *k*K. band, a still greater splitting of the ground (^5E_g) level would in any case be expected. The second possibility may be dismissed largely from an examination of the intensities of the low energy bands: thus, the absorptions are usually slightly weaker than those due to most *d–d* transitions, but are considerably more intense than is customary for spin-forbidden transitions in the 3*d* series. In addition, the intensity of the bands would, on this basis, be expected to decrease as D*q* increased, since the energy gap between the spin-forbidden transition and the quintet — quintet excitation, from which via spin-orbit mixing it derives its intensity, would correspondingly be greater. Since this is not the case, the possibility of the low energy band being a $^5E_g \rightarrow {}^3T_{1g}$ transition may also be eliminated. (It may also be noted that similar bands observed for certain Cu(II) complexes could not possibly be ascribed to spin-forbidden excitations.) Finally, we may consider the assignment of the 9 *k*K. band as a charge-transfer transition. If this were so, such a low energy charge-transfer excitation would seem to be specific for Mn(III) complexes alone, quite apart from the intensities, which are usually much too small to correspond to Laporte allowed transitions. On these grounds only such an assignment is most improbable, and the observation of *Davis et al.* that the position of the low energy band is relatively independent of the nature of the ligand affords convincing evidence for discarding it. Thus, *Davis et al.* report a similar band at 8.3 *k*K. for $MnCl_6^{3-}$ (c.f. 9.0 *k*K. for MnF_6^{3-}).

The origin of the very weak 29 *k*K. peak, observed by *Hatfield* and *Parker* is more obscure, but if real it may correspond to a spin-forbidden transition from the $^5B_{1g}$ ground state to one of the tetragonally split components of the ^3E_g, $^3T_{1g}$, $^3T_{2g}$ ($t_{2g}^3 e_g$) levels. Similarly, *Oelkrug's* (*60*) spectrum shows a very weak sharp inflection at about 24 *k*K. upon which the author does not remark, but which could be due to a component of the ^3E_g ($t_{2g}^3 e_g$) level. Such an assignment would suggest a B value of about 850 cm.$^{-1}$, making Mn(III) comparable in nephelauxetic behaviour to Cr(III) and Fe(III).

For MnF_6^{3-} the charge-transfer bands are reported to begin at a rather lower energy (∼40 *k*K.) than for the MF_6^{3-} ions of Ti, V and Cr, even though the M.O. calculations of *Fenske et al.* (*26*) and the χ_{opt} value, do not predict these bands to lie below about 45—50 *k*K. For $MnCl_6^{3-}$ *Hatfield et al.* (*62*) report the first charge-transfer band to lie at 22.4 *k*K.

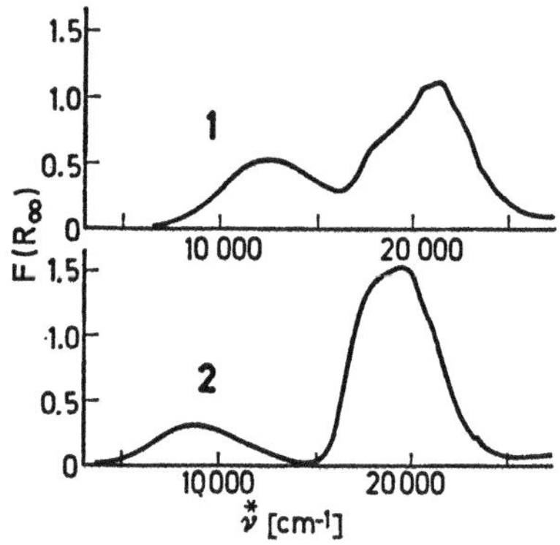

Fig. 7. Electronic spectrum of K_3MnF_6. (1) Na_2MnF_5, (2) K_3MnF_6 (*60*)

(v) Hexafluoroferrate(III), FeF_6^{3-}

Salts of the hexafluoroferrate(III) anion have been obtained in numerous ways. *Peacock* (*49*) obtained K_3FeF_6 by the fusion of $K_2FeF_5.nH_2O$ with KHF_2, and this compound was also prepared by *Klemm* and *Huss* (*63*) by direct fluorination of $KCl:FeCl_3$ mixtures. Similarly the K_2Na salt was prepared (*64*) from FeF_3 by the bifluoride fusion technique, and the Li, Na and NH_4 salts (*39*) by the action of HF on $MCl:FeCl_3$ mixtures. Structural investigations show that K_3FeF_6 possesses a low temperature, tetragonal β-$(NH_4)_3ScF_6$ type lattice as well as the more familiar high temperature, cubic α-K_3FeF_6 type form, and the same result is obtained for the ammonium salt. See, however, *Babel* (*40*) for a detailed discussion of the structural differences between K_3FeF_6 and $(NH_4)_3FeF_6$. As usual, the K_2Na salt has the cubic, K_2NaAlF_6 type lattice.

The d^5 configuration of the Fe^{3+} ion again permits either high- or low-spin behaviour, but, as for Mn(III), low-spin ground states are uncommon, and usually encountered only for cyano complexes. Thus, for FeF_6^{3-} the magnetic moments of all the salts studied lie very close to the spin-only value of 5.92 B.M. expected for a high-spin, $^6A_{1g}$, ground level, e.g. Li, 5.87; Na, 5.95 (*39*); K, 6.0 (*65*); NH_4, 5.80 B.M. (*39*), but despite the abundance of magnetic measurements spectroscopic data is very limited. However, the only generally available spectrum, given by *Jørgensen* (*3*) for the $(enH)_3$ salt (en = ethylenediamine), shows clearly that all the *d*–*d* bands correspond to weak, spin-forbidden, excitations, as required for a $^6A_{1g}$ ground state, and a study of K_3FeF_6, carried out in this laboratory led to the same conclusion (*66*). The general features of both spectra are essentially the same, and, although the bands are found at somewhat lower energies in the enH salt, the difference is mainly reflected in the value of Dq, B remaining almost the same. Both sets of data are collected in Table 2 (v).

The spectrum given by *Jørgensen* does not extend above about 32 kK., but for K_3FeF_6, although the absorption increases gradually above about 33 kK., there is no evidence of any well defined peak, corresponding to a charge-transfer band, below about 50 kK. This is in general agreement with M. O. predictions (*26*), and with the location of the first charge-transfer band in $FeCl_6^{3-}$ at 22.1 kK. (*62, 67*).

Table 2 (v). *Spectroscopic data for* $(enH)_3FeF_6$ *and* K_3FeF_6

$(enH)_3FeF_6$		K_3FeF_6	
Band Position	Assignment	Band Position	Assignment
14.2	$^6A_{1g}\rightarrow{}^4T_{1g}$	16.6	$^6A_{1g}\rightarrow{}^4T_{1g}$
19.7	$^6A_{1g}\rightarrow{}^4T_{2g}$	21.0	$^6A_{1g}\rightarrow{}^4T_{2g}$
25.4	$^6A_{1g}\rightarrow{}^4A_{1g}$, ^4E_g	26.6	$^6A_{1g}\rightarrow{}^4A_{1g}$, ^4E_g
28.8	$^6A_{1g}\rightarrow{}^4T_{2g}$	29.8	$^6A_{1g}\rightarrow{}^4T_{2g}$
30.2	$^6A_{1g}\rightarrow{}^4E_g$	32.0	$^6A_{1g}\rightarrow{}^4E_g$

Dq = 1400 cm.$^{-1}$, B = 845 cm.$^{-1}$ Dq = 1335 cm.$^{-1}$, B = 835 cm.$^{-1}$

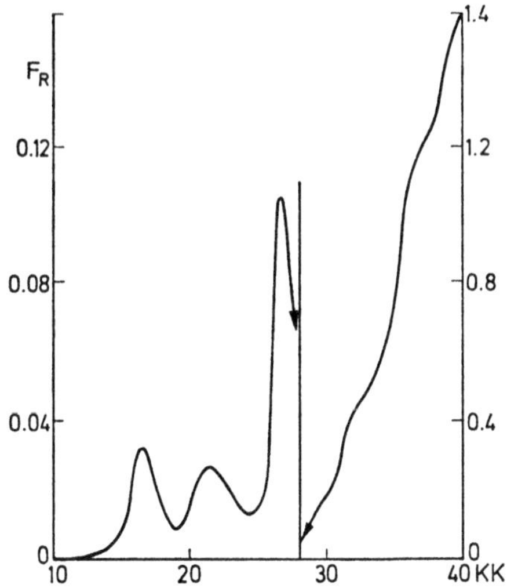

Fig. 8. Electronic spectrum of K_3FeF_6 (*66*)

The band at 25.4 *k*K. in $(enH)_3FeF_6$ is reported (*3*) to show vibrational fine structure with a spacing of the order of 200 $cm.^{-1}$. This separation is considerably too small to represent an α_{1g} vibrational progression (s. Sect. 6), but such structure is not detected for the corresponding band in the potassium salt (*66*).

(vi) Hexafluorocobaltate(III), CoF_6^{3-}

The CoF_6^{3-} ion has been obtained (*68*) as K_3CoF_6 by the action of HF on a mixture of CoF_3 and KF, and also (*69, 70*) by direct fluorination of a 3:1 mixture of KCl and $CoCl_2$, but the salts are probably best prepared by fluorination of the appropriate hexacyanocobaltate compound, according to the method of *Klemm, Brandt,* and *Hoppe* (*70*). In this way the Li, Na, K, Rb and Cs salts were obtained and reported all to have cubic, or nearly cubic, lattices, but *Cotton* and *Meyers* (*71*) have shown that, of the K_2Na, K_3, Na_3 and Li_3 salts, only the first is strictly cubic (K_2NaAlF_6 type lattice). The K_3 salt and the Na_3 salt show respectively small tetragonal and small monoclinic distortions, whilst thc Li_3 salt, and $Ba_3(CoF_6)_2$ are certainly not cubic, the latter being strongly distorted but the CoF_6^{3-} unit in the K_3 and Na_3 salts could well be strictly octahedral.

As in the cases of the d^4 and d^5 ions of Mn^{3+} and Fe^{3+}, the d^6 configuration of Co(III) may constitute either a high- or low-spin situation. Because of the relatively low crossover value of Dq/B, the overwhelming majority of Co(III) complexes are in fact low-spin, but the magnetic moments recorded by *Klemm et al.* (*70*), 5.1—5.3 B.M. at room temperature, clearly indicate high-spin behaviour. *Cotton* and *Meyers* report somewhat higher values, e.g. 5.65 B.M. between 73 and 299 °K for K_3CoF_6, but these are in excellent agreement with the predicted values for a $^5T_{2g}$ ($t_{2g}^4\, e_g^2$) ground state.

For a high-spin ground state only one spin-allowed *d—d* transition, $^5T_{2g} \rightarrow {}^5E_g$, might be expected, and the observation of a double peaked band, centred at about 13 *k*K., is difficult to accommodate. Moreover, the splitting is found to a greater or lesser extent for all the five salts studied (see Table 2 (vi)). Now the $^5T_{2g}$ ground state involves a t_{2g} orbital degeneracy and should not be appreciably split by Jahn-Teller effects, but, although the upper ^5E_g state corresponds to an e_g orbital degeneracy and should therefore show a considerable Jahn-Teller instability, it does not seem possible to postulate a static Jahn-Teller distortion in this system, since the strictly cubic K_2Na salt also shows a double peaked band, and infra-red vibrational data indicate that the CoF_6^{3-} unit is strictly O_h. It was therefore proposed (*71*) that the splitting represented a dynamic Jahn-Teller effect (s. Sect. 7), in contrast to the behaviour of the ^5E_g ground state in MnF_6^{3-}, where the K_2Na salt showed a significant static distortion. It seems likely that for the other salts, especially the Ba derivative, some measure of static distortion is also operative, thus accounting for the greater splitting of the *d–d* band, and this conclusion is supported by the broadening of the $\nu_3(\tau_{1u})$ and $\nu_4(\tau_{1u})$ infra-red active vibrational bands, for which details were also given (*72*).

The spectra of *Cotton* and *Meyers* (*71*) do not extend above 20 *k*K. and it was not therefore possible to check the M.O. predictions of *Fenske et al.* (*26*), which indicated that in CoF_6^{3-} the charge-transfer transitions should lie at somewhat lower energies than for the corresponding ions of Ti, V, Cr, Mn and Fe. In a species identified as $CoCl_6^{3-}$ a charge-transfer band is reported to lie at 32.5 *k*K., but the intensities ascribed to the putative *d–d* bands are so high as to cast doubt on the whole assignment. However, the spectra of *Cotton* and *Meyers* do reveal a weak, rather narrow band, just above the region corresponding to the $^5T_{2g} \rightarrow {}^5E_g$ transition, upon which the authors make no comment. The peak does though appear at 19.3 *k*K. in both the K_2Na and the K_3 salts, and at slightly lower energy in the Li_3 compound, and from its intensity and shape should represent a spin-forbidden excitation. Its identification can clearly only be tentative but the following rationalisation suggests itself.

Since the CoF_6^{3-} ion is virtually the only complex in which Co(III) shows high-spin behaviour, it is reasonable to assume that for this complex the value of Dq/B lies not too far below the high-spin—low-spin crossover point. If we take the average of the split components of the $^5T_{2g} \rightarrow {}^5E_g$ band as 10 Dq, giving $Dq = 1310$ cm.$^{-1}$, this is somewhat lower than predicted by *Jørgensen's* (*3*) *g* and *k* parameters (s. Sect. 4 (i)), (1710 cm.$^{-1}$), but the corresponding prediction of $\beta = 0.72$ ($B = 787$, taking B_{gas} as 1093 cm.$^{-1}$), together with the observed Dq value, yields $Dq/B = 1.66$, i.e. tolerably close to the crossover point which occurs at 2.0 for $C/B = 4.8$. On this assignment the weak peak at 19.3 *k*K.

Table 2 (vi). *Spectroscopic data for the hexafluorocobaltate(III) anion*

Salt	Band Positions	
	$^5T_{2g} \rightarrow {}^5E_g$	$^5T_{2g} \rightarrow {}^3T_{1g}, {}^3T_{2g}$ (?)
K_2NaCoF_6	11.8, 14.4	19.2
K_3CoF_6	11.4, 14.5	19.2
Na_3CoF_6	11.8, 14.4	—
Li_3CoF_6	11.2, 14.6	18.9
$Ba_3(CoF_6)_2$	10.2, 16.7	—

coincides almost exactly with the predicted position of the $^5T_{2g} \rightarrow {}^3T_{1g}$, $^3T_{2g}$ $(t_{2g}^4 e_g^2)$ transition, its relative narrowness being in accordance with an intra-configurational excitation, bearing in mind the juxtaposition of the two triplet levels.

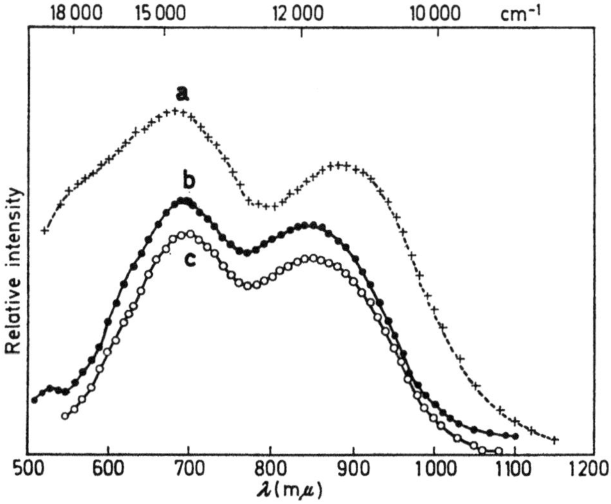

Fig. 9. Electronic spectrum of Li_3CoF_6 (A), K_2NaCoF_6 (B), and Na_3CoF_6 (C) *(71)*

(vii) Hexafluoronickelate(III), NiF_6^{3-}

Complexes containing Ni(III) are very much rarer than those of this oxidation state for the previous members of the first transition series, and the NiF_6^{3-} ion constitutes one of the few well authenticated examples. The potassium salt, K_3NiF_6, was first prepared by *Klemm, Brandt* and *Hoppe* *(70)*, by fluorination of a 3:1 mixture of KCl and $NiSO_4$, an impure sodium salt also being obtained, and was shown to have a cubic, α-K_3FeF_6 type lattice. For the d^7 configuration of Ni(III) there is, once more, the possibility of high- or low-spin ground states, but the magnetic moment *(69, 70)* of K_3NiF_6, rising from 2.12 B.M. at 90 °K to 2.54 B.M. at 294 °K, clearly shows the latter to be the case, although the temperature dependence of the moment is rather large for a ^2E_g $(t_{2g}^6 e_g)$ ground level. In $Co(NO_2)_6^{4-}$ for example it amounts only to about 0.1 B.M. over a similar range *(73)*.

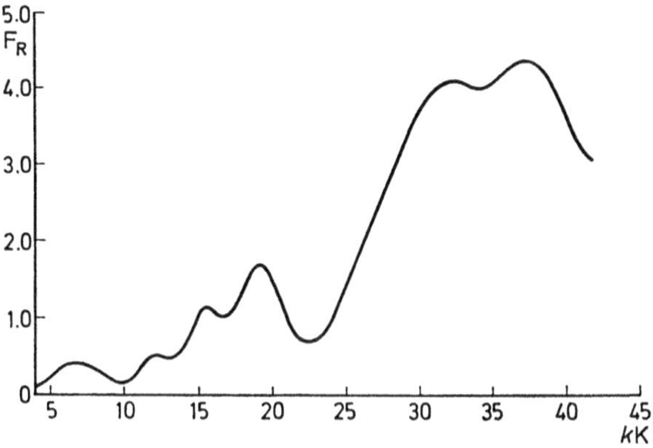

Fig. 10. Electronic spectrum of K_3NiF_6 (*74*)

The only complete treatment of the electronic spectrum of this ion has been given by *Allen* and *Warren* (*74*), but the interpretation of this, together with the magnetic data, is far from straightforward. The spectrum shows the presence of at least four d–d transitions, of comparable intensities, and thus rules out the possibility of a high-spin, $^4T_{1g}$, ground state, for which only two spin-allowed bands are permitted which correspond to one-electron excitations. For an E_g ground state though there is no orbital contribution to the magnetic moment which should therefore normally be essentially temperature independent, in contrast to the observed behaviour. Even at 90 °K the magnetic moment is significantly above the spin-only value, and it is not unreasonable therefore to postulate a thermal equilibrium between the low- and high-spin configurations. However, the results of *Barraclough* (*75*) show that this assumption is not strictly necessary: when spin-orbit coupling is included the ^2E_g ground state becomes Γ_8 in the double group nomenclature, indicating that the magnetic behaviour will be complicated by the spin-orbit matrix elements which connect Γ_8 (^2E_g) with the two Γ_8 components of $^4T_{1g}$. In fact, the observed temperature dependence of the magnetic moment can be reproduced on the assumption that Dq/B lies very close to the crossover point, on the low-spin side. *Allen* and *Warren* (*74*), following *Tanabe* and *Sugano* (*9*), took $C/B = 4.90$ for Ni^{3+}, for which the crossover point is at $Dq/B = 2.26$, and on this basis the magnetic results seem to require Dq/B to lie somewhere in the range 2.30–2.50.

Nevertheless, it was not possible to obtain a satisfactory fit of the d–d bands using the Tanabe-Sugano matrices for d^7 O_h systems. The spectroscopic data for K_3NiF_6 are given in Table 2 (vii), and in Fig. 10, and the band at 6.8 *k*K. is too strong to be assigned as a spin-forbidden $^2E_g \rightarrow {}^4T_{1g}$ transition, which would, moreover, also be inconsistent with the magnetic behaviour. Furthermore, the assignment of the 6.8 *k*K. band as a $^2E_g \rightarrow {}^2T_{1g}$ transition also fails satisfactorily to predict the positions of the other *d—d* transitions, and it was therefore postulated (*74*) that the observed spectrum arose from a tetragonal Jahn-Teller distortion of the ^2E_g ground state. This ground state has the $t_{2g}^6\, e_g$ configuration and the e_g orbital degeneracy should therefore (s. Sect. 7) give rise to a substantial Jahn-Teller instability. (The inclusion of spin-orbit coupling does not affect this conclusion since the Γ_8 ground state is not a Kramers' doublet). The spectrum was analysed using the strong-field repulsion matrices for d^7 systems in D_{4h} symmetry, setting Ds/Dt = 3.0 (c.f. *Davis et al.* (*61*) for MnF_6^{3-}), with Dq/B = 2.30, i.e. just above the $^2E_g \rightarrow {}^4T_{1g}$ crossover point, and the assignments and fitting parameters obtained on this basis are shown in Table 2 (vii). It was found that the most consistent set of parameters was obtained on the assumption of elongation, rather than compression, along the four-fold axis, and it was also noted that the most marked effect of the superimposed

Table 2 (vii). *Spectroscopic data for potassium hexafluoronickelate(III)*

Band Position Observed	Calculated[a]	Assignment[b]
6.8[c]	6.7	$^2A_1 \rightarrow {}^2B_1$
12.3[c]	11.7	$^2A_1 \rightarrow {}^2E$
	14.0	$^2A_1 \rightarrow {}^2A_2$
15.7	16.1	$^2A_1 \rightarrow {}^2E$
	16.8	$^2A_1 \rightarrow {}^2B_2$
19.2[d]	18.6	$^2A_1 \rightarrow {}^2A_2$
	19.4	$^2A_1 \rightarrow {}^2E$
	26.1	$^2A_1 \rightarrow {}^2E$
	26.2	$^2A_1 \rightarrow {}^2B_2$
32.0	—	$\pi \rightarrow e_g$
37.0	—	$\pi \rightarrow e_g$

a) Dq = 1620 cm.$^{-1}$, B = 703 cm.$^{-1}$, Ds/Dt = 3.0, Dt = 423 cm.$^{-1}$.
b) D_4 symmetry for *d—d* bands, O_h symmetry for charge-transfer bands.
c) Found at 6.1 and 12.8 *k*K. respectively in Na_3NiF_6.
d) Also found in HF solution: Ref. (*29*). Reflectance spectrum shows weak vibrational detail at 77 °K.

tetragonal field was in the very substantial splitting of the ground state ^2E_g level into $^2A_1$ and $^2B_1$, the former lying lower (Fig. 11). The lowest $^2T_{1g}$ level is split into $^2A_2$ and ^2E, the latter being considerably depressed, but apart from this the splittings of the excited doublet levels are relatively small compared to that of the ground state. These results are consistent with the observation of the greatest differences between the K_3 and Na_3 salts for the two lowest energy transitions, but in the absence of more detailed structural information (the Na salt is reported to be monoclinic) it is not possible to rationalise the effect further, nor indeed to establish whether the proposed Jahn-Teller effect is static or dynamic in origin.

However, although a fairly satisfactory set of fitting parameters was obtained by assuming a tetragonal Jahn-Teller distortion of the ^2E_g ground state, an alternative interpretation of the spectrum has recently been put forward by *Nelson* (*76*). It is suggested that K_3NiF_6 behaves as though it contained not merely NiF_6^{3-} moieties, but NiF_6^{2-} and NiF_6^{4-} moieties, in equal amounts, as well. In fact, some preparations do contain significant amounts of the latter in separate phases, as K_2NiF_6 and $K_2NiF_4 + 2$ KF respectively, although it is possible to eliminate these (c.f. *Stein, Neil* and *Alms* (*29*)). It may then be argued that the magnetic properties may be interpreted in terms of a mixture of Ni^{2+} plus low-spin Ni^{4+} (in equal amounts), together with *high*-spin Ni^{3+}, thereby removing the apparent anomaly of CoF_6^{3-} being high-spin and NiF_6^{3-} low-spin. In this way the 19.2 *k*K. band of K_3NiF_6, with slight fine-structure at 77 °K, would correspond to the 19.0 *k*K band of K_2NiF_6, and the 6.8 and 12.3 *k*K. bands to the 7.4 and 12.7 *k*K. absorptions (*77*) of $KNiF_3$, thereby leaving the 15.7 *k*K. band to represent the $^4T_{1g} \rightarrow {}^4T_{2g}$ transition of high-spin NiF_6^{3-}.

There are though a number of objections to this rationalisation which compel us to prefer the original interpretation (*74*). In the first place the structural evidence seems to be quite adequate (*40, 78*) to show that K_3NiF_6 has the cubic, K_3FeF_6 lattice, even though the distinction between this and the $(NH_4)_3FeF_6$ type lattice is not completely clear, and if the suggested moieties were in fact present it would be expected to lead to a doubling of the characteristic lines in the powder photographs. Also, as shown above, the magnetic behaviour can quite readily be explained without recourse to any such hypothesis, and the assignment of certain spectral absorptions to $KNiF_3$ leads to difficulties. Thus, $KNiF_3$ has a spin-allowed transition (*77*) at 23.7 *k*K., but this corresponds to a minimum in the K_3NiF_6 spectrum. Furthermore, the $^4T_{1g} \rightarrow {}^4T_{2g}$ transition has, in the first order, an energy of $10\,Dq - 3\,B$, and it is not possible to find reasonable values of D*q* and B, which will both reproduce the energy of 15.7 *k*K. and *also* leave D*q*/B on the high-spin side of the crossover point. It is also worth noting that there is no particular anomaly involved in CoF_6^{3-} being high-spin and NiF_6^{3-} low-spin: the value of the Dq/B crossover point varies appreciably from one d^n configuration to another, and the evidence suggests that both CoF_6^{3-} and NiF_6^{3-} lie close to their respective crossover values. In fact, very small changes in Dq/B can produce quite profound spectroscopic and magnetic consequences without being of themselves of any fundamental significance.

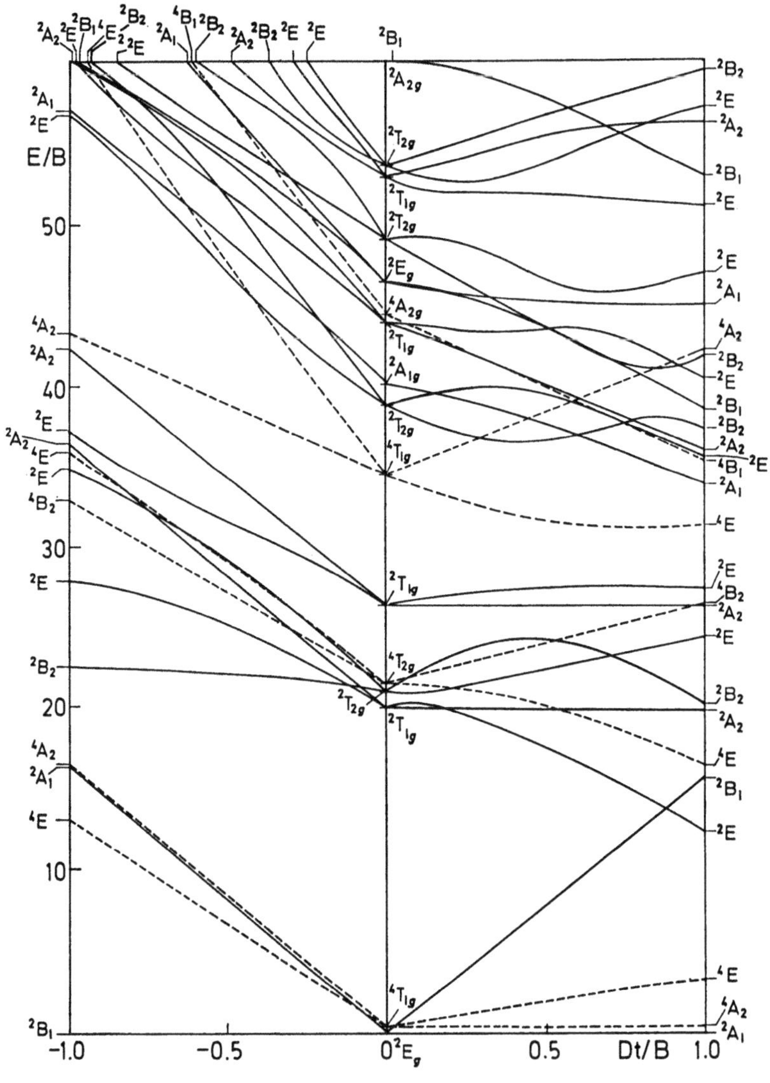

Fig. 11. Energy level diagram for a low-spin d^7 system in O_h and D_4 symmetry (*74*)

Our preference for the explanation in terms of the Jahn-Teller effect is also supported by the characteristics of the 6.8 *k*K. band: it appears (s. Sect. 7) that a band of this nature, rather broad and slightly weaker than most *d*—*d* transitions, is typical for transitions between the split components of a Jahn-Teller unstable ground state. Finally, evidence reinforcing our conclusions has recently been communicated to us by *Dove* (*79*), who, like *Stein et al.* (*29*), experienced some difficulty in obtaining a homogeneous K_3NiF_6 system, free from discrete Ni(II) and Ni(IV) phases. However, his purest samples also give X-ray powder data indicating only one phase, and that with a K_3FeF_6 type lattice, and a ν_3 (τ_{1u}) vibration at 580 cm.$^{-1}$, which he considers too high to be attributed to a high-spin ion. He also (c.f. *Stein et al.*) reports the ν_3 band to be unusually broad, which is consistent with a Jahn-Teller active ^2E_g ground state, and has shown that at 77 °K K_3NiF_6 exhibits a strong e.s.r. resonance ($g \sim 2.26$), diminishing in intensity at lower temperature, as would be expected if the $^2A_1$ component of ^2E_g did indeed constitute the ground level.

For NiF_6^{3-} the $\pi \rightarrow e_g$ charge-transfer transitions are found at 32.0 and 37.0 *k*K. Their location at such relatively low energies is not unexpected in view of the instability of the Ni(III) state, and the great tendency of nickel to attain the Ni(II) condition.

(viii) Hexafluorocuprate(III), CuF_6^{3-}

As for Ni(III) the trivalent state is of very rare occurrence, and the hexafluorocuprates afford one of the very few well established examples of Cu(III). The CuF_6^{3-} anion was first obtained as K_3CuF_6 by *Klemm* and *Huss* (*63*) by fluorination of a 3:1 mixture of KCl and Cu_2Cl_2, and this salt was shown to have a magnetic moment of 2.8 B.M. at room temperature, close to the spin-only value. More recently *Hoppe* and *Homann* (*80*) have prepared the Cs_2K and Cs_2Rb salts, which were shown to have a cubic, K_2NaAlF_6 type lattice, whilst K_3CuF_6 has the cubic, α-K_3FeF_6 type structure.

The only spectroscopic study reported is that of *Allen* and *Warren* (*74*), but unlike the NiF_6^{3-} anion the interpretation of the spectrum presents few problems. The results are shown in Fig. 12 and in Table 2 (viii), together with the assignments and the D*q* and B fitting parameters. The Cu^{3+} ion possesses a d^8 electronic configuration, and the two moderately strong bands at 14.1 and 20.4 *k*K. are readily assigned as the $^3A_{2g}$ $(t_{2g}^6\, e_g^2) \rightarrow {}^3T_{2g}$, and $^3T_{1g}$ $(t_{2g}^5\, e_g^3)$ transitions respectively, whilst the weak peak at 9.6 and the faint shoulder at 16.4 *k*K. correspond to the intra-subshell $(t_{2g}^6\, e_g^2)$ spin-forbidden transitions, $^3A_{2g} \rightarrow {}^1E_g$, and $^1A_{1g}$. From the positions of the spin-allowed bands D*q* and B_{35} are obtained, whilst the spin-forbidden transitions yield B_{33}, which is, as expected, somewhat smaller than B_{35}.

Allen and *Warren* (*74*) also measured the electronic spectrum of the AgF_6^{3-} anion (s. Sect. 4 (iii)), and for both this ion and for CuF_6^{3-}, the spectra were also fitted using the complete spin-orbit matrices for d^8 O_h systems, as well as the simple Tanabe-Sugano formulation. From the eigenvectors resulting from the diagonalisation of these matrices it was

Table 2 (viii). *Spectroscopic data for caesium potassium hexafluorocuprate(III)*

Band Position	Assignment
9.6	$^3A_{2g} \rightarrow ^1E_g$
14.1	$^3A_{2g} \rightarrow ^3T_{2g}$
16.4	$^3A_{2g} \rightarrow ^1A_{1g}$
20.4	$^3A_{2g} \rightarrow ^3T_{1g}$
29.5	$\pi \rightarrow e_g$
37.2	$\pi \rightarrow e_g$

$Dq = 1410$ cm.$^{-1}$, $B_{35} = 641$ cm.$^{-1}$, $B_{33} = 536$ cm.$^{-1}$

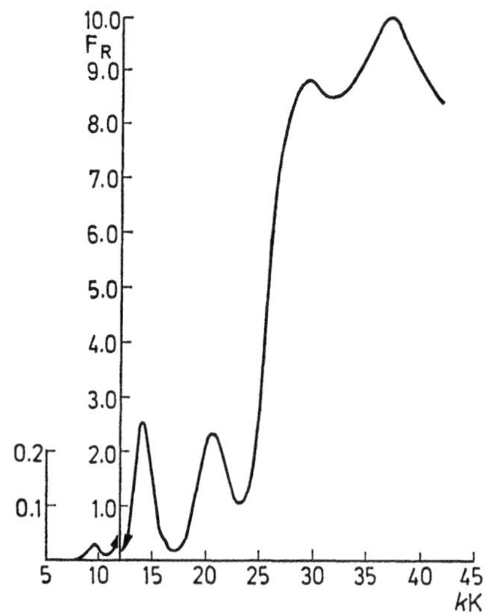

Fig. 12. Electronic spectrum of Cs_2KCuF_6 (*74*)

possible to calculate the proportions of the Γ_1 and Γ_3 components of the triplet states admixed into the ^1E_g and $^1A_{1g}$ levels, and to show that these were in qualitative agreement with the observed singlet: triplet intensity ratios. It may also be noted that the band at 9.6 *k*K. in CuF_6^{3-}, assigned as $^3A_{2g} \rightarrow {}^1E_g$, is relatively narrow, as would be expected for a spin-forbidden intra-subshell excitation.

For Cs_2KCuF_6 the charge-transfer transitions are found at 29.5 and 37.2 *k*K., as compared with 32.0 and 37.0 in K_3NiF_6. The lower position of the first band in the Cu(III) complex is in accordance with the decreasing stability of the 3+ oxidation state, as reflected by the general fall off in β, the nephelauxetic ratio (s. Sect. 4), towards the end of the first transition series, and is discussed further in Sections 4 and 5.

3. Electronic Spectra of Hexafluorometallate(IV) Complexes

For the tetravalent cations of the first transition series only five MF_6^{2-} anions are known. These are the complexes VF_6^{2-}, CrF_6^{2-}, MnF_6^{2-}, CoF_6^{2-}, and NiF_6^{2-}, corresponding to the configurations d^1, d^2, d^3, d^5, and d^6. With the solitary exception of MnF_6^{2-}, which has been known since 1899 (*87*), all these salts have been prepared only within the last twenty years or so, by direct fluorination techniques, and their extreme reactivity has severely hampered spectroscopic studies. Recently however, some spectroscopic data has become available, mostly by reflectance techniques, for all of these anions, but repeated attempts, in the authors' laboratory and elsewhere, to prepare the elusive d^4 FeF_6^{2-} ion have been unsuccessful, possibly because it may simply be thermodynamically unstable with respect to the high-spin d^5 FeF_6^{3-} ion.

In contrast to the MF_6^{3-} series there is abundant evidence that the hexafluoro complexes of Co and Ni, CoF_6^{2-} and NiF_6^{2-}, are both low-spin. This seems to be due to a combination of two factors: in the first place a substantially greater value of D*q* (ca. 2000—2200 cm.$^{-1}$) is found for the tetravalent cations than for the M(III) series, and secondly the much greater degree of covalency which is found for the metal-fluorine linkages, as evidenced by the nephelauxetic ratios observed. These latter are substantially smaller than the corresponding values for the MF_6^{3-} ions (s. Sect. 4) and thus lead to D*q*/B values which lie on the strong field (low-spin) side of the crossover point. As for the trivalent series the trends in the observed β values are fully discussed elsewhere (Section 4).

Unlike the MF_6^{3-} series there is very little indication of any significant Jahn-Teller effects for the M(IV) complexes. For the d^6 NiF_6^{2-} system of course the change in the spin multiplicity of the ground state, as compared with CoF_6^{3-}, leads to a $^1A_{1g}$ level, obviating such complications, whilst the hypothetical d^7 ion CuF_6^{2-} is unknown. The effects were not in any case large in the d^1 and d^2 complexes of the trivalent series, and there is no clear evidence of them in the VF_6^{2-} and CrF_6^{2-} ions.

The MF_6^{2-} ions do however differ markedly from the MF_6^{3-} series in the manifestation of vibrational fine structure. This phenomenon is not commonly observed in absorption or by reflectance, and certainly not in the first transition series, and for the trivalent anions the only, very marginal, indication of it is found for NiF_6^{3-} at liquid nitrogen temperatures. Substantial structure is though found for the MF_6^{2-} ions of Mn, Co, and Ni, and this is discussed under the appropriate anion heading, and further in Section 6.

With the exception of the d^1 VF_6^{2-} ion, for which only D*q* can be derived, both D*q* and B can be estimated for the anions listed, and these parameters are given in the relevant Tables. As before (Section 2) detailed discussion of the charge-transfer bands is reserved until Section 5.

(i) Hexafluorovanadate(IV), VF_6^{2-}

In their early work on this ion *Klemm* and *Huss* considered that the evidence for its formation was inconclusive, as far as the direct fluorination of K_2VF_5 at about 200° was concerned. More recently however *Liebe, Weise,* and *Klemm* (*82*) have succeeded in obtaining the K, Rb, and Cs salts, and these are best prepared by heating the appropriate M_2VF_5 compound with 1% fluorine in CO_2 for about 15 minutes at 500°. Depending on the temperature of preparation different crystalline modifications may be obtained, and both the K and Rb salts give rise to a high temperature hexagonal, K_2MnF_6-type, form, and to a lower temperature trigonal, K_2GeF_6-type structure. For the Cs compound a cubic, K_2PtCl_6-type modification is stable at higher temperatures, in addition to the other two forms. The trigonal form of K_2VF_6 has also been prepared by *Cavell* and *H. C. Clark* (*83*) by the interaction of KF and VF_4 in SeF_4.

Contrary to the implication of *R. J. H. Clark* (*84*), the K, Rb, and Cs salts were all found by *Liebe et al.* (*82*) to have magnetic moments significantly above the spin-only value at 295 °K, and for the three compounds μ was essentially constant between 90 and 295 °K, the average values being K_2VF_6, 1.99; Rb_2VF_6, 1.95; and Cs_2VF_6, 2.03 B.M. Very similar behaviour was found by *Cavel* and *H. C. Clark* for K_2VF_6 — the recorded value being 2.05 B.M. at 295 °K — and the high value of the moment and the large θ values suggest that antiferromagnetic interactions may be operating.

Spectroscopic information concerning the VF_6^{2-} ion is scanty: a spectrum of K_2VF_6, obtained from a KBr disc, has been published by

Bedon, Hatfield, Horner, and *Tyree* (*30*), but the single *d–d* band, ${}^2T_{2g} \rightarrow {}^2E_g$, which is estimated to lie at 20.1 *k*K., is largely obscured by a steadily rising background absorption extending into the charge-transfer region. However, a reexamination of the spectrum of K_2VF_6 has recently been carried out by *Allen* and *El-Sharkarwy* (*85*), who effectively confirmed this finding (see Table 3 (i)), but their spectrum is very much better resolved, and shows a clear separation of the *d–d* band from the charge-transfer peaks (see Fig. 13.). The *d–d* transition shows as a rather broad peak, but the resolution is still not good enough to detect any indication of a Jahn-Teller splitting of the upper ${}^2E_g$ state, such as occurs for the iso-electronic TiF_6^{3-} ion.

The charge-transfer bands are found at 34.0 and 37.4 *k*K., and their relatively low energy reflects the comparative instability of V(IV) with respect to the V(III) state. The molecular orbital calculations of *Bedon*

Table 3 (i). *Spectroscopic data for potassium hexafluorovanadate(IV)*

Band Position	Assignment
20.25	${}^2T_{2g} \rightarrow {}^2E_g$
34.0	$\pi \rightarrow t_{2g}$
37.4	$\pi \rightarrow t_{2g}$

$Dq = 2025$ cm.$^{-1}$.

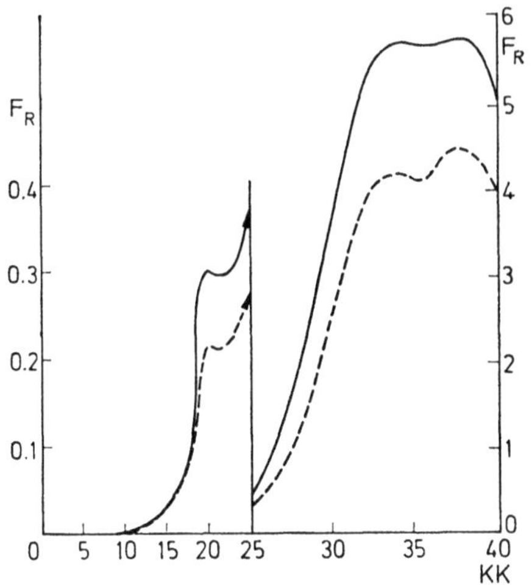

Fig. 13. Electronic spectrum of K_2VF_6 (*85*)

et al. (*30*) suggest that the first charge-transfer band in VF_6^{2-} should lie at around 41 *k*K., as compared with ca. 50 *k*K. for VF_6^{3-}, and the agreement with experimental is very reasonable for a one-electron Wolfsberg-Helmholz type treatment.

The VCl_6^{2-} anion has also been studied by a number of workers, (*86, 87, 88, 89, 90*), and shows charge-transfer bands at considerably lower energies, as would be expected. In acetonitrile solution the first charge-transfer band appears (*86, 87, 90*) at 21.3 *k*K., or thereabouts, and is located at 20.4 *k*K. in the diffuse reflectance spectrum (*86*).

No vibrational fine structure could be detected for the $^2T_{2g} \rightarrow {}^2E_g$ transition of VF_6^{2-}, even at 77 °K. (compare with CrF_6^{2-} and contrast with MnF_6^{2-}, CoF_6^{2-}, and NiF_6^{2-}).

(ii) Hexafluorochromate(IV), CrF_6^{2-}

The hexafluorochromate(IV) anion was first obtained by *Huss* and *Klemm* (*91*) as the potassium salt, K_2CrF_6, by the fluorination of a 3:1 mixture of KCl and $CrCl_3$. It was shown (*92*) to have a hexagonal, K_2MnF_6-type lattice, but *Bode* and *Voss* demonstrated that both K_2CrF_6 and Rb_2CrF_6, although hexagonal as normally prepared, go over to a cubic modification when maintained at about 450° over a period of time. The caesium salt though is always cubic (K_2PtCl_6 type); and only a hexagonal form is known for the Na salt.

For K_2CrF_6 a magnetic moment of 2.8 B.M. at room temperature is reported by *Klemm* (*93*), as compared with a figure of about 2.1 B.M. which would be expected (*51*), assuming a spin-orbit coupling constant of 325 cm.$^{-1}$. It is possible to account for this higher value on the basis of a trigonal Jahn-Teller distortion of the $^3T_{1g}$, d^2 ground state, with a $^3A_2$ level lying lowest, (c.f. *Figgis et al.* (*46*) for a general treatment of $^3T_{1g}$ ground states) but the results of *Murray* and *Machin* (*42*) for the iso-electronic VF_6^{3-} ion (q.v.) (Section 2 (ii)) suggest that only small distortion effects are likely to be found for monodentate ligands.

However, the CrF_6^{2-} ion has also been obtained by *Clark* and *Sadana* (*43*) by the interaction of BrF_3 on the appropriate 2:1 MCl/CrF_4 mixture, but K_2CrF_6 is isolated first as the $\frac{1}{2}$ BrF_3 solvate, which is tetragonal, and on heating loses BrF_3 to give the cubic modification, whilst for caesium both Cs_2CrF_6 and the solvate have the same sized cubic lattice. At 294 °K Cs_2CrF_6 has $\mu = 3.14$ B.M., almost the highest theoretically possible moment for a strong field $^3T_{1g}$ system, but $K_2CrF_6 . \frac{1}{2}$ BrF_3 has a moment of 3.54 B.M. between 123 and 294 °K. These results, together with the high θ values, strongly suggest that antiferromagnetic effects are intervening, and that these are enhanced by the lattice distortion accompanying the uptake of the $\frac{1}{2}$ BrF_3 moiety into the K_2CrF_6 structure.

The only spectroscopic study of the CrF_6^{2-} ion so far reported is a preliminary account by *Allen* and *El-Sharkarwy* (*85*), the results of which are shown in Fig. 14 and in Table 3 (ii). The salts studied were the Rb and Cs compounds, but it is difficult in their preparation to avoid loss of chromium, as CrF_5, and to ensure that all trace of Cr(III) is removed, even though a satisfactory result for the percentage of Cr present as Cr(IV) was obtained. Consequently the peak at 14.8 *k*K is attributed to analytically insignificant amounts of Cr(III), probably as CrF_6^{3-}, rather than to the higher of two ^1E states produced by trigonal

Table 3 (ii). *Spectroscopic data for rubidium hexafluorochromate(IV)*

Band Position	Assignment
11.4	$^3T_{1g}\rightarrow{}^1E_g$, $^1T_{2g}$
14.8	Cr(III), $^4A_{2g}\rightarrow{}^4T_{2g}$
20.2	$^3T_{1g}\rightarrow{}^3T_{2g}$
27.5	$^3T_{1g}\rightarrow{}^3T_{1g}$
29.7	$\pi\rightarrow t_{2g}$
37.3	$\pi\rightarrow t_{2g}$

$Dq = 2200$ cm.$^{-1}$, $B_{35} = 575$ cm.$^{-1}$, $B_{55} = 760$ cm.$^{-1}$.

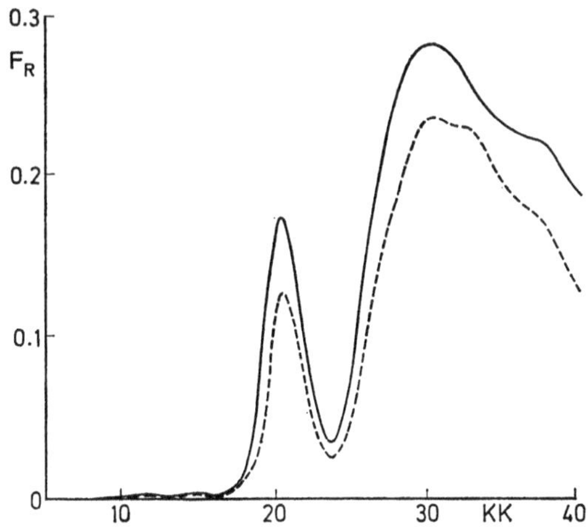

Fig. 14. Electronic spectrum of Rb_2CrF_6 (*85*)

distortion in a manner resembling that once postulated for the iso-electronic VF_6^{3-} ion. The magnitude of such distortion would require a substantial splitting of the $^3T_{1g}$ ground state, and although this could account for some increase in the magnetic moment the results of *Machin* and *Murray* (*42*) for VF_6^{3-} (q.v.) render this explanation improbable. The second spin-allowed *d–d* band is largely obscured by the onset of the charge-transfer transitions, but is tentatively assigned as the shoulder which appears at 27.5 *k*K. for Rb_2CrF_6 at 77 °K. This assignment, together with that of the peak at 20.2 *k*K. as $^3T_{1g} \rightarrow {}^3T_{2g}$, yields $Dq = 2200$ cm.$^{-1}$, and $B_{35} = 608.^{-1}$, whilst the spin-forbidden band at 11.4 *k*K. gives $B_{55} = 750$ cm.$^{-1}$. The appearance of the first charge-transfer band at about 30 *k*K. suggests that Cr(IV) may be somewhat less stable with respect to Cr(III), than is V(IV) with respect to V(III), a conclusion consistent with the general chemical tendencies of these oxidation states. No evidence of vibrational fine structure was observed for the *d–d* bands, even at 77 °K, despite a careful search (s. Sect. 6).

(iii) Hexafluoromanganate(IV), MnF_6^{2-}

Unlike the other hexafluoro complexes of the tetravalent cations of the first transition series, the MnF_6^{2-} ion had been obtained long before the advent of direct fluorination techniques, its preparation being accomplished by *Weinland* and *Lauenstein* (*81*) by the conversion of potassium permanganate to the corresponding manganate by treatment with KOH, followed by reaction with excess HF. Also in contrast to the other MF_6^{2-} salts the golden-brown K_2MnF_6 is quite stable in dry air, and in aqueous HF solutions, from which it may be recrystallised. However, K_2MnF_6 was also obtained by *Huss* and *Klemm* (*63*) by fluorination of a 2:1 $KCl/MnCl_2$ mixture at 375°, and this product, which possessed a hexagonal, K_2MnF_6-type lattice was shown to be identical with that prepared by the procedure of *Weinland* and *Lauenstein*. On prolonged heating of this compound at 350—400° a slightly impure cubic, K_2PtCl_6-type, product was obtained (*94*), and at lower temperatures evidence was also found for a trigonal, K_2GeF_6-type, lattice.

For the rubidium salt the transition temperature appears to be somewhat lower since when crystallised between 70 and 100° the bright yellow cubic form is formed, whereas below 40° the dark yellow hexagonal form is stable. For the caesium salt though only the cubic form can be obtained, whilst Li_2MnF_6 is always hexagonal, and is unstable in aqueous media (*95*).

For a large number of MnF_6^{2-} salts a magnetic moment of between 3.8 and 3.9 B.M. at room temperature has been found, in agreement with the expectation of about the spin-only value. The spectroscopic evidence however, although abundant, is rather more difficult to interpret. Thus, *Jørgensen* (*1*), in an early reflectance study of K_2MnF_6, prepared according to *Weinland* and *Lauenstein*, located the two broad spin-allowed $d-d$ transitions, $^4A_{2g} \rightarrow {}^4T_{2g}$, and $^4A_{2g} \rightarrow {}^4T_{1g}$ at 21.75 and 28.2 kK. respectively, but also found a number of weak narrow bands at around 16.2 kK. which were assigned as $^4A_{2g} \rightarrow {}^2E_g$, $^2T_{1g}$ transitions. The spin-allowed bands yield $Dq = 2175$ cm.$^{-1}$ and a value for B_{35} of about 600 cm.$^{-1}$, but the assignment of the spin-forbidden transition leads to $B_{55} \sim 790$ cm.$^{-1}$ and although B_{55} would be expected to be somewhat larger than B_{35} the magnitude of the effect is surprisingly great. The salt was therefore reinvestigated by *Allen, El-Sharkarwy* and *Warren* (*96*), who also prepared it by the wet method, and a significantly different reflectance spectrum was obtained (see Table 3 (iii)). The two spin-allowed bands were found at 22.2 and 28.6 kK., in essentially the same positions as recorded by *Jørgensen* (*1*), but no trace could be found of the weak sharp bands around 16 kK., whilst a band of this type was evident at 19.3 kK., which was not present in *Jørgensen's* spectrum. On this basis all the observed transitions could be fitted with a single B value of 587 cm.$^{-1}$ if it were assumed that the 19.3 kK. band represented a $^4A_{2g} \rightarrow {}^2T_{2g}$ transition instead of a $^4A_{2g} \rightarrow {}^2E_g$, $^2T_{1g}$ excitation, to which latter a broad, very weak, absorption at about 14 kK. was assigned. The spectrum of *Allen et al.* also differed from that of *Jørgensen* in showing very marked vibrational fine structure on the $^4A_{2g} \rightarrow {}^4T_{2g}$ band. A regular spacing of 475 ± 50 cm.$^{-1}$ was found which was interpreted as an upper state ($^4T_{2g}$) vibrational progression in the ν_1, α_{1g} mode, by analogy with previous findings for the NiF_6^{2-} and CoF_6^{2-} anions (s. Sect. 6 for further discussion).

Similar vibrational detail has also been found by *Nelson* (*76*) for the cubic Cs_2MnF_6, and a full investigation of this salt has recently been completed by *Reisfeld, Matwiyoff,* and *Asprey* (*97*). For the spin-allowed bands their results are extremely similar to those of *Allen et al.* for K_2MnF_6, yielding $Dq = 2180$ cm.$^{-1}$ and $B_{35} = 585$ cm.$^{-1}$, but the positions of the spin-forbidden bands corresponds much more closely with the previous data and assignments of *Jørgensen* (see Table 3 (iii)) and give a value for B_{55} of about 720 cm.$^{-1}$. The $^4A_{2g} \rightarrow {}^4T_{2g}$ band in Cs_2MnF_6 is reported by *Matwiyoff* and *Asprey* (*98*) to lie at about 21.5 kK., and they too observe vibrational fine structure, a regular spacing of 520 cm.$^{-1}$ being recorded, and ascribed to the ν_1, α_{1g} mode of the $^4T_{2g}$ state. These authors also identified the Raman active modes for the $^4A_{2g}$ ground state, and the value of 615 cm.$^{-1}$ for ν_1 is fully consistent with the lower figure obtained from the vibrational spacing since a decreased force constant would be expected for the excited state, by virtue of the excitation of a t_{2g} electron to an e_g anti-bonding level.

Table 3 (iii). *Spectroscopic data for the hexafluoromanganate(IV) anion*

Assignment	Band Position			
	Jørgensen (*1*)	*Allen et al.* (*96*)	*Reisfeld et al.* (*97*)	This work
	K_2MnF_6	K_2MnF_6	Cs_2MnF_6	K_2MnF_6
$^4A_{2g} \rightarrow {}^2E_g$			16.26	15.7[a]
	16.2	14.0 (?)		
$^4A_{2g} \rightarrow {}^2T_{1g}$			17.30	16.15
$^4A_{2g} \rightarrow {}^4T_{2g}$	21.75	22.2	21.80	22.2
$^4A_{2g} \rightarrow {}^2T_{2g}$	—	19.3	25.71	—
$^4A_{2g} \rightarrow {}^4T_{1g}$	28.2	28.6	28.65	28.6
$\pi \rightarrow t_{2g}$	—	38.8	38.0	38.5
Dq	2175	2220	2180	2220 cm.$^{-1}$
B_{35}	600	587	585	585 cm.$^{-1}$
B_{55}	780	587	720	708 cm.$^{-1}$

[a]) Flint gives 16.06 (*99*).

The situation is further complicated by the results of a recent study of crystalline K_2MnF_6 by *Flint* (*99*) which are closely akin to those of *Jørgensen.* Flint carried out both luminescence and absorption studies and detected a number of weak sharp lines in the 16 *k*K. region both by emission and reflectance, assigning the origin of the $^2E_g \rightarrow {}^4A_{2g}$ transition at 16.08 *k*K., from the phosphorescence spectrum. From the data he was able to determine a number of the vibrational frequencies of the MnF_6^{2-} unit, both for the ground state and for the ^2E_g level, for which ν_1 is found to be 570 cm.$^{-1}$, intermediate in value, as would indeed be expected, between the $^4A_{2g}$ figure and the $^4T_{2g}$ result. *Flint* has also measured the infra-red and Raman spectra of K_2MnF_6, all the vibrational frequencies being determined, and the result of 600 cm.$^{-1}$ for ν_1, α_{1g} accords closely with the value of 615 cm.$^{-1}$ found by *Reisfeld et al.* (*97*) for Cs_2MnF_6.

As well as the phosphorescence spectrum in the region of 16 *k*K,. a broad band was also observed in emission at 14.3 *k*K., and this is ascribed to $^4T_{2g} \rightarrow {}^4A_{2g}$ fluorescence. It is probably significant, as pointed out by *Flint,* that in general for Mn^{IV} complexes only fluorescence is observed in emission when Dq is less than about 2000 cm.$^{-1}$. This suggests that until Dq attains this magnitude the lowest lying doublet level (^2E_g) is above the lowest quartet state ($^4T_{2g}$), and the appearance of phosphorescence for MnF_6^{2-}, with Dq ca. 2200 cm.$^{-1}$, argues strongly that the lowest observed spin-forbidden level in the MnF_6^{2-} spectrum must be the ^2E_g, $^2T_{1g}$ state, and not the $^2T_{2g}$ level as previously proposed by *Allen et al.* (*96*). This view is borne out by observations in our laboratory on a sample of K_2MnF_6 prepared by direct fluorination, and kindly supplied to us by Prof. *Hoppe*. This shows the two spin-allowed bands, virtually as before, although the fine structure on the $^4A_{2g} \rightarrow {}^4T_{2g}$ band is not so clear, but there is no trace of a band at 19 *k*K., whilst two weak sharp peaks occur at 15.7 and 16.15 *k*K. which it seems necessary to assign as $^4A_{2g} \rightarrow {}^2E_g$ and $^2T_{1g}$, thus giving a result very similar to that of *Jørgensen* (*1*) (s. Fig. 15).

It is difficult in the extreme to rationalise all these observations, but it is possible that the lattice geometry may play some part in this problem. Thus, the caesium salt is always cubic, whilst the potassium salt may be either cubic, hexagonal, or trigonal, and since none of the investigations of K_2MnF_6 have included an unequivocal determination of the lattice type it is possible that the spectra may not all relate to the same structure. A systematic examination of the relationship between spectrum and structural type would be most welcome, but, in the interim, it seems safe to conclude that (i) the existence of the 16 *k*K. bands in MnF_6^{2-} is adequately established, and (ii) that the lowest doublet level lies only just slightly below the lowest quartet state.

The nature of the 19.3 *k*K. peak observed by *Allen et al.* therefore remains in doubt, but it cannot be reasonably be ascribed to a $^2T_{2g}$ state, and it is clear that B_{55} must indeed be substantially greater than B_{35}.

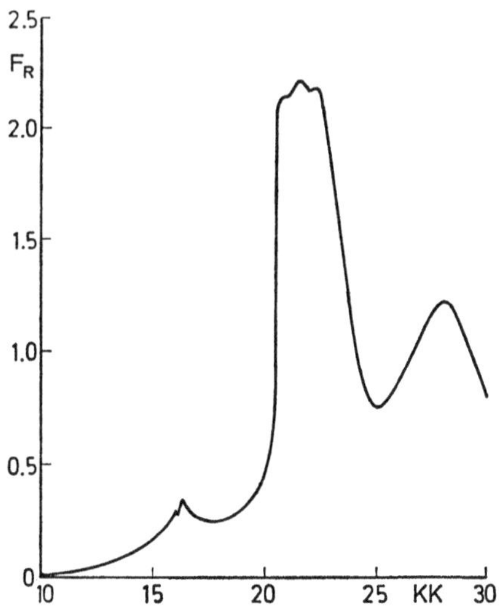

Fig. 15. Electronic spectrum of K_2MnF_6

Possibly this effect is associated with the particular stability of the half-filled t_{2g}^3 shell, but it seems evident that the criterion of the comparability of the group overlap integrals relating to the metal e_g and t_{2g} orbitals (which the authors have used when speculating that the two B values should be about equal) is too approximate to be of any validity.

Very recently another study of the K, Rb, and Cs salts of the MnF_6^{2-} anion has been reported by *Novotny* and *Sturgeon* (*100*). When K_2MnF_6 was prepared by fluorination of a KCl:$MnCl_2$ mixture at 375° the product was hexagonal, showed no vibrational fine structure at ca. 22 *k*K. and no peak at 19.3 *k*K., although the weak sharp bands at ca. 16 *k*K. were observed. A sample prepared at 250° by fluorination of $K_2MnF_5 \cdot H_2O$ showed similar behaviour, except that a faint peak at ca. 19 *k*K. was detected, but a sample obtained by fluorination of $K_2MnF_5 \cdot H_2O$ at 150° could no longer be cleaıly indexed as hexagonal, and showed a definite peak at 19.3 *k*K. with only a faint absorption around 16 *k*K. For the cubic Cs_2MnF_6 *Novotny* and *Sturgeon* found a very similar spectrum to that reported by *Reisfeld et al.*, with well marked fine structure on the 22 *k*K. band, but for the potassium and rubidium salts vibrational detail was only detected for the cubic modifications. They therefore suggest that the spectrum recorded by *Allen et al.* relates to the cubic form of K_2MnF_6 and that the 19.3 *k*K. peak is due to some unidentified impurity which also serves to obscure the 16 *k*K. peaks. *Novotny* and *Sturgeon* also considered the possibility that the 38.8 *k*K. peak might correspond to the $^4A_{2g} \rightarrow {}^4T_{1g}$ (^{4}P) transition, rather than to a $\pi \rightarrow \gamma_5$ excitation, since the former would be predicted to lie in the region of 45 *k*K., and the band at 38.8 *k*K. was of comparable intensity to the two *d*–*d* bands at ca. 22 and 28 *k*K. The agreement is not however very satisfactory, and such an assignment appears improbable: in any case the intensities derived from diffuse reflectance spectra need to be expressed in F_R units to have any quantitative significance, and even then the strengths of bands in the charge-transfer region tend to be considerably underestimated relative to the *d*–*d* transitions because of the usual increase in the scattering coefficient above about 25 *k*K. (s. Sect. 1 (iv)).

Potassium hexachloromanganate(IV), which is the only MCl_6^{2-} ion known, apart from VCl_6^{2-}, in the first transition series, has also been studied spectroscopically. *Moews* (*101*) found strong absorptions at 15.4, 27.4, and 33.3 *k*K., which he ascribed to charge-transfer transitions, and the much lower energy of these bands, completely obscuring the *d*–*d* transitions, is consistent with the beginning of the charge-transfer bands at about 38 *k*K. in the hexafluoro complex. For the hexafluoro complex the first charge-transfer band occurs at somewhat higher energies than in VF_6^{2-} and CrF_6^{2-}, which accords with the rather greater stability of the Mn(IV) state in these compounds (s. Sect. 5).

(iv) Hexafluorocobaltate(IV), CoF_6^{2-}

Complexes containing Co(IV) are exceedingly rare and indeed the only well characterised entity is the CoF_6^{2-} anion, which itself is known only as the caesium salt, Cs_2CoF_6. This was prepared by *Hoppe* (*69*), and by *Klemm, Brandt,* and *Hoppe* (*70*), by the fluorination of $Cs_2Co(SO_4)_2 \cdot 2\,H_2O$ or of Cs_2CoCl_4, and was shown to have a cubic, K_2PtCl_6-type lattice, with a magnetic moment rising from 2.46 B.M. at 90 °K to 2.97 B.M. at 294 °K. Here again for a d^5 system we have the possibility of either high- or low-spin behaviour, but *Cotton* and *Wilkinson* (*102*) felt that, as the observed moment did not clearly correspond either to one or to five unpaired spins, the compound required further characterisation. However, the lattice constant and molecular volume for Cs_2CoF_6 fit in smoothly with the data for other Cs_2MF_6 complexes of the first transition series (*40*), and there seems no adequate reason to doubt its identity. Nevertheless, *Figgis* and *Lewis* (*51*) calculated that for a low-spin ${}^2T_{2g}$ (t_{2g}^5) ground state a magnetic moment of about 2.1 B.M. would be expected for Co^{4+} at room temperature, compared with a high-spin value of 5.92 B.M., and concluded that the complex was more likely to be high-spin, with some unusual magnetic interaction. On the other hand the actual values of the moment lie much closer to those predicted for the low-spin situation, and exhibit the sort of temperature dependence which would be expected for a ${}^2T_{2g}$ ground state, even though they are about 0.8—0.9 B.M. too high. Furthermore, the magnetic moments recorded for the d^1 and d^2 systems, VF_6^{2-} and CrF_6^{2-}, both lie significantly above the predicted values, with high θ constants, and since the data for CoF_6^{2-} can be fitted with a Weiss constant of about 60°, it seems reasonable to conclude that the ground state is in fact ${}^2T_{2g}$, and that antiferromagnetic interactions may be responsible for the rather high observed moments.

This view is borne out by the spectroscopic study of *Allen* and *Warren* (*103*). For a high-spin d^5 system the ground state is a ${}^6A_{1g}$ level, and all d—d transitions would be spin-forbidden, but the intensities of the bands observed below about 25 kK. are comparable with those found for the spin-allowed bands of low-spin NiF_6^{2-}. Whilst it is not possible to place too much reliance on intensity values obtained by reflectance (s. Sect. 1 (iv)), it is usually easy to distinguish between spin-allowed and spin-forbidden transitions, and on this basis the bands lying between 17 and 25 kK. (Table 3 (iv)) are assigned as spin-allowed doublet-doublet excitations, whilst the much weaker absorptions at 6.4 and 10.3 kK. represent transitions to quartet states. Resolution of the absorption region above 15 kK., by Gaussian analysis, yields the peak positions shown in the Table, and the bands may be fitted by the parameters $Dq = 2030$ cm.$^{-1}$,

Table 3 (iv). *Spectroscopic data for caesium hexafluorocobaltate(IV)*

Band Position		Assignment
Obsd.	Calcd.[a])	
6.4	6.3	$^2T_{2g} \rightarrow {}^4T_{1g}$
10.3	10.9	$^2T_{2g} \rightarrow {}^4T_{2g}$
17.3	17.3	$^2T_{2g} \rightarrow {}^2A_{2g}, {}^2T_{1g}$
18.9	18.9	$^2T_{2g} \rightarrow {}^2T_{2g}$
21.4	20.5	$^2T_{2g} \rightarrow {}^2E_g$
24.4	24.8	$^2T_{2g} \rightarrow {}^2T_{1g}$
28.3	....	$\pi \rightarrow t_{2g}$
32.2	32.1	$^2T_{2g} \rightarrow {}^2E_g(?)$
37.1	....	$\pi \rightarrow t_{2g}$

[a]) $Dq = 2030$ cm.$^{-1}$, $B = 635$ cm.$^{-1}$.

and $B = 635$ cm.$^{-1}$. The very satisfactory agreement with experiment thereby obtained on the basis of a low-spin, $^2T_{2g}$ ground state thus supports our earlier interpretation of the magnetic data.

Although an alternative explanation of the temperature dependence of the magnetic moment in terms of a thermal high-spin — low-spin equilibrium is a possibility, it can in this case be shown to be improbable. Using the assumption (*9*) of $C/B = 4.9$, the fitting parameters require $Dq/B = 3.20$, whereas the $^6A_{1g}$—$^2T_{2g}$ crossover lies at $Dq/B = 3.05$, and since the crossover position is sensitive to the C/B ratio, a thermal equilibrium cannot be ruled out. However, the magnetic behaviour in the region of the crossover point will not be complicated as it is in the case of d^7 systems (s. Sect. 2 (vii), NiF_6^{3-}) since there are no spin-orbit matrix elements connecting the $^6A_{1g}$ and $^2T_{2g}$ states, and if such an explanation were valid the temperature variation of the magnetic moment could be due only to changes in the position of equilibrium. If this were the case a noticable variation of the electronic spectrum with temperature would be anticipated, but in fact, as shown in Fig. 20, the spectra at 77 °K and at room temperature are virtually identical except for a rather better resolution at the lower temperature.

The measurements at liquid nitrogen temperature also illustrate well the vibrational fine structure found for this complex. At room temperature this vibrational detail is observed between about 17 and 22 *k*K., but at 77 °K it is possible to distinguish an over-lapping pair of regular progressions on the $^2T_{1g} \rightarrow {}^2T_{2g}$ and $^2T_{2g} \rightarrow {}^2E_g$ bands. As for MnF_6^{2-} and NiF_6^{2-} (q.v.) these are assigned to the α_{1g} vibrational mode, and the spacing of 520 ± 50 cm.$^{-1}$ is fully consistent with the observation of the

ν_1, α_{1g} mode at 560 cm.$^{-1}$ in the Raman spectrum of Cs_2CoF_6. As before (MnF_6^{2-}) a somewhat lower frequency would be expected for the vibrational mode relating to the excited states.

The first charge-transfer band for CoF_6^{2-} is located at 28.3 *k*K., that is some 10 *k*K. lower than for MnF_6^{2-}, and this feature is in accordance with the markedly greater oxidising character of the Co(IV) state, as compared with Mn(IV). The assignment of this band is reasonably secure but there is some uncertainty over the bands at 32.2 and 24.4 *k*K. The former may be a further charge-transfer transition and not a *d–d* band, although the analysed intensity supports the latter view, and *Jørgensen* has suggested to us that the 24.4 *k*K. band may constitute a Laporte-forbidden charge transfer transition, possibly of the $t_{1g} \rightarrow t_{2g}$ type.

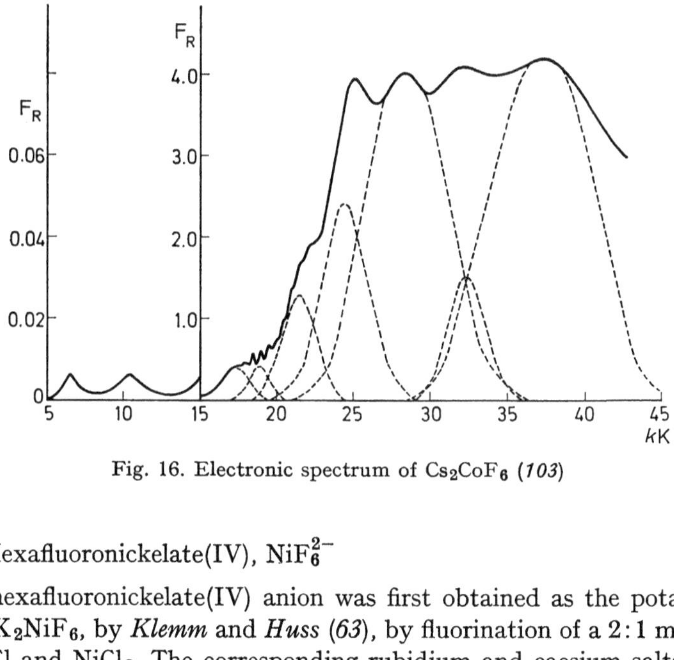

Fig. 16. Electronic spectrum of Cs_2CoF_6 (*103*)

(v) Hexafluoronickelate(IV), NiF_6^{2-}

The hexafluoronickelate(IV) anion was first obtained as the potassium salt, K_2NiF_6, by *Klemm* and *Huss* (*63*), by fluorination of a 2:1 mixture of KCl and $NiCl_2$. The corresponding rubidium and caesium salts were prepared by *Bode* and *Voss* (*92*), and found to have a cubic, K_2PtCl_6 lattice, whilst *Hoppe* (*104*) showed the magnetic moment of K_2NiF_6 to be zero, and the ion therefore to have a low-spin t_{2g}^6 configuration. More recently *Henkel, Hoppe,* and *Allen* (*105*) have reported the preparation and characterisation of the sodium salt, and have shown that whereas the K, Rb, and Cs salts exist only in the cubic form, both cubic and hexagonal species are indicated for the Na compound.

Over the last few years several spectroscopic investigations of the NiF_6^{2-} system have been reported. As early as 1965 *Westland, Hoppe,* and *Kaseno* (*106*) recorded, without discussion, a rough spectrum of K_2NiF_6, whilst in 1968 *Bougon* (*107*) prepared several salts of the NiF_6^{2-} anion, and reported the ν_3 and ν_4 (τ_{1u}) vibrational frequencies, and noted the presence of a band at about 18.5 *k*K. in the electronic spectrum. In 1969 two studies of the NiF_6^{2-} spectrum were published, the first by *Reisfeld, Asprey,* and *Penneman* (*31*), who used a mull technique, and the other by *Allen* and *Warren* (*108*), who made measurements by diffuse reflectance, using samples sealed in silica cells.

Reisfeld et al. prepared K_2NiF_6 by fluorination of $K_2Ni(CN)_4$ and obtained the Rb and Cs salts in a similar way. Spectroscopic measurements were made between 4 and 53 *k*K. on these products and on samples recrystallised from liquid anhydrous HF, at both room and liquid nitrogen temperatures, using compounds ground into mulls with Fluorolube oil and placed between KBr discs. For K_2NiF_6 at 80 °K two *d*–*d* transitions centred at 19.2 and 25.8 *k*K. were observed and assigned respectively to transitions from a low-spin $^1A_{1g}$ (t_{2g}^6) ground level to the $^1T_{1g}$ and $^1T_{2g}$ states respectively, the parameters $Dq = 2090$ cm.$^{-1}$ and $B = 485$ cm.$^{-1}$ being derived. In the charge-transfer region a broad band with a maximum at 32.8 *k*K. was found, assigned as $t_{2u} \rightarrow e_g$, and an indication of a further peak at 51 *k*K. which was not further analysed because of instrumental difficulties. The spin-forbidden transitions to the $^3T_{1g}$ and $^3T_{2g}$ states, which should also lie in the low energy region, were not detected.

The band at 19.2 *k*K. contained extensive well defined fine structure, both at room and liquid nitrogen temperatures, the intensity of the vibrational series being essentially independent of temperature, indicating that it was connected with vibrational modes of the excited state. The structure showed a regular spacing of 486 ± 30 cm.$^{-1}$, and *Reisfeld et al.*, after consideration of a number of possible interpretations, concluded that this interval corresponded to the ν_2, ε_g, Raman active vibrational mode, which was found (*109*) for the ground state at 520 cm.$^{-1}$. Since the observation of a uniquantal progression in a degenerate vibrational mode may in certain cases be indicative of a lowering of the effective symmetry of the upper state by a Jahn-Teller distortion (s. Sect. 6), it was concluded that the vibrational fine structure was due to the operation of such an effect for the excited $^1T_{1g}$ level.

Allen and *Warren*, on the other hand, studied K_2NiF_6 and Na_2NiF_6 by diffuse reflectance measurements between 4 and 45 *k*K., and located the $^1A_{1g} \rightarrow {}^1T_{1g}$ and $^1A_{1g} \rightarrow {}^1T_{2g}$ transitions at virtually the same energies as reported by *Reisfeld et al.* As found by these latter workers the spectra appeared to be almost independent of the nature of the cation. However,

Allen and *Warren* did succeed in locating the two spin-forbidden *d–d* transitions, $^1A_{1g} \rightarrow {}^3T_{1g}$ and $^1A_{1g} \rightarrow {}^3T_{2g}$, at 12.9 and 16.0 *k*K. respectively, (see Table 3 (v)) and these, together with the spin-allowed bands, could all be very satisfactorily fitted with the parameters $Dq = 2010$ cm.$^{-1}$, and $B = 515$ cm.$^{-1}$, taking $C/B = 5.0$.

Allen and *Warren* also observed the extensive vibrational fine structure associated with the 19.0 *k*K. peak, reporting a spacing of 490 ± 50 cm.$^{-1}$, but interpreted this as being due to the superposition of an α_{1g} progression on the vibronically allowed (probably τ_{1u}) transition. They pointed out that the assumption of ν_1, $\alpha_{1g} = 490$ cm.$^{-1}$ for the excited $^1T_{1g}$ state would be fully consistent with the observation of the Raman active ν_1 mode at 560 cm.$^{-1}$ for the ground state, since an appreciable reduction in the force constant would be expected for the excited state by virtue of the increased occupation of the e_g anti-bonding levels. In a more recent communication the authors (*110*) have pointed out some inconsistencies in the discussion of results put forward by *Reisfeld et al.*, and have also adduced further reasoning and experimental evidence to support the assignment as an α_{1g} progression. (s. Sect. 6 for further details.)

Since for a low-spin d^6 configuration the spin-allowed transitions take place between singlet levels, the $^1T_{1g}$ and $^1T_{2g}$ levels contain only a single Γ_4 and Γ_5 component respectively in the octahedral double group, O*. It is therefore possible using the spin-orbit matrix elements which connect Γ_4 of $^3T_{1g}$ with $^1T_{1g}$, and Γ_4 of $^3T_{2g}$ with $^1T_{1g}$, to estimate the relative intensities of the spin-forbidden bands. Using a spin-orbit coupling constant of 750 cm.$^{-1}$ for Ni^{4+} the intensity ratios $^3T_{1g}$:$^1T_{1g}$ and $^3T_{2g}$:$^1T_{1g}$ are respectively calculated to be 1/560 and 1/30, whilst

Table 3 (v). *Spectroscopic data for the hexafluoronickelate(IV) anion*

Band Position (K_2NiF_6)		Assignment
Reisfeld et al. (*31*)	*Allen and Warren* (*108*)	
—	12.9	$^1A_{1g} \rightarrow {}^3T_{1g}$
—	16.0	$^1A_{1g} \rightarrow {}^3T_{2g}$
19.2	19.0	$^1A_{1g} \rightarrow {}^1T_{1g}$
25.8	25.3	$^1A_{1g} \rightarrow {}^1T_{2g}$
32.8	30.8	$\pi \rightarrow e_g$
—	38.9	$\pi \rightarrow e_g$
51.0	—	$\sigma \rightarrow e_g$
$Dq = 2090$ cm.$^{-1}$, $B = 485$ cm.$^{-1}$.	$Dq = 2010$ cm.$^{-1}$, $B = 515$ cm.$^{-1}$.	

the experimental values are 1/315 and 1/100. The correlations are both quite satisfactory and support the assignments given, and the agreement can in fact be substantially improved if the origin of the $^1T_{1g}$ band lies below the value of 18.4 *k*K. then assumed. In fact, the origin of the band appears to lie at about 17 *k*K., according to the later note of *Allen* and *Warren* (*110*). Thus, the fitting of all the *d*–*d* transitions, assuming a low-spin ground state, affords ample confirmation of the diamagnetic situation indicated by the magnetic measurements.

The two peaks at 30.8 and 38.9 *k*K. in the broad high intensity region of the reflectance spectrum were assigned by *Allen* and *Warren* as $\pi \rightarrow e_g$ charge-transfer transitions, t_{2u} and $t_{1u} \rightarrow e_g$ in a molecular orbital interpretation, but only one peak was resolved in this region by *Reisfeld et al.*, at 32.8 *k*K., which they similarly interpreted as a charge-transfer excitation. It is possible that this disagreement is a result of the different experimental techniques adopted (s. Sect. 1 (iv)), and since the data of *Allen* and *Warren* do not extend beyond 45 *k*K. it is not possible to confirm the presence of the 51 *k*K. peak. In any case, the two charge-transfer bands recorded by *Reisfeld et al.* are 18 *k*K. apart, which is much greater than would be expected if they represented the t_{1u} and $t_{2u}(\pi) \rightarrow e_g$ transitions, and we are therefore inclined to believe that if real the 51 *k*K. band correspond to a $t_{1u}(\sigma) \rightarrow e_g$ excitation. Nevertheless the first charge-transfer band is, as for Co(IV), at rather a low energy for a fluoride complex, and is consistent with the strongly oxidising character of Ni(IV). The whole problem of the optical electronegativities derived from the charge-transfer transitions, and their relationship to nephelauxetic character and covalency, is discussed in greater detail in Section 5.

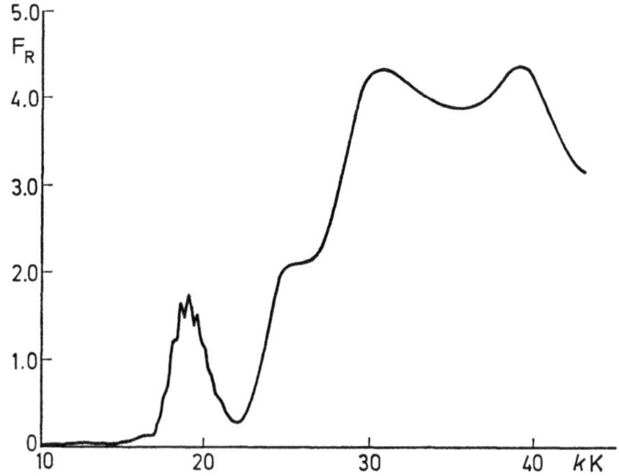

Fig. 17. Electronic spectrum of K_2NiF_6 (*105*)

4. Nephelauxetic Effects

(i) General Considerations

In Section 1 (ii) we have considered some features of the information which may be obtained from the evaluation of the nephelauxetic ratio, β, for a given complex. For this purpose the determination of the free-ion value of the Racah parameter, B_{gas}, is obviously of prime importance, since $\beta = B_{complex}/B_{gas}$, and *Tanabe* and *Sugano* (*9*) utilised a number of extrapolation techniques with spectroscopic data to obtain the B and C integrals for various ionisation states of the elements of the 3*d* series. These values have since then remained in general usage for finding β, although a number of other compilations of the F_k (and hence B and C) integrals have subsequently appeared. (See, for example, *Hinze* and *Jaffe* (*111*), and *Tondello et al.* (*112*).)

It is of course also possible to calculate the F_k integrals theoretically from Hartree-Fock radial functions, but these values exceed those obtained empirically by a factor of roughly $(z+3)/(z+2)$ where z is the ionic charge of the cation (*113*). Nevertheless, for fairly high ionic charges, there exists a strong proportionality between the spectroscopic B values and the calculated average reciprocal radii, $<r^{-1}>$, of the Hartree-Fock function, as well as the dependence on z, and *Jørgensen* (*113*), modifying a suggestion of *Racah* (*114*), has shown that for the 3*d* group the variation of B (in cm.$^{-1}$) with z and with q (the occupation number of the d^q shell) is well expressed by the relationship:

$$B = 384 + 58q + 124\,(z+1) - 540/(z+1)$$

The results thus derived for integer values of z and q agree closely with those of *Tanabe* and *Sugano* (*9*) (see Table 4 (i)), and also allow the determination, by interpolation, of the effective cationic charges for various complexes (*vide infra*).

As regards the magnitude of the ratio of the Racah parameters, C/B, the approximation C = 4 B, adopted by *Jørgensen* (*3, 115*), is in many cases quite satisfactory. The authors have in general prefered to adopt the values listed by *Tanabe* and *Sugano* (*9*) for a given free cation, assuming the same ratio to hold for the complex, but small changes in the defined value of C/B do not usually affect a spectroscopic fitting significantly, except in those cases in which Dq/B lies close to a spin crossover point. The C/B data of *Tanabe* and *Sugano*, and values extrapolated therefrom, are also given in Table 4 (i).

In Section 1 (ii) we also explained the origin of the three nephelauxetic parameters, β_{33}, β_{55} and β_{35}. This whole topic has been exhaustively treated by *Jørgensen* (*3, 115*), but briefly, for octahedral systems,

Table 4 (i). *Values of the Racah parameter B from spectroscopic data* (*9*)

Element	M^0	M^+	M^{2+}	M^{3+}	M^{4+}
V	—(258)	—(594)	755(750)	862(861)	—(954)
Cr	—(316)	—(652)	810(808)	918(919)	1014(1012)
Mn	—(374)	—(710)	860(866)	965(977)	1065(1070)
Fe	—(432)	—(768)	917(924)	1015(1035)	—(1128)
Co	—(490)	—(826)	971(982)	1065(1093)	—(1186)
Ni	—(548)	—(884)	1030(1040)	1115(1151)	—(1244)
Cu	—(—)	—(942)	—(1098)	—(1209)	—(1302)

Values in parentheses from the Jørgensen-Racah relationship.

Values of the Racah parameter ratio C/B from spectroscopic data (*9*)

Element	M^{2+}	M^{3+}	M^{4+}
V	4.31	4.43	—
Cr	4.40	4.50	4.55
Mn	4.48	4.61	4.62
Fe	4.41	4.73	—
Co	4.63	4.81	—
Ni	4.71	4.89	—

Note: The Slater-Condon parameters of *Tondello et al.* (*112*) lead to B values in reasonable agreement with those listed for the M^+ and M^{2+} ions, but give appreciably larger values for M^0. Fitting of the data of Table 4 (vi) on this basis would not substantially alter the conclusions where the listed z_{eff} lies above about $+1$, but would lead to rather smaller values of z_{eff} than the listed figures where these are in the 0—$+1$ range.

it is usually only possible to evaluate β_{55} for d^3 complexes, and β_{33} is derived solely for d^8 configurations. In both cases these parameters are obtained from the spin-forbidden, intra sub-shell, transitions, whilst the spin-allowed bands yield the value of β_{35}. This restriction to octahedral d^3 and d^8 systems is largely a result of the Jahn-Teller effect: both these configurations have orbitally non-degenerate ground states ($^4A_{2g}$ and $^3A_{2g}$), but other situations which could in principle yield values of β_{55} — d^2 and low-spin d^4 complexes — give rise to orbitally degenerate ground levels for which Jahn-Teller distortions might be sufficient to nullify any conclusions concerning, say, the relative magnitudes of β_{55} and β_{35}. Thus, for systems other than d^3 and d^8, the nephelauxetic effect is generally expressed in terms of the single parameter β_{35}.

The position of the fluoride ion ligand in both the spectrochemical and the nephelauxetic series was also discussed earlier. In fact, both series can be adequately approximated (*3*) as the product of two functions, one characteristic of the ligand and the other of the central cation. Thus we write $10\,Dq = f$ (ligands) . g (central ion) and $(1 - \beta_{35}) = h$ (ligands) . k (central ion) representative values of these parameters being listed in Table 4 (ii). Clearly the ordering of the g and f values corresponds to the respective spectrochemical series for central cations and for ligands, whilst that of the k and h quantities similarly represents the corresponding nephelauxetic series. Whilst there exists no clear cut theoretical reason for the above relationships, it is noteworthy that the Dq and β_{35} values are most successfully predicted for the simplest ligands such as the halides, H_2O, and NH_3, whilst the greatest discrepancies arise for ligands with large π-bonding propensities such as NO_2^- and CN^-. The expressions are therefore of potential use for the approximate prediction of β values for hexafluoro complexes which show too few d–d bands for the value of $B_{complex}$ to be determined.

Table 4 (ii). *Parameters for the empirical estimation of 10 Dq and the nephelauxetic ratio* (*3, 116*)

Metal Ion	g	k	Ligand Group	f	h
V^{2+}	12.3	0.08	6 F^-	0.9	0.8
Cr^{2+}	14.1	—	6 H_2O	1.00	1.00
Mn^{2+}	8.0	0.07	6 NH_3	1.25	1.4
Fe^{2+}	10.0	—	6 Cl^-	0.80	2.0
Co^{2+}	9.3	0.24 [a]	6 Br^-	0.76	2.3
Ni^{2+}	8.9	0.12	6 I^-	0.7	2.7
Cu^{2+}	12.0	—			
V^{3+}	18.6	—			
Cr^{3+}	17.4	0.21			
Mn^{3+}	21	—			
Fe^{3+}	14.0	0.24			
Co^{3+}	19.0	0.35			
Mn^{4+}	23	0.5			

[a]) This probably represents an upper limit: *Jørgensen* (private communication) suggests a value closer to 0.1.

(ii) Hexafluorometallate(III) and Hexafluorometallate(IV) Complexes

In Sections 2 and 3 the electronic spectra of the MF_6^{3-} and MF_6^{2-} complexes of the first transition series have been surveyed in some detail, and values of Dq and B deduced. For the d^1 systems of TiF_6^{3-} and VF_6^{2-} there are of course no d–d transitions involving the parameter B, even allowing for Jahn-Teller effects, and accordingly no g and k parameters are known which might allow the indirect estimation of the B values. However, for the remainder of the MF_6^{2-} series, which shows low-spin behaviour for the d^5 and d^6 configurations, both Dq and B are readily derived, but in the MF_6^{3-} series Dq is sufficiently smaller to lead to high-spin behaviour for all but the NiF_6^{3-} ion. As a consequence of this essentially only one spin-allowed transition $^5E_g \rightarrow {}^5T_{2g}$, and $^5T_{2g} \rightarrow {}^5E_g$ respectively would be expected for the d^4 and d^6 configurations, thereby preventing the determination of B for MnF_6^{3-} and for CoF_6^{3-}.

Nevertheless, the availability of g and k parameters for Co^{3+} permits a reasonable estimate to be made of $B_{complex}$ for CoF_6^{3-} of 787 cm.$^{-1}$ ($\beta_{35} = 0.72$). The experimental Dq value of 1310 cm.$^{-1}$ is subject to some uncertainty by virtue of the dynamic Jahn-Teller effect associated with the upper ^5E_g state, and is somewhat lower than the value of 1710 cm.$^{-1}$ predicted from 10 D$q = f \cdot g$, but the experimental and theoretical Dq values lead respectively to Dq/B equal to 1.66 and 2.17, i.e. on either side of the crossover point, and on balance are in accordance with the expectation that Dq/B should lie slightly on the high-spin side of the crossover. It is also noteworthy that the calculated β value agrees extremely closely with that deduced from the tentative assignment of the spin-forbidden transition. (c.f. Section 2 (vi)).

On the other hand, for Mn^{3+}, only the g parameter is known, from which we calculate a Dq of 1890 cm.$^{-1}$, as compared with the experimental value of 1740 cm.$^{-1}$. The difficulty here is that a k value for Mn^{3+} can only be derived from a low-spin d^4 complex, and as yet there exists no fully satisfactory analysis of the spectrum of the only well established low-spin system, $Mn(CN)_6^{3-}$. Moreover, data derived from g and k parameters for CN^- ligands are in any case considerably less reliable than that obtained from more weakly π-bonding ligands. However, if, making these reservations, we accept *Ballhausen's* (*4*) tentative parameters of D$q = 3000$ and B $= 500$ cm.$^{-1}$ for $Mn(CN)_6^{3-}$, we obtain k (Mn^{3+}) $= 0.24$ and thence β_{35} (MnF_6^{3-}) $= 0.81$. This value is in tolerably good agreement with that deduced from the tiny inflection at 24 kK. in *Oelkrug's* (*60*) spectrum of MnF_6^{3-} ($\beta_{35} = 0.86$), but must clearly be regarded as a very qualitative figure.

On the above basis, and from the values compiled in Sections 2 and 3, Tables 4 (iii) and 4 (iv) have been drawn up in which are shown respectively the Dq and β_{35}, and where appropriate the β_{55} and β_{33}, values for the MF_6^{3-} and MF_6^{2-} anions of the 3d series. Similarly, in Table 4 (v) are shown the g and k parameters calculated from the collected data, assuming $f = 0.9$ and $h = 0.8$ for the 6 F^- group. It is apparent that the values of β reflect quite closely the general chemical stability of the particular hexafluoro complexes, and the parallel between β and the overall chemical behaviour of the cations of the first transition series is quite striking.

Table 4 (iii). *Crystal field splitting constants and nephelauxetic ratios for hexafluoro complexes of trivalent transition metal ions (MF_6^{3-})*

	V	Cr	Mn	Fe	Co	Ni	Cu
Dq	1610	1510	1740 [a]	1335	1310 [a]	1620 [a]	1410
β_{35}	0.75	0.89	(0.81) [b]	0.81	(0.72) [b]	0.61	0.53
β_{55}	0.80	0.93	—	—	—	—	—
β_{33}	—	—	—	—	—	—	0.44

[a] Estimated after allowance for Jahn-Teller effects.
[b] Approximate values calculated using g and k parameters.

Thus, if we examine the properties of the trivalent state for the ions V^{3+} to Cu^{3+}, we find initially that V(III) although not as stable as V(IV), is still an important oxidation state, whilst for chromium the trivalent state is easily the most stable. In turn Mn(III) and Fe(III) are slightly less stable with respect to their divalent states, Co(III) is more stable than Co(II) only when complexed, whilst Ni(III) and Cu(III) are very unstable oxidation states and their best characterised compounds are the hexafluoro anions in which lattice energy stabilisation is probably all important. Thus, the β values obtained for a constant ligand, F^-, for a constant oxidation state (3+), markedly reflect the decreasing stability of the higher oxidation states towards the end of the first transition series. This parallel between β and oxidation state stability is readily understandable since *Jørgensen* (*115*) has shown that decreasing values of β are associated with a reduction in the effective positive charge on the cation, and thus with greater tendency to go down to the next lowest oxidation state, i.e. $M^{3+} \rightarrow M^{2+}$.

For the MF_6^{2-} series the situation is not so clear cut since the M(IV) state constitutes throughout a fairly unstable (oxidising) condition, and the β values are, for fluoro complexes, all extremely low. Here too though

Table 4 (iv). *Crystal field splitting constants and nephelauxetic ratios for hexafluoro complexes of tetravalent transition metal ions (MF_6^{2-})*

	V	Cr	Mn	Co	Ni
Dq	2025	2150	2180	2030	2010
β_{35}	—	0.57	0.55	0.53	0.41
β_{55}	—	0.75	0.79	—	—

β decreases noticably towards the end of the series, in accordance with the general expectation of decreasing stability of the high oxidation states. As would be anticipated β (M^{4+}) is always less than the corresponding β (M^{3+}).

Table 4 (v). *Empirical g and k parameters from spectroscopic data for hexafluoro complexes of M^{3+} and M^{4+} ions*

	g			*k*		
	M^{2+}	M^{3+}	M^{4+}	M^{2+}	M^{3+}	M^{4+}
V	(12.3)	17.9(18.6)	22.5 —	(0.08)	0.31 —	— —
Cr	(14.1) [a]	16.9(17.4)	23.3 —	—	0.14(0.21)	0.54 —
Mn	(8.0)	19.3(21)	24.7(23)	(0.07)	0.24 —	0.56(0.5)
Fe	(10.0) [a]	14.8(14.0)	— —	—	0.24(0.24)	— —
Co	(9.3) [a]	14.6(19.0)	22.6 —	(0.24)	0.35(0.35)	0.59 —
Ni	(8.9)	18.0 —	22.3 —	(0.12)	0.49 —	0.74 —
Cu	(12.0)	15.7 —	— —	—	0.59 —	— —

[a]) Subject to correction for Jahn-Teller effects.
Values in parentheses from Table 4 (ii).

In Section 1 (iii) we have already considered briefly the contributions which both central-field and symmetry restricted covalency may make to the nephelauxetic effect. The former is taken as arising from the reduction of the effective cationic charge by the screening of the d^n configuration by the ligands, whereas the latter is due to the respective involvment of the e_g and t_{2g} metal orbitals in molecular orbital formation with appropriate symmetry adapted ligand combinations. In an LCAO description symmetry restricted covalency may be represented as $\psi_{MO} = a\psi_d + b\psi_{ligands}$ where ψ_d is a metal d orbital (e_g or t_{2g}), $\psi_{ligands}$ is an appropriate matching ligand orbital combination, and a and b are their respective contributing coefficients, usually known as the *Stevens'* (*117*) delocalisation coefficients.

Although it is possible to argue that central-field covalency arises only as a necessary consequence of the bonding of the ligands, i.e. as a result of symmetry restricted covalency, it is nevertheless highly desirable that we should be able to make some attempt to assess the relative magnitudes of their respective contributions. Unfortunately, this is far from being a simple matter. *Jørgensen* (*115*) has shown that for a gaseous ion the F_k parameters, and hence B, show an approximate proportionality to $(z+Z)$, where z is the ionic charge and Z a small constant, whilst in a complex the corresponding proportionality is to $a^4(z_{eff}+Z)$, where a is the Stevens' delocalisation coefficient, and z_{eff} is the effective cationic charge in the complex. Since z_{eff} will depend upon the extent of central-field covalency it is clear that we cannot ascribe all the reduction in B, below the gas phase value, to symmetry restricted covalency, yet, on the other hand, we must not assume that all the nephelauxetic effect arises from the reduction of z to z_{eff} occasioned by the central-field effect.

However, these two extreme assumptions do enable us to obtain the limiting possibilities for any given complex. Thus, if we interpret β as arising solely from central-field effects we can derive the minimum value of z_{eff} for the complex, whilst, conversely, if β is assumed to be due only to symmetry restricted covalency we shall obtain the minimum value for the coefficient of presence, a^2, for the metal atom wave function. In this connection it is well to remember that the *Stevens'* coefficients a and b are not strictly normalised by the relationship $a^2+b^2=1$, since the metal and ligand orbitals are not orthogonal, even though S is usually quite small.

Jørgensen (*115*) has though proposed that a reasonable solution to the problem is to assume that central-field and symmetry restricted covalency are of equal importance and to assign $(\beta)^{\frac{1}{2}}$ to each of the two effects, thus obtaining intermediate values of z_{eff} and a^2 as compared with the two extreme cases cited above. For this purpose we may make use of the more accurate relation between z_{eff} and B contained in the Jørgensen-Racah expression for the 3*d* series, and the values of z_{min} (all central-field), a^2_{min} (all symmetry restricted) and of z_{root} and a^2_{root} listed in Table 4 (vi) are derived in this fashion. It is worth noting here that a rough justification of the $(\beta)^{\frac{1}{2}}$ approximation can in fact be derived from results for octahedral d^3 systems. If, for systems such as CrF_6^{3-}, $CrCl_6^{3-}$, and MnF_6^{2-} one assumes that β_{55} arises solely from central-field covalency, i.e. one neglects the probably small π-anti bonding effects, one obtains (*113*) values of z_{min} which agree quite closely with those of z_{root} obtained from β_{35}, thus suggesting that for these cases at least the $(\beta)^{\frac{1}{2}}$ approximation is not too far removed from the truth. Finally, it should be noted that the values of a^2 calculated from this treatment are in fact averaged figures, since, using β_{35} the proportionality strictly involves $a_3^2 \cdot a_5^2$, whilst for β_{33} and β_{55} respectively the proportionality would be to a_3^4 and a_5^4.

The results obtained by use of these various assumptions are collected in Table 4 (vi) and it is seen that for both the MF_6^{3-} and the MF_6^{2-} series negative values of z_{eff} are encountered when it is assumed that the nephelauxetic effect is due only to central-field covalency. This physically unrealistic result supports the contention that it is unreasonable to ascribe all of β to central-field effects, but the fairly satisfactory parallel between the z_{min} values derived from β_{55} and the z_{root} quantities rein-

Table 4 (vi). *Interpretations of the nephelauxetic effect*

	β_{35} (β_{55})	$z_{\min}$	$a^2_{\min}$	z_{root}	a^2_{root}
MF_6^{3-} Complexes					
V	0.75(0.80)	1.25(1.55)	0.87	1.9	0.93
Cr	0.89(0.93)	2.1(2.4)	0.94	2.5	0.97
Mn	[0.81] [a]	1.4	0.90	2.1	0.95
Fe	0.81	1.25	0.90	2.3	0.95
Co	[0.72] [a]	0.75	0.85	1.6	0.92
Ni	0.61	0.3	0.78	1.1	0.88
Cu	0.53(0.64) [b]	—0.1(0.4)	0.73	0.7	0.85
MF_6^{2-} Complexes					
Cr	0.57(0.75)	0.7(1.6)	0.75	1.65	0.87
Mn	0.55(0.79)	0.55(1.8)	0.74	1.5	0.86
Co	0.53	0.35	0.73	1.2	0.85
Ni	0.41	—0.1	0.64	0.65	0.80

[a] Obtained from *g* and *k* parameters.
[b] Derived from the β_{33} value assuming $\beta_{55}/\beta_{35} = \beta_{35}/\beta_{33}$.

forces the view that the $(\beta)^{\frac{1}{2}}$ approximation is tolerably adequate. It may be noted that eight of the eleven z_{root} values listed in the Table lie between 0 and $+2$, most of them in the range $+1$ to $+2$, and this bears out *Jørgensen's* (*113*) conclusion that the effective central atom charge, as defined above, is always somewhat below the oxidation state, but usually above $+1$ and frequently above $+1.5$.

This conclusion is in fact based on the assumption of a regular relationship between $\langle r^{-1} \rangle$ and the interelectronic repulsion parameters, but although other interpretations of the nephelauxetic effect have been put forward, *Jørgensen* (*113, 115*) has demonstrated that the physical cause is the invasion of the partly filled metal shell by the electronic density of the ligands, and it therefore follows that the methods described above do constitute a consistent approach.

Finally, the results given in Table 4 (vi) show that it is highly hazardous to draw conclusions about the magnitude of the effective cationic charge in a given complex merely by considering the formal oxidation state. Thus, *Jørgensen* (*113, 115*) found that for Ni(II) in an octahedral fluoride environment the effective central atom charge, as given by z_{root}, was $+1.75$, i.e. not too far below the formal $+2$ value, but for Ni(III) and Ni(IV) in the NiF_6^{3-} and NiF_6^{2-} complexes the corresponding z_{root} values are $+1.1$ and $+0.65$, showing that the effective cationic charge may in fact actually *decrease* as the formal oxidation number increases. Similarly it may be seen from the Table that all the z_{root} values for the M(III) oxidation state (MF_6^{3-} series) are appreciably *greater* than the values for the same metal in the M(IV) state, confirming the expected increase in covalency for the higher oxidation states in the 3*d* series. It may of course also be noted that the parameters a_{root}^2 also decrease as β_{35} decreases, but it is not possible to draw any reliable quantitative conclusions about the extent of ligand participation in the $2\,e_g^*$ and $2\,t_{2g}^*$ molecular orbitals since in the first place the normalisation condition $a^2+b^2=1$ does not really apply, and secondly a_{root}^2 properly relates to an averaged $(a_3^2 \cdot a_5^2)^{\frac{1}{2}}$ value. Qualitatively though the trends are as would be anticipated.

(iii) Comparison with Hexachloro Compounds and with Hexahalo Complexes of the 4*d* and 5*d* Series

In this Section we compare the nephelauxetic and related phenomena associated with the MF_6^{3-} and MF_6^{2-} series of the 3*d* elements with those exhibited by the corresponding hexachloro (and hexabromo) complexes, and with similar results for the hexafluoro compounds of the 4*d* and 5*d* series. There exists in the literature of course a vast amount of data concerning the electronic spectra of other hexahalo (—Cl, —Br and —I) compound of the second and third transition series, but in order to restrict this Section to managable proportions we have limited our coverage to the extent indicated.

a) *Comparison with other complexes of the 3d series.* In general the stability of the hexahalo complexes of the 3*d* series decreases quite markedly with increasing atomic number and with increasing size of the halide ligand. Thus hexachloro anions of the M(III) series are known only for the elements Ti to Fe (and possibly Co), the corresponding bromo ions are limited to Ti and V, whilst for the M(IV) series only the hexachloro anions of V and Mn are known. Similarly it appears necessary in the MCl_6^{3-} series for a fairly large cation to be used, such as pyH or enH, in order to stabilise the complex, and the simple alkali metal salts are restricted to Ti and V.

In Table 4 (vii) the available data for D*q* and β for the hexachloro and hexabromo complexes of the M(III) and M(IV) series are given. The spectrum of $(pyH)_3\,TiCl_6$ shows (*38*) a broad band at 12.75 *k*K. with a shoulder at 11.0 *k*K. closely resembling the behaviour of the TiF_6^{3-} anion, and similarly ascribed to Jahn-Teller splitting of the upper ^2E_g state, whilst the $TiBr_6^{3-}$ salt gives a broad band at 11.75 *k*K. In the hexachloro compound the charge-transfer bands begin at about 25 *k*K. (too low to be due to pyH).

Table 4 (vii). *Interpretations of the nephelauxetic effect*

	Dq	β_{35}	z_{min}	a^2_{min}	z_{root}	a^2_{root}
MCl_6^{3-} Complexes						
Ti	1190	—	—	—	—	—
V	1265	0.60	0.7	0.77	1.4	0.88
Cr	1320	0.61	0.65	0.78	1.35	0.88
Mn	1540 [a]	0.52 [a]	0.35	0.72	1.0	0.85
Fe	1110	0.58	0.45	0.76	1.1	0.87
Co(?)	1730	0.34	—0.2	0.58	0.35	0.76
MBr_6^{3-} Complexes						
Ti	1175	—	—	—	—	—
V	1200	0.51	0.45	0.71	1.1	0.84
MCl_6^{2-} Complexes						
V	1540	—	—	—	—	—
Mn	[b]	[b]	—	—	—	—

[a] $MnCl_6^{3-}$ ion experiences strong Jahn-Teller distortion of ^5E_g ground state: values calculated from *g* and *k* parameters.
[b] All *d–d* transitions obscured by strong charge-transfer bands.

Fowles and *Russ* (*38*) also investigated the pyH salts of VCl_6^{3-} and VBr_6^{3-}, reporting bands at 11.4 and 18.0 and at 11.1 and 16.8 *k*K. respectively, which are in turn readily assigned as the $^3T_{1g} \rightarrow {}^3T_{2g}$ and $^3T_{1g} \rightarrow {}^3T_{1g}$ transitions. The data for VCl_6^{3-} are in good agreement with those of *Gruen* and *McBeth* (*47*) for melt systems, which show the charge-transfer bands to start at around 25 *k*K., whilst *Fowles* and *Russ* (*38*) report a charge-transfer band at 23.25 *k*K. for VBr_6^{3-}.

The hexachlorometallate(III) complexes of Cr, Mn and Fe have all been studied by *Hatfield, Fay, Pfluger* and *Piper* (*62*), using the tris-(1,2 propanediamine)-cobalt(III) cation. For $CrCl_6^{3-}$ the bands at 13.2 and 18.7 *k*K. were assigned as $^4A_{2g} \rightarrow {}^4T_{2g}$ and $^4A_{2g} \rightarrow {}^4T_{1g}$, yielding $Dq = 1318$ cm.$^{-1}$ and $B = 575$ cm.$^{-1}$, whilst for the pyH salt *Fowles* and *Russ* found the bands at 12.7 and 18.2 *k*K., leading to $Dq = 1266$ cm.$^{-1}$ and $B = 571$ cm.$^{-1}$. The parameters given in the Table are from *Jørgensen's* (*113*) fitting.

Hatfield et al. (*62*) also report the spectrum of the $MnCl_6^{3-}$ ion, quoting bands at 17.5 and 22.4 *k*K., the latter being ascribed to charge-transfer. They point out that 1750 cm.$^{-1}$ may well be too high a value for Dq because of the substantial tetragonal Jahn-Teller distortion which would be expected for the ^5E_g ground state of this d^4 system, and this view is supported by the observation of a band at 8.3 *k*K. by *Davies, Fackler* and *Weeks* (*61*), which most probably arises from a transition between the Jahn-Teller split components of the ground state (c.f. MnF_6^{3-}, Section 2 (iv)). However, it is not possible to determine B from this information, and, as the value of D*q* cannot be found without resolving the band at 17.5 *k*K., both D*q* and β (Table 4 (vii)) were found using the *g* and *k* parameters.

For the hexachloroferrate anion several investigations have been reported (*62, 67, 118*), with substantially different assignments. On the whole the authors tend to prefer the view of *Yamamata* and *Kato* (*67*), who reported bands at 8.9, 12.8 and 18.0(sh) *k*K., (with the charge-transfer region starting at 23 *k*K.), ascribed respectively to $^6A_{1g} \rightarrow {}^4T_{1g}$, $^6A_{1g} \rightarrow {}^4T_{2g}$, and $^6A_{1g} \rightarrow {}^4A_{1g}$, ^4E_g transitions and leading to the parameters $Dq = 1110$ cm.$^{-1}$ and $B = 605$ cm.$^{-1}$. *Hatfield et al.* (*62*) assign a band at 22.1 *k*K. as $^6A_{1g} \rightarrow {}^4A_{1g}$, ^4E_g, which seems more likely to be charge-transfer, whilst *Balt* and *Verwey's* (*118*) assignments of their spectra lead to very low values of both B and D*q*. The situation though is not at all clear and our inclination to accept *Yamamata* and *Kato's* assignment is due in some measure to the fact that the fitting parameters thus derived seem somewhat the more reasonable.

The $CoCl_6^{3-}$ anion is of rather dubious existence. It is reported (*119*) to be prepared by the action of chloride ions on Co(III) acetate in acetic acid, but the species has only been obtained in solution. This shows bands at 16.5, 21.8 and 32.5 *k*K., the latter attributed to a charge-transfer excitation, but the former two have extinction coefficients of about 2×10^3, which makes their assignment as *d*–*d* transitions most improbable. If, however, such an attribution is accepted, the spectrum can be analysed in terms of a low-spin, t_{2g}^6, ground state, the transitions being respectively $^1A_{1g} \rightarrow {}^1T_{1g}$, $^1A_{1g} \rightarrow {}^1T_{2g}$ and $\pi \rightarrow e_g$, and yielding $Dq = 1730$ cm.$^{-1}$ and $B = 374$ cm.$^{-1}$.

For the M(IV) series only two MCl_6^{2-} anions are known — VCl_6^{2-} and $MnCl_6^{2-}$ — although for vanadium the VBr_6^{2-} species has also been obtained (*20*). A considerable number of investigations, using various large cations, have been carried out on the VCl_6^{2-} ion, but these are all in essential agreement. Thus, for the NEt_2H_2 and NEt_3H salts in acetonitrile *Fowles* and *Walton* (*86*) report the $^2T_{2g} \rightarrow {}^2E_g$ band to lie at 14.9 *k*K., with the charge-transfer region beginning at 21.3 *k*K., whilst by diffuse reflectance the bands are found at 15.1 (broad and asymmetric) and 20.4 *k*K. respectively. *Kilty* and *Nicholls* (*87*) similarly studied the pyH salt, finding these same bands at 15.4 and 21.1 *k*K., both by diffuse reflectance and in MeCN, and *Bereman* and *Brubaker* (*90*), for the Et_4N

salt, found the ${}^2T_{2g} \rightarrow {}^2E_g$ band at 15.4 *k*K., with charge-transfer transitions at 21.2, 37.3 and 41 *k*K., in MeCN. For the Me_4N salt the Jahn-Teller distortion was sufficient to split the ${}^2T_{2g} \rightarrow {}^2E_g$ band, giving a shoulder at 13.8 *k*K., and a peak at 15.3 *k*K.

For the $MnCl_6^{2-}$ anion only the unstable black solid, K_2MnCl_6, has been clearly established. *Moews* (*101*) found strong bands at 15.4, 27.4 and 33.3 *k*K. by absorbance, the former appearing at 16.6 *k*K. by diffuse reflectance, all of these being assigned as charge-transfer excitations on the basis of their intensities, and from the expectation that the $\pi \rightarrow t_{2g}$ band found at about 38 *k*K. in MnF_6^{2-} should lie some 20—25 *k*K. lower in the hexachloro compound. *Moews* (*101*) rejects the quite different spectrum reported by *Sartori et al.* (*121*) which he attributes to hydrolysis of the unstable $MnCl_6^{2-}$ species to give Mn(III) salts.

From the data collected in Table 4 (vii) it may be seen that the Dq and β parameters for the hexachloro and hexabromo complexes follow very much the pattern predicted from the trends embodied in the spectrochemical and nephelauxetic series (s. Sect. 1), and with the exception of the dubious $CoCl_6^{3-}$ the effective cationic charges all lie between + 1 and + 1.5, again following the tendency noted by *Jørgensen* (*113*). It is perhaps noteworthy that in both the MF_6^{3-} and MCl_6^{3-} series the values of β actually decrease very little from V to Fe, but the non-existence of the MCl_6^{3-} ions for Ni and Cu is consistent with the expectation of even smaller nephelauxetic ratios for these anions than those actually found for the hexafluoro compounds.

b) *Comparison with hexafluoro complexes of the 4d and 5d series.* Whilst there exists a considerable body of spectroscopic data for the MCl_6^{n-} complexes of the second and third transition series, the reliable information available for the corresponding hexafluoro anions is rather limited. To facilitate a direct comparison with the results of the MF_6^{3-} and MF_6^{2-} series of the $3d$ elements, we have considered only the hexafluoro anions of the M(III) and M(IV) cations, and have limited the data presented to those for which adequate allowance is made for the now appreciable effects of spin-orbit coupling. (In the $4d$ series ζ is substantially greater than for the $3d$ series, and for the $5d$ elements it becomes comparable with Dq.)

In addition to these restrictions, the significantly greater Dq values encountered in the second and third transition series result in a greater tendency for the d—d bands to be obscured by the charge-transfer transitions, and consequently the data given in Table 4 (viii) relate to only a small number of configurations — $4d^3$, $4d^6$, $4d^8$, $5d^3$ and $5d^6$. When ζ is large the analysis of the d—d bands is often complicated, and we therefore list in the Table only the Dq and β values obtained, together with the $z_{\min}$, $a^2_{\min}$, z_{root} and a^2_{root} parameters derived as before.

Table 4 (viii). *Interpretation of the nephelauxetic effect*

	Dq	$\beta_{35}(\beta_{55})$	z_{min}	a^2_{min}	z_{root}	a^2_{root}
4d Series						
MF_6^{3-} Complexes						
$4d^3$ MoF_6^{3-}	2350	0.90	2.05	0.95	2.35	0.97
$4d^6$ RhF_6^{3-}	2230	0.64	0.60	0.80	1.20	0.89
$4d^8$ AgF_6^{3-}	1840	0.61(0.77) [a]	0.40(1.0)	0.78	1.0	0.88
MF_6^{2-} Complexes						
$4d^3$ TcF_6^{2-}	2840	0.75(0.79)	1.35(1.60)	0.87	2.25	0.93
$4d^6$ PdF_6^{2-}	2590	0.43	0.1	0.66	0.7	0.81
5d Series						
MF_6^{2-} Complexes						
$5d^3ReF_6^{2-}$	3280	—(0.83)	—(2.20)	—	—	—
$5d^6PtF_6^{2-}$	3300	0.53	0.30	0.73	1.15	0.85

[a]) Derived from the β_{33} value assuming $\beta_{55}/\beta_{35}=\beta_{35}/\beta_{33}$.

For the *4d* series the variation of B with *z* and with *q* is given by *Jørgensen* (*113*) as

$$B = 472 + 28\,q + 50\,(z+1) - 500/(z+1)$$

which, for the M^{n+} ($n=1$ to 4) ions gives values of B closely similar to those obtained using the assumption B, M^{n+} (*4d*) $=0.66$B, M^{n+} (*3d*), and is also in reasonable agreement with the values derived from the F_k parameters of *Di Sipio et al.* (*122*). These latter authors also find appreciably greater values of B for M^0 than suggested by the above equation (c.f. *3d* series, *Tondello et al.* (*112*)), and for the *4d* complexes the same reservations may apply concerning the z_{eff} values as were indicated in the footnote to Table 4 (i). For the *5d* series, the values of B_{gas} were obtained assuming B, M^{n+} (*5d*) $=0.60$ B, M^{n+} (*3d*), and the a^2 and z_{eff} parameters derived as before.

An inspection of the results given in Table 4 (viii) shows that very much the same conclusions may be drawn from the data as were advanced on the basis of the information collected in Table 4 (vi), especially as regards the magnitude of the effective cationic charge. The data are too fragmentary for any other firm deductions to be made, but do suggest a more definite decrease in Dq towards the end of the *4d* series than is found for the *3d* complexes, and also support the view that for a given MF_6^{n-} anion the nephelauxetic ratio tends to increase in the order

$3d < 4d < 5d$ thus reinforcing the generally accepted view that the higher oxidation states become relatively more stable in the second and third transition series.

(iv) Miscellaneous Considerations

It is, in principle, possible, from magnetic moment data, to obtain some independent estimate of the extent of the delocalisation of the electrons of the t_{2g} metal orbitals onto the ligands. The usual effect of such delocalisation is in fact to reduce the moment below that predicted by the *Kotani* (*123*) treatment, bringing it closer to the spin-only value, and from the extent of such deviations a semi-quantitative parameter, k, the orbital reduction parameter, can be deduced which affords some measure of the extent of metal-ligand mixing. (See, for example, *Figgis* and *Lewis* (*51*) for full details.)

Unfortunately the magnitude of such effects is usually small in the first transition series, because of the low values of the spin-orbit coupling constants, and in a number of cases deviations from the Kotani predictions also occur by virtue of the presence of low symmetry ligand field components (*124, 125*). Moreover, very few hexafluoro complexes appear to have been studied over a sufficiently wide temperature range and even in these instances there is evidence to suggest, especially in the MF_6^{2-} series, that antiferromagnetic interactions may be producing anomalously high μ values. Thus, for example, Cs_2CoF_6 obeys the *Kotani* predictions in a qualitative sense, but the actual values of the moment are some 0.8 B.M. too high over the measured temperature range as compared to the expected low-spin value.

It was therefore concluded that because of all these limitations it was not possible to obtain meaningful correlations between magnetic data and the nephelauxetic parameters. On the other hand though it is possible to deduce some qualitative connection between the M—F bond lengths (and the implied M^{n+} radii) listed in Table 4 (ix) and the increasing tendency towards covalency on approaching the end of the transition series. For the MF_6^{3-} ions the data (*40*) is very sparse, but for the high-spin situation one would normally expect a *larger* ionic radius for the d^5 Fe^{3+} ion than for the d^3 Cr^{3+}, because of the greater screening power of the e_g electrons, and the reduction of the M—F distance from 1.93 to 1.85 Å is consistent with a shortening due to increased covalency.

For the MF_6^{2-} series of course we are dealing throughout with low-spin ground states and a general reduction of the M—F distance towards the end of the series would be expected in any case. However, for the d^6 NiF_6^{2-} ion the Cs, Rb and K salts exist only in the cubic modification (*40*) and the well known cubic-hexagonal polymorphism of the MF_6^{2-}

series begins only at the Na salt (*105*). This has been attributed to the particularly small size of the NiF_6^{2-} unit, this being fully consistent with the high degree of covalency indicated by the nephelauxetic ratio.

Table 4 (ix). *Metal-fluorine bond lengths for hexafluoro complexes*

	V	Cr	Mn	Fe	Co	Ni
MF_6^{3-} Complexes						
M—F (Å)	—	1.933	[a]	1.85	—	—
MF_6^{2-} Complexes						
M—F (Å)	1.90	1.725	1.73	—	—	1.705

[a]) Jahn-Teller distorted: 4×1.86 Å, 2×2.06 Å.

5. Charge-Transfer Bands and Optical Electronegativities

In the previous Sections we have dealt with the relatively weak, Laporte-forbidden transitions of MF_6^{n-} complexes which arise from excitations within the d^n configurations. Such transitions can be treated quite successfully by the most general application of ligand field theory, but for an adequate description of the strong Laporte-allowed bands it is necessary to utilise the full molecular orbital scheme, as outlined in Section 1 (ii) b. For the present situation, that is for ligands such as fluorine and the other halogens, the surrounding atoms do not possess their own π-bonding system, as for example in CN^-, and it is generally accepted that the Laporte allowed transitions normally represent the transfer of charge from the ligands towards the central metal atom. For simplicity it is usually assumed that one is dealing with well defined configurations, but even so it must be remembered that all the metal orbitals involved, and the ligand orbitals except the 1 t_{1g} and 1 t_{2u} sets, are to some extend delocalised, and that the excited electron cannot then be regarded as being completely transferred from the ligands to the metal. In fact normally there is an appreciable probability of it being in either place, both in the ground and in the excited state.

Nevertheless, for any given type of charge-transfer transition, the positions of the bands may be qualitatively described on the assumption that their energy decreases the more oxidising the central metal and the more reducing the ligands. This expectation arises from the fact that the electron transfer from ligand to metal, which accompanies the transition, effectively constitutes a redox process in which the oxidation number of the central atom is reduced by one unit, whilst the *d—d* bands represent transitions between states characterised by the same oxidation number.

The most obvious distinction between the charge-transfer and the *d—d* bands lies in their intensities however, since the former relate to the excitation of an electron from a level of odd parity to the even t_{2g} or e_g, predominantly metal orbitals, whilst for the latter both the ground and the excited states are formally of even parity, the upper levels being only slightly modified by vibronic interactions. It is seen from Fig. 2 that the highest lying filled molecular orbitals of odd parity are the predominantly ligand $1\,t_{2u}$ and $2\,t_{1u}$ levels, which are respectively purely and predominantly π-bonding, whilst at lower energy the $1\,t_{1u}$ orbital is mainly σ-bonding. In the general case therefore, in which both the even $t_{2g}(\gamma_5)$ and the even $e_g(\gamma_3)$ levels are only partly filled, we can anticipate four main groups of Laporte-allowed charge-transfer transitions:

(i) $\pi \rightarrow \gamma_5$
(ii) $\pi \rightarrow \gamma_3$
(iii) $\sigma \rightarrow \gamma_5$
(iv) $\sigma \rightarrow \gamma_3$

Of these $\pi \rightarrow \gamma_5$ will always lie lowest and $\sigma \rightarrow \gamma_3$ highest whilst the order of $\pi \rightarrow \gamma_3$ and $\sigma \rightarrow \gamma_5$ will depend on whether 10 Dq is greater or less than the energy difference between the π- and σ-levels.

In fact, for the MF_6^{3-} and MF_6^{2-} anions of the first transition series, there is no difficulty in assigning the charge-transfer bands where these have been determined. For the MF_6^{3-} series there is a paucity of information and only for NiF_6^{3-} and CuF_6^{3-} has the charge-transfer region been properly investigated, although for many of the other MF_6^{3-} ions it seems likely that the charge-transfer bands may well lie above 45—50 *k*K. in any case[1]. For both NiF_6^{3-} and CuF_6^{3-} two charge-transfer bands are found quite close together (8—10 *k*K.), which are both assigned as $\pi \rightarrow \gamma_3$ ($t_{2u} \rightarrow e_g$ and $t_{1u} \rightarrow e_g$), since the $\sigma \rightarrow \gamma_3$ transition would be expected to lie about 10 Dq higher.

For the MF_6^{2-} series the positions of the charge-transfer bands have been determined for all the species known, and here each complex shows a rather broad high intensity region with two maxima, again fairly close together (4—9 *k*K. apart) which, for the d^1, d^2, d^3, and d^5 anions, are assigned as $\pi \rightarrow \gamma_5$ ($t_{2u} \rightarrow t_{2g}$ and $t_{1u} \rightarrow t_{2g}$). For these complexes the bands all lie within the range 28—39 *k*K., and no further transitions were detected

[1] For the MF_6^{3-} complexes of Ti, V, Cr, Mn, Fe, and Co the $\pi \rightarrow \gamma_5$ transitions are found respectively at 47.7, 43.3, ca. 55—60, ca. 50—55, 48.8, and 40.6 *k*K., *G. C. Allen, G. A. M. El-Sharkarwy,* and *K. D. Warren,* submitted for publication.

below 45 *k*K [2]). This is consistent with the value of 10 D*q* in the MF_6^{2-} series — 20—22 *k*K. — and furthermore suggests that the $\sigma \rightarrow \gamma_5$ transition also lies at least some 15 *k*K. above the $\pi \rightarrow \gamma_5$ excitation. For the d^6 species, NiF_6^{2-}, only transitions to the γ_3 level are of course available, but the observation (*31*) of a band at 51 *k*K., assigned as $\sigma \rightarrow \gamma_3$, i.e. 20 *k*K. above the first $\pi \rightarrow \gamma_3$ transition, is in accordance with the above conclusions concerning the relative positions of the π- and σ-levels.

For convenience the positions of the charge-transfer bands in the MF_6^{3-} and MF_6^{2-} series are summarised in Table 5 (i). It may be noted here that since all these values relate to diffuse reflectance measurements it is not possible to draw any useful conclusions from intensity data: above about 25 *k*K. the scattering coefficient usually increases sharply so that F_R does not increase in proportion to the extinction coefficient, and is apparently not large enough for the charge-transfer bands as opposed to the *d*—*d* transitions (s. Sect. 1 (iv)).

Table 5 (i). *Charge-transfer transitions of the MF_6^{3-} and MF_6^{2-} anions of the 3d series*

MF_6^{3-} Complexes	$\pi \rightarrow \gamma_3$		
NiF_6^{3-}	32.0, 37.0		
CuF_6^{3-}	29.5, 37.2		
MF_6^{2-} Complexes	$\pi \rightarrow \gamma_5$	$\pi \rightarrow \gamma_3$	$\sigma \rightarrow \gamma_3$
VF_6^{2-}	34.0, 37.4	—	—
CrF_6^{2-}	29.7, 37.3	—	—
MnF_6^{2-}	38.5	—	—
CoF_6^{2-}	28.3, 37.1	—	—
NiF_6^{2-}	—	30.8, 38.9	51.0

For the MF_6^{3-} series there is too little data for any coherent rationalisation of the band positions to be made, but for the MF_6^{2-} complexes the results may be satisfactorily interpreted following the treatment advanced by *Jørgensen* (*3, 126*) for the hexahalo complexes of the 4*d* and 5*d* series. Thus it is assumed that the lowest energy group of electron-transfer bands in each complex results from the transfer of electrons from the mainly ligand π-type orbitals to the partly filled γ_5 sub-shell for the d^1, d^2, d^3, and d^5 systems, and to the empty γ_3 sub-shell for the d^6

[2]) For MnF_6^{2-} a shoulder is observed at 44 *k*K., followed by a rising absorption towards 50 *k*K. (*66*).

system. It is postulated that the orbital energy of the γ_5 electrons varies linearly with the occupation number, q, and is expressed as $W - qE$, whilst that of the γ_3 electrons is then $W - qE + 10\,Dq$. The contribution of the inter-electronic repulsion terms for the central metal, assuming a dependence on the total spin S for the d^q configuration, is then found to be $q(q-1)\,A^*/2 - D\,S\,(S+1)$ where A^* is expressed in terms of the Racah parameters as $A - 8/5\,B + 7/9\,C$ and D is the spin-pairing energy equal to $7/6\,(5/2\,B + C)$. (See *Jørgensen* (*3, 28, 116, 127, 128*) for a full discussion of these parameters and their significance.)

If the energy of the π-electrons is assumed constant (or its variation to be linear in q and thus comprehended in E), the variation of the wave number with q within the series can be expressed as:

$$
\begin{array}{ll}
d^0 & V - 3/4\,D \\
d^1 & V - 5/4\,D + A - E \\
d^2 & V - 7/4\,D + 2\,(A - E) \\
d^3 & V + 7/4\,D + 3\,(A - E) \\
d^5 & V + 3/4\,D + 5\,(A - E) \\
d^6 & V - 3/4\,D + 6\,(A - E) + 10\,Dq
\end{array}
$$

on the assumption of pure low-spin configurations in O_h symmetry. The parameter V is a constant characteristic of the particular transition series, the nature of the ligand, and the oxidation state of the metal. Strictly speaking, the spin-pairing parameter, D, is not constant throughout the series, but in practice it is found, using C/B values estimated from the data of *Tanabe* and *Sugano* (*9*), that it is approximately constant throughout at about 5.5 kK. This value, together with $V = 48.5$ kK., $(E - A) =$ 4.5 kK. and $10\,Dq = 20$ kK., gives a very acceptable prediction of the positions of the barycentres of the $\pi \rightarrow \gamma_5$ and $\pi \rightarrow \gamma_3$ transitions, as is shown by a comparison of the calculated figures (Table 5 (ii)) with the experimental data of Table 5 (i).

The figures in the third column indicate the discrepancy between the calculated and observed values: for TiF_6^{2-} there appears to be no informa-

Table 5 (ii). *Barycentre values of charge-transfer transitions in the MF_6^{2-} series, calculated according to Jørgensen* (*28, 126*)

d^0	TiF_6^{2-}	44.4	(?)	$\pi \rightarrow \gamma_5$
d^1	VF_6^{2-}	37.1	(+1.4)	$\pi \rightarrow \gamma_5$
d^2	CrF_6^{2-}	29.9	(—3.6)	$\pi \rightarrow \gamma_5$
d^3	MnF_6^{2-}	44.6	(?)	$\pi \rightarrow \gamma_5$
d^5	CoF_6^{2-}	30.1	(—2.6)	$\pi \rightarrow \gamma_5$
d^6	NiF_6^{2-}	37.4	(+2.5)	$\pi \rightarrow \gamma_3$

tion about the position of the charge-transfer bands[3]), but since the Laporte-allowed bands for $TiCl_6^{2-}$ begin (*38*) between 20 and 25 *k*K., it seems likely that in TiF_6^{2-} the corresponding transitions would be located between about 45 and 50 *k*K., in reasonable agreement with the calculated value, whilst for MnF_6^{2-} only one $\pi \rightarrow t_{2g}$ transition is found at about 38.5 *k*K., and it is quite possible that a higher energy transition of the same type might have been overlooked simply because few spectrometers extend far above 45 *k*K. For the remaining four cases the average deviation of the predicted value from the experimental is only 2.5 *k*K., which compares quite well with the accuracy of *Jørgensen's* (*126*) treatment of the 4*d* and 5*d* hexahalo complexes.

Although information is in general abundant for the charge-transfer bands of hexahalo complexes of the 4*d* and 5*d* series, relatively little data in fact refers to hexafluoro species — certainly not enough to attempt a rationalisation similar to that given above for the 3*d* MF_6^{2-} ions. Overall there is no doubt that for a given charge-transfer excitation for an nd^q, MF_6^{x-} species, the band energy tends to increase in the sense $3d < 4d < 5d$, but the data for all the hexafluoro anions of the three transition series, and indeed for all the hexahalo species, are probably best interpreted in terms of another concept due to *Jørgensen* (*28, 127, 128*) — that of optical electronegativity.

We have already commented upon the fact that a charge-transfer excitation, in which electron density is transferred from ligand to metal, does effectively correspond to a redox process, and it is a logical, albeit considerable, step to proceed to the assumption that the energy of the charge-transfer band should be directly proportional to the electronegativity difference between the ligand and the central atom. *Jørgensen* (*28*) has discussed at length why this assumption appears inherently somewhat improbable, and why in fact it actually works, and to this the reader is referred for further details.

For the charge-transfer bands of the hexahalo complexes, MX_6^{n-}, it is well known that the bands move to lower energies very regularly as a function of X, the shift from F to Cl being 28 *k*K., from Cl to Br 6 *k*K., and from Br to I 10 *k*K., these being in turn proportional to the differences between the electronegativity values, χ, on the Pauling scale: F, 3.9; Cl, 3.0; Br, 2.8; I, 2.5 if one unit on this scale corresponds to 30 *k*K. It is therefore postulated that the suitably corrected energy of the lowest lying Laporte-allowed charge-transfer band should be proportional to the electronegativity difference between the ligand and the central atom according to the relationship;

$$\sigma_{corr} = (\chi_{opt}(X) - \chi_{opt}(M)) \cdot 30\ kK.$$

[3]) See also footnote p. 119.

We make however two corrections to the experimental value for the lowest charge-transfer band. We allow for the change in the spin-pairing energy accompanying the $d^q \rightarrow d^{q+1}$ transition on the assumption that the average energy of terms with a given S value for the d^q configuration is: $D(<S(S+1)> - S(S+1))$, where the average S term, $<S(S+1)>$, is given by $<S(S+1)> = 3q/4 - 3q(q-1)/16l+4$. The values of the corresponding energies have been tabulated by *Jørgensen* (*128*), and from these may be calculated the appropriate corrections. These will depend upon whether the complex is of high- or low-spin behaviour, but for the 3*d* elements the MF_6^{2-} series shows low-spin character throughout, as do all the hexahalo complexes of the 4*d* and 5*d* elements. Secondly, when dealing with $\pi \rightarrow \gamma_3$ transitions, the experimental value is corrected by subtracting 10 D*q*, in order to make a direct comparison with data derived from $\pi \rightarrow \gamma_5$ transitions. The values thus obtained from the data for the MF_6^{3-} and MF_6^{2-} complexes of the 3*d* series are shown in Table 5 (iii).

Table 5 (iii). *Optical electronegativities for hexafluoro complexes of the 3d series*

MF_6^{3-} Complexes	$\pi \rightarrow \gamma_3$	σ_{corr}	χ_{opt}
NiF_6^{3-}	32.0	26.0	3.05
CuF_6^{3-}	29.5	12.3	3.50
MF_6^{2-} Complexes	$\pi \rightarrow \gamma_5$	σ_{corr}	χ_{opt}
VF_6^{2-}	34.0	38.0 [a]	2.6
CrF_6^{2-}	29.7	37.8	2.65
MnF_6^{2-}	38.5	25.0	3.05
CoF_6^{2-}	28.3	24.6	3.1
	$\pi \rightarrow \gamma_3$	σ_{corr}	χ_{opt}
NiF_6^{2-}	30.8	15.0	3.4

[a]) Using estimated value of D (based on $\beta_{55} \sim 0.8$).

The values of σ_{corr} listed in Table 5 (iii) are obtained by applying the appropriate corrections to the value of σ_{obs} for the $\pi \rightarrow \gamma_5$ or $\pi \rightarrow \gamma_3$ transitions, as listed in Table 5 (iv), and it is also possible, using this Table, to estimate the positions of the lowest charge-transfer bands for the MF_6^{3-} series, for which information is largely lacking. Thus, from *Jørgensen's* (*28*, *127*) χ_{opt} values for V(III), Cr(III), Mn(III), Fe(III), and Co(III), deduced from other halo complex data, (s. Table 5) (v)) we calculate values of 46, 75, 73, 61 and 58 kK. respectively for the lowest $\pi \rightarrow \gamma_5$ transitions, but for Mn and Co these figures are subject to a greater than usual uncertainty since χ_{opt} values relating to low-spin situations have been used in the absence of high-spin values[4]).

Table 5 (iv). *Corrections for changes in the spin-pairing energy parameter associated with charge-transfer transitions*

d^0		0	
d^1		+2/3 D	
d^2		+4/3 D	
d^3		—2 D	
d^4	—4/3 D (low-spin)		—7/3 D (high-spin)
d^5	—2/3 D (low-spin)		—8/3 D (high-spin)
d^6	+D—Δ (low-spin)		—2 D (high-spin)
d^7	+5/3 D—Δ (low-spin)		—4/3 D (high-spin)
d^8		—2/3 D—Δ	
d^9		—Δ	

All d^1 to d^5 configurations refer to $\pi \rightarrow \gamma_5$ transitions, as do the high-spin values for d^6 and d^7, whilst all remaining configurations refer to $\pi \rightarrow \gamma_3$ excitations. $\Delta = 10$ Dq.

From the large amount of data relating to the hexahalo complexes of the 4d and 5d series optical electronegativities have been derived for many of the corresponding oxidation states of these metals. These values, together with a collection of all those now available for the 3d series, are shown in Table 5 (v), from which it may be seen that the χ_{opt} values reflect quite faithfully the general trends of decreasing stability of the high oxidation states towards the end of a transition series, and increasing stability of such states in the order 3d, 4d, 5d. Note that the high values of χ_{opt} represent relatively unstable, strongly oxidising, situations, and low values the more stable conditions.

From the arguments advanced above and in Section 4 it would be expected that some relationship should exist between χ_{opt} and β, the nephelauxetic ratio. Thus, χ_{opt} constitutes a measure of the stability of a given oxidation state, i. e. the ease or otherwise of reducing it, and β, since it assesses the tendency towards covalency, essentially parallels the same trend, i. e. the extent of electron accession from the ligand, orbitals towards the metal in any given complex. The optical electronegativity has of course been shown to be virtually independent of the

4) See also footnote p. 118. The derived χ_{opt} values for trivalent Ti, V, Cr, Mn, Fe, and Co are therefore appreciably higher than those given in Table 5 (v).

ligand, as far as the halides are concerned, so that we should expect a correlation to exist if we consider β values for a series of complexes of the *same* ligand — here fluorine[5]).

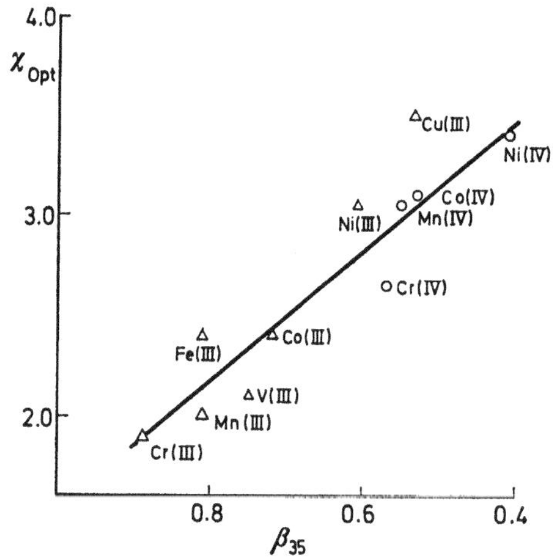

Fig. 18. Correlation between β_{35} and χ_{opt}

In fact, the available results do support this view, especially for the *3d* series. For the MF_6^{2-} series both β and χ_{opt} have been deduced from the same measurements on the hexafluoro complexes, but for the MF_6^{3-} series the charge-transfer bands lie above about 45 *k*K. in general, except for Ni(III) and Cu(III), and it is necessary to use χ_{opt} values derived (*28, 127*) from other systems in order to make the comparison with the β values for the hexafluoro anions. The rather limited amount of data for the β values of the 4*d* and 5*d* hexafluoro complexes somewhat curtails the treatment of the corresponding complexes of these series, but the indications are that a similar situation obtains. The position for the 3*d* series is illustrated in Fig. 18 in which χ_{opt} is plotted against β_{35} for the MF_6^{3-} and MF_6^{2-} series.

As noted previously, reliable spectroscopic data for other hexahalo complexes of the 3*d* series are rather scarce, but where it is to hand the optical electronegativity values deduced agree quite reasonably well with the results derived from the hexafluoro compounds, although for

[5]) For a theoretical justification of the connection between χ_{opt} and β see *G. C. Allen* and *K. D. Warren*, Mol. Phys., *20*, 379 (1971).

V(IV), since it is a d^1 system, it is necessary to assume a rough value of B, which may play some part in the discrepancy in the χ_{opt} values obtained from VF_6^{2-} and VCl_6^{2-}. For $MnCl_6^{2-}$ it is also necessary to assume a B value, since all the *d–d* bands are obscured, but the χ_{opt} value of 2.7 is tolerably close to the 3.05 deduced from MnF_6^{2-}. For VBr_6^{3-} the value obtained is the same as is listed in Table 5 (v), and good agreement is also found for $CoCl_6^{3-}$. For $FeCl_6^{3-}$ the χ_{opt} value is rather larger than listed before, but not unacceptably so, which serves to underline the uncertainty involved in attempting to *estimate* the position of the charge-transfer bands for the MF_6^{3-} anions on the basis of χ_{opt} values.

Table 5 (v). *Optical electronegativities*

3*d* Series	Ti	V	Cr	Mn	Fe	Co	Ni	Cu
M (II)	—	—	—	—	—	1.9	2.1	2.3
M (III)	—	2.1	1.9	2.0 [a]	2.1 [a] 2.4 [b]	2.4 [a]	3.05	3.5
M (IV)	1.8	2.6	2.65	3.05	—	3.1	3.4	—
4*d* Series			**Mo**	**Tc**	**Ru**	**Rh**	**Pd**	**Ag**
M (III)			1.7	—	2.05	2.3	—	3.65
M (IV)			—	2.2	2.4	2.6	2.7	—
5*d* Series			**W**	**Re**	**Os**	**Ir**	**Pt**	**Au**
M (III)			—	—	1.95	2.25	—	2.9
M (IV)			—	2.05	2.2	2.35	2.7	—
M (VI)			2.0	—	2.6	2.9	3.2	—

[a] low-spin.
[b] high-spin.

Table 5 (vi). *Optical electronegativities for hexachloro and hexabromo complexes*

	$\pi \to \gamma_5$	χ_{opt}		$\pi \to \gamma_5$	χ_{opt}
VBr_6^{3-}	23.25	1.9	VCl_6^{2-}	21.2	2.25
$FeCl_6^{3-}$	23.0	2.6	$MnCl_6^{2-}$	16.6	2.7
$CoCl_6^{3-}$	32.5 ($\pi \to \gamma_3$)	2.3			

6. Vibrational Fine Structure of *d—d* Bands

For measurements carried out by the ordinary absorption spectrophotometric methods vibrational fine structure is only rarely reported, and seems to be largely confined to the elements of the 5*d* series. Thus it is reported (*129*) for ReF_6 and PtF_6, in the gas phase, and for $ReCl_6^{2-}$ and $OsCl_6^{2-}$ in solution (*130*), but more usually spectrophotographic techniques are required for its detection, even in the third transition series. It is therefore noteworthy that such vibrational detail has been found by diffuse reflectance (*85, 103, 108*) for some of the *d—d* bands of NiF_6^{2-}, CoF_6^{2-}, and MnF_6^{2-}, especially in view of the fact that the quality of resolution attainable by this technique is usually lower than that achieved by absorption spectrophotometry. (Figs. 19 and 20).

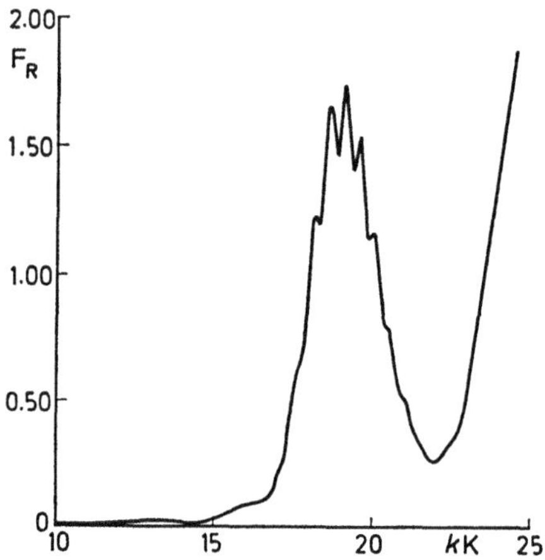

Fig. 19. Vibrational fine structure in the spectrum of K_2NiF_6 (*108*)

For the NiF_6^{2-} ion vibrational fine structure was also detected by *Reisfeld et al.* (*31*), using a fluorolube mull technique, both investigations revealing a well marked regular vibrational progression on the $^1A_{1g} \rightarrow {}^1T_{1g}$ band. Originally *Reisfeld et al.* interpreted this feature as indicating the operation of a significant Jahn-Teller distortion, via the ν_2, ε_g mode, for the excited state, but have, we understand (*97*), recently revised their views on this as a result of studies on K_2MnF_6, and are now in agreement with the authors in assigning the progression to the α_{1g} mode. We therefore now summarise briefly the evidence and reasoning leading to this conclusion as a general explanation of the vibrational fine structure in this series.

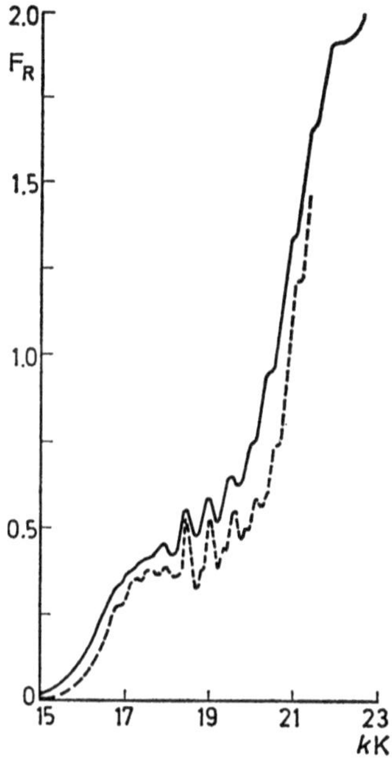

Fig. 20. Vibrational fine structure in the spectrum of Cs_2CoF_6 (*103*)
——— Room temperature, --------- Liquid nitrogen temperature

It is generally accepted that the *d–d* bands shown by transition metal complexes, although formally forbidden, acquire their intensities by a vibronic coupling mechanism in which one quantum of an odd vibrational mode, usually τ_{1u}, is co-excited. One can then rule out the possibility that the observed structure corresponds to a uni-quantal progression in an odd mode since all even overtones of an odd vibration are necessarily of even parity, and therefore still constitute Laporte-forbidden transitions, except in the improbable case that the centre of inversion is destroyed in the excited state. We have thus to consider the even vibrational modes, α_{1g}, ε_g, and τ_{2g}, as possible participants in the observed effect, but it is clear that the transition will only take place with any appreciable intensity at distances in the region of the equilibrium configuration of the $^1A_{1g}$ ground state, and that by the Franck-Condon principle it will take place without change in configuration or in the Ni–F distance. This would probably limit the vibrational changes of all the modes to one quantum, except for the fully symmetric α_{1g} vibration or for the ε_g and τ_{2g} modes should the excited state show a large enough Jahn-Teller distortion.

In this latter case the distortion would have to be very large for the excited state vibrational levels to be evenly spaced, as required by the regularity of the progression, and in such cases it has been shown by *Dunn* (*131*) that a rather distorted irregular band contour would normally be expected to result. Conversely, the α_{1g} mode would be expected to give rise to a perfectly regular spacing — anharmonicity at higher vibrational quantum numbers would be inhibited by the O_h site symmetry — and since the excited state, $^1T_{1g}$ ($t_{2g}^5\, e_g$), has a increased occupation of the e_g anti-bonding level the equilibrium inter-nuclear distance should be appreciably greater than in the ground state, thereby decreasing the force constant for the Ni—F bond. It thus follows, from the Franck-Condon principle, that transitions to high vibrational α_{1g} levels will occur (see Fig. 21), which corresponds to the maximum intensity representing the (4,0) transition, at 19.08 *k*K., with the origin (0,0), of the α_{1g} progression at 17.09 *k*K. Our results (*110*) suggest that the band as a whole is τ_{1u} enabled, although the intensities are not accurate enough to distinguish between ν_3 and ν_4, and thus the true origin of the whole band, which is of course absent, must lie one τ_{1u} quantum below 17.09 *k*K.

For the NiF_6^{2-} anion the ground state values for the ν_1, α_{1g} and ν_2, ε_g modes are 562 and 520 cm.$^{-1}$ respectively, and since the progression spacing is about 490 cm.$^{-1}$, *Reisfeld et al.* (*31*) originally favoured its assignment as an ε_g mode. However, a reduction of the α_{1g} frequency from 562 cm.$^{-1}$ in the ground state to 490 cm.$^{-1}$ in the excited state is entirely reasonable in view of the expected concomitant reduction of the force constant for the Ni—F bonds, and the observations of *Reisfeld, Matwiyof*, and *Asprey* (*97, 98*) on K_2MnF_6 support this interpretation. Here the ground state Raman data yield ν_1, $\alpha_{1g} = 612$ cm.$^{-1}$, and ν_2, $\varepsilon_g = 488$ cm.$^{-1}$, whilst for the $^4A_{2g} \rightarrow {}^4T_{2g}$ band a vibrational spacing of 520 cm.$^{-1}$ is found: this latter cannot then represent an ε_g progression, but the order of magnitude of the reduction in wave number suggests strongly that it too represents an excited state α_{1g} progression. For CoF_6^{2-} the authors (*103*) have observed vibrational fine structure for the $^2T_{2g} \rightarrow {}^2T_{2g}$ and $^2T_{2g} \rightarrow {}^2E_g$ bands, the progressions overlapping to some extent. This makes the estimation of the spacing rather difficult but the spectrum obtained at 77 °K can be analysed in terms of progressions of 520 ± 50 cm.$^{-1}$, which is in good agreement with the ground state ν_1, α_{1g} value of 560 cm.$^{-1}$.

Whilst however the mechanism by which the vibrational detail is produced seems to be reasonably well established, it is not at all clear just why it is actually observed in some cases and not in others. It has been suggested by *Nelson* (*76*) that the actual site symmetry of the MF_6 unit, as opposed to the micro-symmetry about the central metal, may play some part, and in support of this we may quote the observation of fine structure for K_2NiF_6, Cs_2CoF_6, and Cs_2MnF_6, all of which have cubic lattices, and in which both the site and the micro symmetry is O_h. On the other hand *Allen et al.* (*96*) found vibrational detail for K_2MnF_6, prepared by the wet method and therefore most probably hexagonal, and this result has also been recorded by *Reisfeld* and co-workers (*97*). Although *Novotny* and *Sturgeon* (*100*) considered that the K_2MnF_6 sample of *Allen et al.* might well have been cubic, there is no conclusive evidence to support this, and the vibrational detail is also observed for K_2MnF_6 prepared by direct fluorination which would be expected to be either hexagonal or trigonal. Moreover, the authors (*108*) also found vibrational fine structure for Na_2NiF_6, which is a (*105*) cubic-hexagonal mixture, and we are therefore inclined to think that site symmetry is not a critical factor in the production of fine structure.

However it is noteworthy that the observation of vibrational detail appears, for the 3*d* series, to be restricted to complexes of metals in the tetravalent state. Thus it is detected for Ni(IV), Co(IV), and Mn(IV), although not found for Cr(IV) and V(IV), but absent for V(III), Cr(III), Mn(III), Co(III), and Cu(III). The structure reported (3) by *Jørgensen* for FeF_6^{3-} also shows too small a spacing for an α_{1g} progression, and for NiF_6^{3-} only traces of structure are found at 77 °K. Moreover, such detail appears to be more pronounced for complexes showing low values of the nephelauxetic ratio, β. Thus, the effect is extremely well marked for NiF_6^{2-} ($\beta = 0.41$), less clear for CoF_6^{2-} (0.53), and rather indistinct for most samples of MnF_6^{2-} salts (0.55). Similarly, for 5*d* compounds, the effect is well marked for PtF_6 ($\beta_{55} = 0.30$), for $ReCl_6^{2-}$ ($\beta_{55} = 0.62$), and for $OsCl_6^{2-}$ ($\beta_{55} = 0.52$).

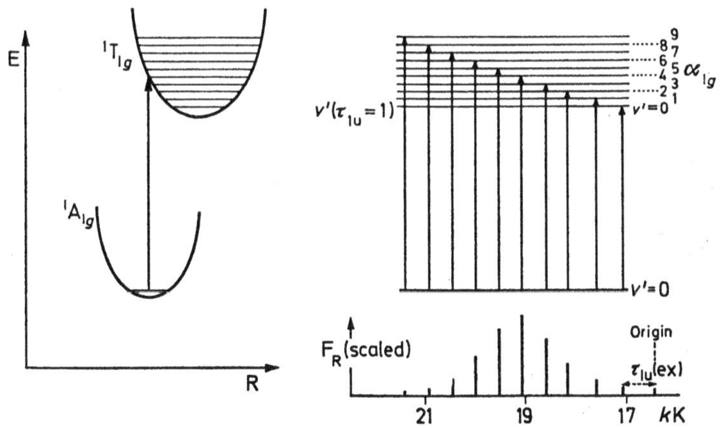

Fig. 21. Origin of the vibrational fine structure in the $^1A_{1g} \rightarrow {}^1T_{1g}$ band of K_2NiF_6

Consequently, these results can be rationalised on the assumption that the superposition of the α_{1g} progression will result in changes in polarisability which may lead to significant changes in the transition moment of the τ_{1u} enabled vibronic excitation. Such an effect should be most marked in complexes of metals bearing high formal positive charges, since, in general these show rather low values of β — i.e. a substantial measure of covalency — and are therefore more polarisable than the more highly ionic complexes of metals in the lower oxidation states. It cannot of course be maintained that these are the only factors which may contribute to the effect, but whilst it is always dangerous to place too much relicance on negative information, the available information is at least consistent with this hypothesis.

As regards solution measurements, aimed at detecting vibrational fine structure, very little work has been attempted, probably because of the extreme experimental difficulties which arise. It is however reported *(29)* that K_2NiF_6 shows no vibrational fine structure in the spectrum obtained from solutions in anhydrous HF, but the conclusion that this is a consequence of the removal of the NiF_6^{2-} unit from its octahedral lattice environment appears premature: it is certainly not possible to rule out any of the usual solvent broadening effects as being responsible for the disappearance of the vibrational detail.

7. Jahn-Teller Effects

In this Section we consider briefly the spectroscopic consequences of the operation of the Jahn-Teller effect in the hexafluoro complexes of the $3d$ series. In this context we shall neglect the influence of spin-orbit coupling, which is small in the first transition series, and treat only the orbital degeneracies involved.

For the MF_6^{2-} complexes, VF_6^{2-} (d^1), CrF_6^{2-} (d^2), MnF_6^{2-} (d^3), CoF_6^{2-} (d^5) and NiF_6^{2-} (d^6), the corresponding ground states are: ${}^2T_{2g}$ (t_{2g}^1), ${}^3T_{1g}$ (t_{2g}^2), ${}^4A_{2g}$ (t_{2g}^3), ${}^2T_{2g}$ (t_{2g}^5) and ${}^1A_{1g}$ (t_{2g}^6), and it therefore follows that ground state Jahn-Teller effects will not be operative for MnF_6^{2-} and NiF_6^{2-}. For all the other species the ground states exhibit t_{2g} orbital degeneracies and would thus not be expected to show very significant Jahn-Teller distortions because the t_{2g} orbitals are involved only in the usually rather small π-bonding. The individual ions are discussed in more detail in Section 3, but overall there appears no compelling evidence for any appreciable ground state splittings.

There remains though the possibility of excited state Jahn-Teller effects; thus, in a number of d^1 complexes the rather distorted band shape which is often observed for the $^2T_{2g} \rightarrow {}^2E_g$ transition has been ascribed to Jahn-Teller splitting of the upper ^2E_g level. Such an effect is not inherently unreasonable since Jahn-Teller splittings would be expected to be substantially larger for e_g orbital degeneracies since the e_g level is involved in the normally dominant σ-bonding. On the whole though there are few indications of such an effect for the MF_6^{2-} series: even for the d^1 system, VF_6^{2-}, no splitting of the single $^2T_{2g} \rightarrow {}^2E_g$ band is found, although the band width is slightly greater than normal, and the beginning of the charge-transfer region may perhaps serve to prevent the observation of any band distortion.

On the other hand the MF_6^{3-} series provides four quite clear cut examples of the effect of Jahn-Teller distortions on the electronic spectrum. For the MF_6^{3-} anions, TiF_6^{3-} (t_{2g}^1), VF_6^{3-} (t_{2g}^2), CrF_6^{3-} t_{2g}^3), MnF_6^{3-} ($t_{2g}^3\, e_g$), FeF_6^{3-} ($t_{2g}^3\, e_g^2$), CoF_6^{3-} ($t_{2g}^4\, e_g^2$), NiF_6^{3-} ($t_{2g}^6\, e_g$) and CuF_6^{3-} ($t_{2g}^6\, e_g^2$), the respective ground states are: $^2T_{2g}$, $^3T_{1g}$, $^4A_{2g}$, ^5E_g, $^6A_{1g}$, $^5T_{2g}$, ^2E_g and $^3A_{2g}$, from which we deduce that CrF_6^{3-}, FeF_6^{3-} and CuF_6^{3-} will be free from ground state Jahn-Teller effects. However, excited state effects involving e_g orbital degeneracies would be anticipated for TiF_6^{3-} and CoF_6^{3-} and ground state splittings, also involving e_g orbital degeneracies, should occur for MnF_6^{3-} and NiF_6^{3-}. In fact all these expectations are realised and the four systems concerned are now discussed in greater detail.

a) *Excited State Jahn-Teller Effects.* The results of *Bedon, Horner* and *Tyree* (*36*) for the TiF_6^{3-} anion have already been surveyed in Section 2 (i). As indicated there it seems most probable that the splitting of the single $^2T_{2g} \rightarrow {}^2E_g$ d–d band is due to a tetragonal Jahn-Teller effect operating for the excited state, since the splitting is observed even for the strictly cubic K_2NaTiF_6. The rather larger separation found for the NH_4 salt is consistent with the assumption of deviations from O_h symmetry in the crystal lattice since it is likely that the ammonium salt is not in fact cubic.

A rather similar, but rather more marked, effect arises for CoF_6^{3-} (s. Sect. 2 (vi)). The single spin-allowed band, $^5T_{2g} \rightarrow {}^5E_g$ is split into two components separated by 2.6 kK. or more, depending on the cation, and the data again do not indicate any static distortion for the K_2Na salt, although once more the salts showing the greater deviations from cubic symmetry exhibit the larger splitting of the two bands. Thus the Li and Ba salts which are definitely non-cubic show splittings of 3.4 and 6.5 kK. respectively. It thus seems necessary here also to postulate an excited state Jahn-Teller effect, but this can be rationalised on the grounds that the ^5E_g level corresponds to a $t_{2g}^3\, e_g^3$ configuration — i.e. a configuration with a stable half-filled shell for the t_{2g} level, but an unstable orbital degeneracy for the e_g level.

b) *Ground State Jahn-Teller Effects.* For the MnF_6^{3-} ion there is definite crystallographic evidence for a ground state tetragonal distortion in the K_2Na salt, but from a spectroscopic standpoint it has been known for some time that high-spin Mn(III) complexes behaved somewhat anomalously. Thus only one spin-allowed transition, $^5E_g \rightarrow {}^5T_{2g}$ would normally be expected for this system, but very many Mn(III) complexes show at least two absorptions below about 20—25 *k*K. which seem of the right intensity to correspond to spin-allowed *d—d* transitions. In particular the origin of a low energy band, frequently found at around 10 *k*K., remained a mystery, but the situation has been analysed in detail by *Davis*, *Fackler*, and *Weeks* (*61*), who adduced convincing arguments that both this band, and a higher energy transition found between 15 and 20 *k*K., resulted from a tetragonal Jahn-Teller splitting of the ^5E_g ground state. Such a conclusion is not altogether unexpected: the ^5E_g ground level has a $t_{2g}^3\,e_g$ configuration, once more constituting an unstable e_g orbital degeneracy for which a substantial tetragonal Jahn-Teller splitting would be predicted, but the excited state $^5T_{2g}$ level corresponds to a $t_{2g}^2\,e_g^2$ configuration, with a stable half-filled e_g level, and only a small splitting of the $^5T_{2g}$ level would be expected to arise from the t_{2g} orbital degeneracy. Thus *Davis et al.* assigned the 10 *k*K. band to a transition between the two Jahn-Teller split components of ^5E_g, and the band at 15 to 20 *k*K., which is often broadened and sometimes split, to two transitions to the split components of the $^5T_{2g}$ state. The situation is illustrated diagramatically in Fig. 6 (Sect. 2 (iv)).

Davis et al. also carried out calculations of the potential energy surfaces for the Jahn-Teller split ^5E_g and $^5T_{2g}$ states, on the basis of the treatment given by *Liehr* and *Ballhausen* (*132*). Their findings are reproduced in Fig. 22, and show that a separation of some 8 to 10 *k*K. is indicated between the two potential energy sheets which result for the ground state, but a much smaller splitting is presumed for the $^5T_{2g}$ state. This calculation was made on the assumption of a dynamic Jahn-Teller effect since, as *Davis et al.* point out, it is by no means certain that all Mn(III) complexes do in fact show a static Jahn-Teller splitting. For MnF_6^{3-} there is of course direct evidence of tetragonal elongation from crystallographic measurements but *Davis et al.* found that for tris (acetylacetonate) manganese(III) all the Mn—O bond lengths were identical, and since this complex also showed the characteristic low energy band at 9.5 *k*K (and higher energy bands at 17.9 and 21.5 *k*K.) they concluded that it constituted an example of a dynamic Jahn-Teller effect, or pseudorotation. However, it has recently been shown by *Bartlett* and *Palenik* (*133*) that although all the Mn—O bond lengths are identical for $Mn(acac)_3$ the structure does in fact show angular distortion, so that the postulation of a dynamic effect in this case may be unnecessary.

Davis, Fackler, and *Weeks* (*61*) also pointed out that since the minima of the components of ^5E_g may be quite widely separated (0.2—0.5 Å), the band corresponding to a transition between the two components of ^5E_g should be exceptionally broad, and in the numerous cases in which a low energy band is found for Mn(III) complexes, including MnF_6^{3-}, its width is certainly substantially greater than usual. It might also be expected that for a transition between the two split potential sheets of the ^5E_g state the vibrational overlap integral might be rather smaller than normal, leading, to a band intensity rather less than that of a normal spin-allowed *d—d* transition, and this expectation is also fulfilled, both for MnF_6^{3-} and for Mn(III) compounds in general.

On the basis of the preceeding arguments a substantial Jahn-Teller effect would also be expected for the NiF_6^{3-} ion: the spectroscopic and magnetic data (Sect. 2 (vii)) clearly indicate a low-spin ^2E_g ground state, with a $t_{2g}^6\, e_g$ configuration, and once again the orbital degeneracy in the σ-bonding e_g level should produce an appreciable ground state splitting. The situation is also somewhat similar to the MnF_6^{3-} case in that the $^2T_{1g}$ and $^2T_{2g}$ levels, which correspond to spin-allowed one-electron excitations, possess the $t_{2g}^5\, e_g^2$ configuration in which the e_g shell is now half-filled and the orbital degeneracy transferred to the π-bonding t_{2g} orbitals, where its effect should be much smaller. We can therefore anticipate that the tetragonal Jahn-Teller splitting of the ^2E_g ground state will be quite considerable, but that only small separations will result for the excited levels. The situation is illustrated in Fig. 10 (Sect. 2 (vii)).

Once again it is noteworthy that the first *d—d* band, at 6—7 *k*K. is both significantly broader, and appreciably weaker, than the other *d—d* bands, in accordance with the reasoning outlined above, and it is therefore of interest to examine the evidence available to determine whether this is a general characteristic of bands which represent transitions between the two potential sheets of a Jahn-Teller split ground state. In fact we have already noted that this behaviour seems to be typical for ^5E_g, d^4 complexes of high-spin Mn(III), and *Fackler* and *Holah* (*134*) found a similar result for the isoelectronic $Cr(H_2O)_6^{2+}$ ion. The d^9 Cu(II) species, with a ^2E_g ground state, would also appear capable of producing a band corresponding to a transition between the split ground state components, but the position is here less straightforward; for most ligands for which Cu(II) forms formally octahedral complexes the value of 10 D*q* is such that the transition thought to arise from excitation between the ground level components seems to overlap that representing transitions to the two components of the upper $^2T_{2g}$ level. Consequently only one very broad, rather asymmetric, band is observed, and although it has been maintained (*135*), on the basis of Gaussian analysis, that this contains three distinct transitions, the evidence is less conclusive than in the other cases, even though it has gained fairly general acceptance.

Nevertheless, there are other systems which may be investigated to test the hypothesis further. For example, the Co(II) complex, $Co(NO_2)_6^{4-}$, is known from magnetic studies (*73*) to have a low-spin d^7, ^2E_g ground state, and its reflectance spectrum has been measured by *Elliott, Hathaway*, and *Slade* (*136*) and by the authors (*137*), together with those of the analogous Fe(II), Ni(II), and Cu(II) species. It is found that both the complexes possessing a ^2E_g ground state, that is $Co(NO_2)_6^{4-}$ and $Cu(NO_2)_6^{4-}$, exhibit a very broad, rather weak band at about 7.5 *k*K., and a stronger band, of normal *d–d* intensity, at around 16 *k*K. For the Co(II) compound the low energy band is strikingly similar to that shown by NiF_6^{3-} and provides further evidence for the correctness of the assignment proposed in Section 2 (vii), whilst for the Cu complex the observation of *two* bands which apparently represent *d–d* transitions also supports the views described above for Cu(II) species as a whole. In this latter case it appears that the rather large D*q* value associated with the NO_2^- ligand ensures that the ^2E_g and $^2B_{2g}$ levels, arising from the tetragonal splitting of the upper $^2T_{2g}$ state, lie so far above the upper split component ($^2A_{1g}$) of the split ^2E_g ground state that two distinct bands are observed. Here again the low energy band is very broad, and quite noticably weaker than usual for a *d–d* transition, thereby supporting the view that the behaviour observed for the MnF_6^{3-} and NiF_6^{3-} anions is typical for excitations taking place between the Jahn-Teller split potential sheets arising from the ground state. It may here be noted that

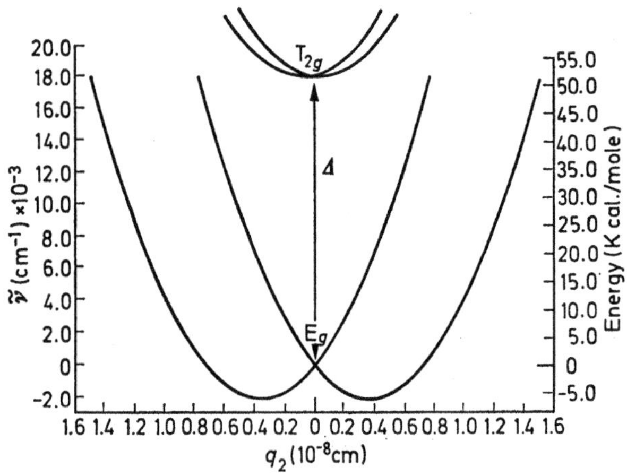

Fig. 22. Potential energy surfaces for Jahn-Teller split E_g and T_{2g} states (*61*)

Elliott et al. (*136*) were unable to make a clear cut decision between a static and a pseudorotational dynamic Jahn-Teller effect for the hexanitrito complexes, as is the case also for NiF_6^{3-}, but this distinction would not in principle affect the spectroscopic consequences, since such measurements reflect the instantaneous rather than the time averaged situation.

Acknowledgements

We wish to thank Mr. *G. A. M. El-Sharkarwy*, of this College, for making available to us a number of hitherto unpublished spectra, and Dr. *P. G. Nelson*, of the University of Hull, for a number of private communications and a stimulating correspondence. We also would like to thank Dr. *M. F. A. Dove*, of the University of Nottingham, for supplying us with his latest data on K_3NiF_6, and Dr. *M. J. Reisfeld*, of the University of California, Los Alamos, for permission to use his unpublished material, and for results on the MnF_6^{2-} ion.

8. References

1. *Jørgensen, C. K.:* Acta Chem. Scand. *12*, 1539 (1958).
2. *von Schnering, H. G.:* Z. Anorg. Allgem. Chem. *353*, 1 (1967).
3. *Jørgensen, C. K.:* Absorption Spectra and Chemical Bonding in Complexes. London: Pergamon Press 1962.
4. *Ballhausen, C. J.:* Introduction to Ligand Field Theory. New York: McGraw-Hill Book Co., Inc. 1962.
5. *Figgis, B. N.:* Introduction to Ligand Fields. New York: Interscience Publishers 1966.
6. *Bethe, H.:* Ann. Physik *3*, 133 (1929).
7. *Condon, E. U., Shortley, G. H.:* The Theory of Atomic Spectra. London: Cambridge University Press 1963.
8. *Orgel, L. E.:* J. Chem. Phys. *23*, 1004 (1955).
9. *Tanabe, Y., Sugano, S.:* J. Phys. Soc. Japan *9*, 753, 766 (1954).
10. *Griffith, J. S.:* The Theory of Transition Metal Ions. London: Cambridge University Press 1964.
11. *Dunn, T. M., McClure, D. S., Pearson, R. G.:* Some Aspects of Crystal Field Theory. London: Harper Row 1965.
12. *Kettle, S. F. A.:* Coordination Compounds. London: Nelson 1969.
13. *Cotton, F. A.:* Chemical Applications of Group Theory. New York: Interscience Publishers 1963.
14. *Liehr, A. D., Ballhausen, C. J.:* Ann. Phys. (N. Y.) *6*, 134 (1959).
15. *Eisenstein, J. C.:* J. Chem. Phys. *34*, 1628 (1961).
16. *Schroeder, K. A.:* J. Chem. Phys. *37*, 1587, 2553 (1962).
17. *Jørgensen, C. K.:* Acta Chem. Scand. *12*, 903 (1958).
18. *Orgel, L. E.:* Quart. Rev. *11*, 381 (1957).
19. *Offenhartz, P. O'D.:* J. Am. Chem. Soc. *91*, 5699 (1969).
20. *Wolfsberg, M., Helmholz, L.:* J. Chem. Phys. *20*, 837 (1952).
21. *Ballhausen, C. J., Gray, H. B.:* Molecular Orbital Theory. New York: W. A. Benjamin, Inc. 1964.

22. *Basch, H., Viste, A., Gray, H. B.:* J. Chem. Phys. *44*, 10 (1966).
23. *Roothaan, C. C. J.:* Rev. Mod. Phys. *23*, 69 (1951).
24. *Dahl, J. P., Ballhausen, C. J.:* Advan. Quantum Chem. *4*, 170 (1968).
25. *Pople, J. A., Santry, D. P., Segal, G. A.:* J. Chem. Phys. *43*, S129 (1965).
26. *Fenske, R. F., Caulton, K. G., Radtke, D. D., Sweeney, C. C.:* Inorg. Chem. *5*, 951 (1966).
27. *Brown, R. D., Roby, K. R.:* Theoret. Chim. Acta (Berlin) *16*, 175, 194 (1970).
28. *Jørgensen, C. K.:* Orbitals in Atoms and Molecules. New York: Academic Press 1962.
29. *Stein, L., Neil, J. M., Alms, G. R.:* Inorg. Chem. *8*, 2472 (1969).
30. *Bedon, H. D., Hatfield, W. E., Horner, S. M., Tyree, S. Y.:* Inorg. Chem. *4*, 743 (1965).
31. *Reisfeld, M. J., Asprey, L. B., Penneman, R. A.:* J. Mol. Spectry. *29*, 109 (1969).
32. *Kortüm, G.:* Reflectance Spectroscopy. Berlin-Heidelberg-New York: Springer 1969.
33. — *Schottler, H.:* Z. Elektrochem. *57*, 353 (1953).
34. *Ehrlich, P., Pietzka, G.:* Naturwissenschaften *40*, 509 (1953).
35. *Bright, N. F. H., Wurm, J. G.:* Can. J. Chem. *36*, 615 (1958).
36. *Bedon, H. D., Horner, S. M., Tyree, S. Y.:* Inorg. Chem. *3*, 647 (1964).
37. *Van Vleck, J. H.:* J. Chem. Phys. *7*, 61 (1939).
38. *Fowles, G. W. A., Russ, B. T.:* J. Chem. Soc. (A) 517 (1967).
39. *Nyholm, R. S., Sharpe, A. G.:* J. Chem. Soc. 3579 (1952).
40. *Babel, D.:* Struct. Bonding *3*, 1 (1967).
41. *Ballhausen, C. J., Winther, F.:* Acta Chem. Scand. *13*, 1729 (1959).
42. *Machin, D. J., Murray, K. S.:* J. Chem. Soc. (A) 1498 (1967). — See also *Clark, R. J. H.:* J. Chem. Soc. 417 (1964).
43. *Clark, H. C., Sadana, Y. N.:* Can. J. Chem. *42*, 50 (1964).
44. *Pryce, M. H. L., Runciman, W. A.:* Discussions Faraday Soc. *26*, 34 (1958).
45. *Warren, K. D.:* unpublished matrices.
46. *Figgis, B. N., Lewis, J., Mabbs, F. E., Webb, G. A.:* J. Chem. Soc. (A) 1411 (1966).
47. *Gruen, D. M., McBeth, R. L.:* J. Phys. Chem. *66*, 57 (1962).
48. *Christensen, O. T.:* J. Prakt. Chem. *35*, 163 (1887).
49. *Peacock, R. D.:* J. Chem. Soc. 4684 (1957).
50. — Progr. Inorg. Chem. *2*, 193 (1960).
51. *Figgis, B. N., Lewis, J.:* Progr. Inorg. Chem. *6*, 37 (1964).
52. *Wood, D. L., Ferguson, J., Knox, K., Dillon, J. F.:* J. Chem. Phys. *39*, 890 (1963).
53. *Ferguson, J., Knox, K., Wood, D. L.:* J. Chem. Phys. *35*, 2236 (1961) and *37*, 193 (1962).
54. *Schläfer, H. L., Gausmann, H., Zander, H. U.:* Inorg. Chem. *6*, 1528 (1967).
55. — — *Witzke, H.:* J. Chem. Phys. *46*, 1423 (1967).
56. *Furlani, C., Ciana, A.:* Ann. Chim. (Rome) *48*, 286 (1958).
57. *Dingle, R.:* J. Mol. Spectry. *9*, 426 (1962).
58. — Inorg. Chem. *4*, 1287 (1965).
59. *Hatfield, W. E., Parker, W. E.:* Inorg. Nucl. Chem. Letters *1*, 7 (1965).
60. *Oelkrug, D.:* Ber. Bunsenges. Physik. Chem. *70*, 736 (1966).
61. *Davis, T. S., Fackler, J. P., Weeks, M. J.:* Inorg. Chem. *7*, 1894 (1968).
62. *Hatfield, W. E., Fay, R. C., Pfluger, C. E., Piper, T. S.:* J. Am. Chem. Soc. *85*, 265 (1963).
63. *Klemm, W., Huss, E.:* Z. Anorg. Allgem. Chem. *258*, 221 (1949).

64. *Knox, K., Mitchell, D. W.:* J. Inorg. Nucl. Chem. *21*, 253 (1961).
65. *Klemm, W., Krause, J., Wahl, K., Weise, E., Brandt, W.:* Forschungsber. Wirtsch.- u. Verkehrsministeriums Nordrhein-Westfalen No. 160 (1955).
66. Unpublished measurements by *G. A. M. El-Sharkarwy.*
67. *Yamata, H., Kato, A.:* Bull. Chem. Soc. Japan *41*, 2220 (1968).
68. *Grey, J. T.:* J. Am. Chem. Soc. *68*, 605 (1946).
69. *Hoppe, R.:* Rec. Trav. Chim. *75*, 569 (1956).
70. *Klemm, W., Brandt, W., Hoppe, R.:* Z. Anorg. Allgem. Chem. *308*, 179 (1961).
71. *Cotton, F. A., Meyers, M. D.:* J. Am. Chem. Soc. *82*, 5023 (1960).
72. — — J. Am. Chem. Soc. *82*, 5027 (1960).
73. *Figgis, B. N., Nyholm, R. S.:* J. Chem. Soc. 338 (1959).
74. *Allen, G. C., Warren, K. D.:* Inorg. Chem. *8*, 1895 (1969).
75. *Barraclough, C. G.:* Trans. Faraday Soc. *62*, 1033 (1966).
76. *Nelson, P. G.:* private communication.
77. *Knox, K., Shulman, R. G., Sugano, S.:* Bull. Am. Phys. Soc. *5*, 415 (1960).
78. *Bode, H., Voss, E.:* Z. Anorg. Allgem. Chem. *290*, 1 (1957).
79. *Dove, M. F. A.:* private communication.
80. *Hoppe, R., Homann, R.:* Naturwissenschaften *53*, 501 (1966).
81. *Weinland, R. F., Lauenstein, O.:* Z. Anorg. Allgem. Chem. *20*, 40 (1899).
82. *Liebe, W., Weise, E., Klemm, W.:* Z. Anorg. Allgem. Chem. *311*, 281 (1961).
83. *Cavell, R. G., Clark, H. C.:* J. Chem. Soc. 2692 (1962).
84. *Clark, R. J. H.:* The Chemistry of Titanium and Vanadium, pp. 180—1. Amsterdam: Elsevier 1968.
85. *Allen, G. C., El-Sharkarwy, G. A. M.:* Inorg. Nucl. Chem. Letters *6*, 493 (1970).
86. *Fowles, G. W. A., Walton, R. A.:* J. Inorg. Nucl. Chem. *27*, 735 (1965).
87. *Kilty, P. A., Nicholls, D.:* J. Chem. Soc. 4915 (1965).
88. *Machin, D. J., Murray, K. S.:* J. Chem. Soc. (A) 1330 (1967).
89. *Brisdon, B. J., Lester, T. E., Walton, R. A.:* Spectrochim. Acta (A) *23*, 1969 (1967).
90. *Bereman, R. D., Brubaker, C. H.:* Inorg. Chem. *8*, 2480 (1969).
91. *Huss, E., Klemm, W.:* Z. Anorg. Allgem. Chem. *262*, 25 (1950).
92. *Bode, H., Voss, E.:* Z. Anorg. Allgem. Chem. *286*, 136 (1956).
93. *Klemm, W.:* Bull. Soc. Chim. France *10*, 1325 (1956).
94. *Bode, H., Wendt, W.:* Z. Anorg. Allgem. Chem. *269*, 165 (1952).
95. *Hoppe, R., Liebe, W., Dähne, W.:* Z. Anorg. Allgem. Chem. *307*, 276 (1961).
96. *Allen, G. C., El-Sharkarwy, G. A. M., Warren, K. D.:* Inorg. Nucl. Chem. Letters *5*, 725 (1969).
97. *Reisfeld, M. J., Matwiyof, N. A., Asprey, L. B.:* private communication (from M. J. R.).
98. *Matwiyof, N. A., Asprey, L. B.:* Chem. Commun. 75 (1970).
99. *Flint, C. D.:* Chem. Commun. 482 (1970). See also *Pfeil, A.:* Spectrochim. Acta (A) *26*, 1341 (1970).
100. *Novotny, D. S., Sturgeon, G. D.:* Inorg. Nucl. Chem. Letters *6*, 455 (1970).
101. *Moews, P. C.:* Inorg. Chem. *5*, 5 (1966).
102. *Cotton, F. A., Wilkinson, G.:* Advanced Inorganic Chemistry. New York: Interscience Publishers 1962.
103. *Allen, G. C., Warren, K. D.:* Inorg. Chem. *8*, 1902 (1969).
104. *Hoppe, R.:* Angew. Chem. *62*, 339 (1950).
105. *Henkel, H., Hoppe, R., Allen, G. C.:* J. Inorg. Nucl. Chem. *31*, 3855 (1969).
106. *Westland, A. D., Hoppe, R., Kaseno, S. S. J.:* Z. Anorg. Allgem. Chem. *338*, 319 (1965).
107. *Bougon, R.:* Compt. Rend. (C) *267*, 681 (1968).

108. *Allen, G. C., Warren, K. D.:* Inorg. Chem. *8*, 753 (1969).
109. *Reisfeld, M. J.:* J. Mol. Spectry. *29*, 120 (1969).
110. *Allen, G. C., Warren, K. D.:* J. Mol. Spectry. *33*, 180 (1970).
111. *Hinze, J., Jaffe, H. H.:* J. Chem. Phys. *38*, 1834 (1963).
112. *Tondello, E., De Michelis, G., Oleari, L., Di Sipio, L.:* Coord. Chem. Rev., *2*, 65 (1967).
113. *Jørgensen, C. K.:* Helv. Chim. Acta, Fasciculus Extraordinarius Alfred Werner 131 (1967).
114. *Racah, G.:* Lunds Universitets Årsskrift N. F. 50, No. 21, 31 (1954).
115. *Jørgensen, C. K.:* Progr. Inorg. Chem. *4*, 73 (1962).
116. — Struct. Bonding *1*, 3 (1966).
117. *Stevens, K. W. H.:* Proc. Roy. Soc. (London) *A219*, 542 (1953).
118. *Balt, S., Verwey, A. M. A.:* Spectrochim. Acta (A) *23*, 2069 (1967).
119. *Chester, A. W., Heiba, El-A., Dessau, R. M., Koehl, W. J.:* Inorg. Nucl. Chem. Letters *5*, 277 (1969).
120. *Walton, R. A., Brisdon, B. J.:* Spectrochim. Acta (A) *23*, 2222 (1967).
121. *Sartori, G., Cervone, E., Cancellari, P.:* Rend. Accad. Lincei *35*, 226 (1963).
122. *Di Sipio, L., Tondello, E., De Michelis, G., Oleari, L.:* Inorg. Chem. *9*, 927 (1970).
123. *Kotani, M.:* J. Phys. Soc. Japan *4*, 293 (1949).
124. *Figgis, B. N.:* Trans. Faraday Soc. *56*, 1553 (1960).
125. — Trans. Faraday Soc. *57*, 198, 204 (1961).
126. *Jørgensen, C. K.:* Mol. Phys. *2*, 309 (1959).
127. — Mol. Phys. *6*, 43 (1963).
128. — Solid State Phys. *13*, 375 (1962).
129. *Moffitt, W., Goodman, G. L., Fred, M., Weinstock, B.:* Mol. Phys. *2*, 109 (1959).
130. *Jørgensen, C. K.:* Acta Chem. Scand. *16*, 793 (1962).
131. *Dunn, T. M.:* In: Modern Coordination Chemistry, edited by *J. Lewis* and *R. G. Wilkins*. New York: Interscience Publishers 1960.
132. *Liehr, A. D., Ballhausen, C. J.:* Ann. Phys. (N. Y.) *3*, 304 (1958).
133. *Bartlett, M., Palenik, C. J.:* Chem. Commun. 416 (1970).
134. *Fackler, J. P., Holah, D. G.:* Inorg. Chem. *4*, 954 (1965).
135. *Holmes, O. G., McClure, D. S.:* J. Chem. Phys. *26*, 1686 (1957).
136. *Elliott, H., Hathaway, B. J., Slade, R. C.:* Inorg. Chem. *5*, 669 (1966).
137. *Allen, G. C., Warren, K. D.:* unpublished observations.

Received September 10, 1970

Iron(II) Diimine and Related Complexes

Prof. Dr. P. Krumholz

Instituto de Quimica, Universidade de São Paulo, Brasil

Table of Contents

1. Introduction

The Austrian chemist *Fritz Blau* who, at the end of the 19th century, discovered (*1*) the ability of the heterocyclic bases, 2,2′-bipyridine (I; bipy) and 1,10-phenanthroline (II; phen) to form very stable complex-

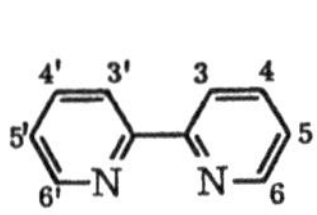

I

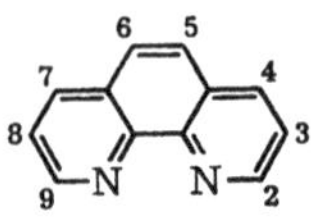

II

es with a variety of metal ions, could hardly have foreseen the important role which these ligands were to play in the future development of coordination chemistry. Among the compounds prepared by *Blau*, those containing the complex cations $Fe(bipy)_3^{2+}$ and $Fe(phen)_3^{2+}$ showed, rather unexpectedly, an intense red color. Strong oxidizing agents transformed those complexes into the corresponding blue and less intensely colored ferric compounds. Very similar complexes of ruthenium(II) (*2, 3*) and osmium(II) (*4*) have since been prepared. Many other transition metal ions in a lower oxidation state are known to form colored bipy and phen complexes. Of particular interest are the complexes of metals with zero or negative formal oxidation number, discovered and studied extensively by *Herzog* and his co-workers ((*5*) and references therein). Among many others, $[Fe(bipy)_3]^0$ (*6*) and $[Fe(bipy)_3]^-$ (*7, 8*) have been prepared. It is now conceded that such complexes involve bipy anions and the metal in a higher than the formal oxidation state (*5, 9*).

Based on a comparative study of colored iron(II) complexes of various organic nitrogen ligands, such as bipy, phen and dimethylglyoxime, *Kozo Sone* (*10*) associated their visible spectra with the presence of a common structural element, viz. the five membered unsaturated chelate ring III, considered as a kind of chromophore.

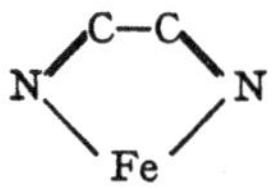

III

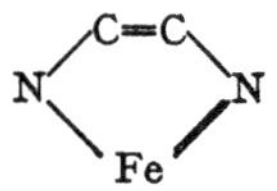

IIIa

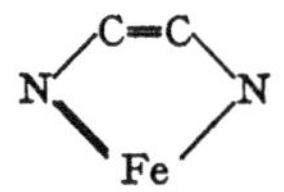

IIIb

The later discovery of the iron(II) complexes of aliphatic α-diimines (IV), e. g. glyoxalbismethylimine (GMI) and biacetylbismethylimine (BMI) (*11*) and of 2-pyridinalimines (V) (*12–16*) with spectral characteristics very similar to those of the bipy and phen complexes, corroborated Sone's chromophore concept. Electron delocalization within

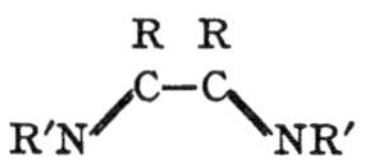

IV

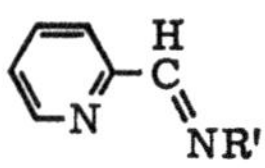

V

the "iron(II)-diimine" grouping (or chromophore) III, suggested by the spectral and other properties, was pictured (*11*) as resonance with structures such as IIIa and IIIb, involving metal-ligand double bonds. Such delocalization may, perhaps, be described more properly in terms of a "back-donation" of electrons from filled metal *d*-orbitals into vacant π^*-orbitals on the ligands.

The afore-mentioned types of ligands, which contain or can be thought to contain the butadiene-like grouping N=C—C=N, are now commonly classified as "diimine" ligands. For convenience, ligands such as 2,2′,2″-terpyridine (terpy) (*17*) or 2-pyridinal-(2′-aminoethyl) imine (V, $R' = CH_2 \cdot CH_2NH_2$) (*18, 19*), which contain the grouping N=C–C=N–C$\doteq$C$\doteq$N ($\doteq$ denotes a single or a double bond) will be referred to as "terimine" (= terdentate diimine) ligands.

The iron(II) complexes of polyfunctional hydrazones, e. g. biacetyl-dihydrazone (IV, $R' = NH_2$) (*16*) or of the "flexidendate" 2-pyridineal-dazine (*20*), are further examples of stable and intensely colored iron(II) compounds containing the chromophoric grouping III. Following the proposition by *Busch* and *Bailar* (*16*), this grouping will be referred to as the "iron(II)-methine" chromophore whenever a *generic* classification of iron(II) complexes of conjugated nitrogen ligands is desired. The terms "dimethine" and "trimethine" ligands are self-explaining.

Beginning with the classic work of *Brandt, Dwyer,* and *Gyarfas* (*13*), several more recent review articles deal with the chemistry of metal diimine and related complexes (*21–24*). This review is chiefly concerned with a discussion of the properties of the complexes of divalent d^6 metal ions, in particular iron(II), which may throw some light on the electronic structure and on the nature and strength of the chemical bonds in those complexes.

2. Magnetic Properties

In the great majority of iron(II) tris — diimine and bis — terimine complexes, the central metal ion is, at room temperature, in the spin-paired fully symmetric ground state, $^1A_1$. Steric hindrance to a close-enough approach of metal and ligand-nitrogens (*25*), or a lower bonding ability of the ligands, may reduce the ligand field parameter $10\,Dq$ below the critical value necessary to overcome the exchange interaction of the metal *d*-electrons. In this case, a spin-free ground state, $^5T_2$ in cubic symmetry, will result. If the energy difference between the spin-free and the spin-paired state is not too large, both forms may coexist in a temperature dependent equilibrium. In solution, *Williams et al.* (*26*) have presented evidence for such an equilibrium for the iron(II) complex of 5,5′-dicarbethoxy-bipy. The phenomenon of a magnetic changeover is more frequent in solid complexes and was observed in some iron(II) tris — diimine (*27–30*) and bis — terimine (*31*) complexes. The behaviour of the iron(II) complex salts of 2-(2′-pyridyl) imidazole (*27, 29*) comes close to what one might expect for a true spin-state equilibrium, the magnetic moment, μ_{eff}, decreasing (or increasing) smoothly and reversibly on cooling (or heating). In some cases, changes of μ_{eff} are accompanied by rather drastic color changes. At a given temperature, μ_{eff}

depends strongly on the nature of the anion and the degree of hydration indicating that crystal lattice effects perturb the rather subtle ballance between 10 D*q* and the interelectronic repulsion. In the case of tris-(2-(2'-pyridyl) imidazoline) iron(II) perchlorate, a blue spin-paired and a purple spin-free form have been separately obtained (*30*). The two forms are likely to represent the vicinal and meridial isomers of the complex. The magnetic behaviour of the spin-free form is different from that of the afore-mentioned complexes. On cooling to ca. 100 °K, μ_{eff} drops rather suddenly to a low value.

A magnetic behaviour of the latter kind has been earlier observed (*32*) in solid bis — diimine complexes such as $Fe(bipy)_2(NCS)_2$ and $Fe(phen)_2(NCSe)_2$ and was further investigated by *König* and *Madeja* (*22, 33, 34*). The nature of the spin changeover as a $^5T_2 \rightarrow {}^1A_1$ transition was confirmed by Mössbauer spectral measurements (*35*). Close to the transition temperature, changes of μ_{eff} are slow, indicating a significant rearrangement of the environment of the metal ion (*36*). Similar spin-state transitions were observed in iron(II) poly(1-pyrazolyl) borates (*37*). It has been concluded that the energy difference between the two spin-states depends strongly on the temperature (*38*).

Bis-diimine complexes $Fe(bipy)_2X_2$ and $Fe(phen)_2X_2$, with X=Cl, Br, I, N_3, NCO, HCOO or CH_3COO, have a $^5T_2$ ground state and show only minor variations of μ_{eff} with temperature (*32, 39–43*). When X is one of the strong-field ligands CN (*44*), CNO (fulminate) (*45*) or NO_2 (*41*), the ground state is, as expected, $^1A_1$. Finally, when X is F, oxalate/2 or malonate/2, a $^3T_1$ state with half-quenched spin becomes, apparently, the ground state (*46*). *Baker* and *Garland* (*47*) prepared a series of similar mixed ligand complexes with phen and substituted phens, having intermediate magnetic moments of 2.2–4.6 BM with little temperature dependence. The authors conclude that all of these systems represent mixtures of singlet and quintet states. Anomalous magnetic moments have also been observed in mixed bis-dihydrazone complexes (*48*).

Mössbauer and IR spectral data (see later) indicate a cis-configuration of the $Fe(diimine)_2X_2$ complexes. A possible trans isomer of $Fe(phen)_2(NCS)_2$ with a spin-paired ground state was reported by *Spacu et al.* (*49*).

3. Electronic Spectra

a) *d–d* Transitions

Spin-paired d^6 metal ions in fields of octehedral symmetry are expected to have four *d–d* transitions in the near infrared-visible region which, in order of increasing energy, are $^1A_{1g} \rightarrow {}^3T_{1g}$, $^3T_{2g}$, $^1T_{1g}$, $^1T_{2g}$. In fields of lower symmetry, the degeneracy of the excited levels will be partially or totally lifted. However, in tris-diimine complexes (point group D_3), any splitting of the octahedral levels is expected to be small as compared with the width of the $d-d$ bands and is, generally, neglected. *König* and *Schläfer* (50) observed weak bands or shoulders on the long wave length tail of the visible band of $Fe(bipy)_3^{2+}$ and $Fe(phen)_3^{2+}$ and associated them with the $d-d$ transitions of the metal ion. In the spectrum of a single crystal of $[Fe(bipy)_3]Br_2 \cdot 6\,H_2O$, there is a weak band at 11.5 kK, completely polarized perpendicularly to the threefold axis of the complex. *Palmer* and *Piper* (*51*) present strong arguments for assigning this band to the spin-forbidden $^1A_1 \rightarrow {}^3T_1$ transition. Using reasonable values for the Racah parameter B, $0.42\ k\mathrm{K} < \mathrm{B} < 0.75\ k\mathrm{K}$, one obtains for the ligand field parameter $16.5\ k\mathrm{K} < 10\ \mathrm{D}q < 19\ k\mathrm{K}$. *Ito* and *Tanaka* (*52*) assign a medium intensity ($\varepsilon_{max} \approx 300$) band at ca. 28 kK, and a weak shoulder at ca. 11 kK in the spectrum of $Fe(BMI)_3^{2+}$, to the $^1A_1 \rightarrow {}^1T_2$ and $^1A_1 \rightarrow {}^3T_1$ transitions. This assignment leads to $10\ \mathrm{D}q \approx 19\ k\mathrm{K}$ and $\mathrm{B} \approx 0.7$.

Spin-free $Fe(diimine)_2X_2$ complexes show a weak broad band at 8.5—12.5 kK with two more or less clearly resolved components separated by ca. 2 kK. *König et al.* (*33, 34, 40*) associate this band with the $^5T_2 \rightarrow {}^5E$ transition, the excited ^5E state being split by low symmetry field components. For spin-free $Fe(bipy)_2(NCS)_2$ and $Fe(phen)_2(NCS)_2$, 10 Dq is estimated at 11.2 and 11.8 kK, respectively.

Since these complexes become spin-paired on cooling, the listed 10 Dq values are believed to be close to the critical value for spin-pairing. On this basis, the values of the electron repulsion parameter B are estimated at ca. 0.6 and 0.64 kK, respectively. A similar estimate of B in spin-paired $Fe(phen)_2(NCS)_2$, yields a value of ca. 0.58 kK.

Robinson, Curry and *Busch* (*53*), very ingeniously, use the 10 Dq values of nickel(II) tris-dimethine complexes as a probe for the critical value for spin-pairing in the respective iron(II) complexes. The 10 Dq value of the tris-chelated iron(II) complex of the strongest ligand known to produce spin-free iron(II) (2-pyridinaldimethylhydrazone, $10\ \mathrm{D}q = 11.63\ k\mathrm{K}$), is taken as the lower limit. The 10 Dq value of the nickel(II) complex of the weakest ligand known to produce spin-paired iron(II), times the high estimate of the ratio $\mathrm{D}q(Fe^{2+})/\mathrm{D}q(Ni^{2+})$ (ca. 1.08), is taken as the upper limit. From the data in Table 1 and the lower limit obtained before, one arrives at an estimate for the critical 10 Dq value for spin-pairing in iron(II) tris-dimethine complexes of $12.1 \pm 0.5\ k\mathrm{K}$. It is, evidently, not expected that this value will be actually independent of the specific nature of the ligand. This is borne out by the data for the terdentate ligands in Table 1, which seem to lead to an upper limit for the critical 10 Dq value of only ca. 11.5 kK.

Table 1. *Values of 10 Dq for nickel(II) tris-dimethine and bis-trimethine complexes*

Ligand	10 D*q* *k*K	Ref.	Ligand	10 D*q* *k*K	Ref.
en[a]	11.20	(*53*)	BdH	12.80	(*53*)
PIC[a]	11.63	(*54*)	GdH	13.20	(*48*)
PPA	11.70	(*54*)	DPEI	12.33	(*55*)
PII	12.38	(*53*)	PDH	12.20	(*55*)
bipy	12.65	(*53*)	DPH	12.50	(*55*)
phen	12.70	(*53*)	DPMH	11.17	(*55*)
GMI	12.65	(*53*)	DPPH	10.63	(*55*)
PMI	12.67	(*53*)	DPMA	12.55	(*54*)
PH	12.72	(*53*)	TPI	12.82	(*56*)

Abbreviations: en = ethylenediamine; PIC = 2-picolylamine; PPA = 2-picolylpicolinamidine; PII = 2-pyridinalisopropylimine; PMI = 2-pyridinalmethylimine; PH = 2-pyridinalhydrazone; BdH = biacetyldihydrazone; GdH = glyoxalbishydrazone; DPEI = 2,6-diacetylpyridinebisethylimine; PDH = 2,6-pyridinedialdihydrazone; DPH = 2,6-diacetylpyridinedihydrazone; DPMH = 2,6-diacetylpyridinebis(methylhydrazone); DPPH = 2,6-diacetylpyridinebis(phenylhydrazone); DPMA = di-(2-pyridylmethyl)amine; TPI = tri-2-pyridylamine.

[a]) Spin free iron(II) complex.

As soon as spin-pairing occurs, 10 D*q* is expected to jump by some 40% (*57, 58*) because of the reduction of the metal-ligand distance on passing from the $^5T_2$ ($t_2^4 e^2$) to the $^1A_1$ (t_2^6) state (*59*)[1].

Comparison of *Palmer's* and *Piper's* estimate of 10 D*q* (16.5—19 *k*K) for *spin-paired* $Fe(bipy)_3^{2+}$, with the value of ca. 13.5 *k*K obtained from Table 1 for the hypothetical *spin-free* complex agrees reasonably with this expectation.

In the spectrum of $Ru(bipy)_3^{2+}$, two medium intensity bands at ca. 29 and 31 *k*K have been tentatively assigned to spin-allowed *d—d* transitions, split under the actual D_3 symmetry of the molecule (*60*). Since electron transfer bands are likely to be present in this spectral region, there seems no compelling reason for this assignment.

b) Internal Ligand Transitions

It is generally agreed that the intense absorption bands in the near UV spectra of metal diimine complexes, first recorded by *Yamasaki* (*61*), arise from $\pi \rightarrow \pi^*$ transitions of the ligands slightly perturbed by the

[1]) Preliminary X-ray results for low- and high-spin $Fe(bipy)_2(CNS)_2$ indicate a difference of ca. 7% in the Fe—N(bipy) distance (*Watson* and *König* quoted in ref. (*22*)).

presence of the metal ion. Just like the lowest energy transitions of the parent hydrocarbons, the transitions of the corresponding aza compounds are believed to involve the two highest filled and the two lowest vacant π-orbitals (π_1, π_2, π_1^* and π_2^* in order of increasing energies).

On the basis of theoretical calculations (*62*), the first (i.e. lowest energy) intense absorption band of bipy at 280 mμ (aqueous solution) can be rather safely assigned to a transition originating mainly from $\pi_2 \rightarrow \pi_1^*$ and polarized along the long molecular axis. The second band at 233 mμ is expected to have the same polarization and to originate mainly from $\pi_1 \rightarrow \pi_2^*$. For phen, *Mason* (*23*) assigned the main peak of the first band at 265 mμ to the long axis polarized $\pi_2 \rightarrow \pi_1^*$ transition (corresponding to the *p*-band in *Clar's* (*63*) classification). However, *Ferguson et al.* (*64*) present arguments for assigning the long wavelength shoulder of the band at ca. 290 mμ, to the *p*-band transition, associating the main peak with *Clar's* β' band ($\pi_1 \rightarrow \pi_2^*$ in the simple HMO picture), having the same polarization as the *p*-band. The absorption band at ca. 209 mμ, in the spectrum of biacetylbis-n-butylimine is likely to represent the long axis polarized transition $\pi_2 \rightarrow \pi_1^*$ (*65*).

Protonation of bipy causes a bathochromic shift of the first and second absorption band of ca. 2.5 and 1 *k*K respectively (*66*). For phen, the shift of the first band is only about one half of that for bipy, while the second band moves to shorter wavelength. Spectral changes of a very similar kind and magnitude occur in bipy and phen complexes of many divalent, high spin transition metal, and post-transition metal ions (*61, 67*). Except for their intensity, the spectra of mono- and tris-diimine complexes are very similar. In all cases the first band of bipy is split by ca. 1.2 *k*K into two components which were associated by *Stammreich et al.* (*67*) with a vibrational structure. Low temperature (77 °K) spectra of $Zn(bipy)_3^{2+}$ and $Ni(bipy)_3^{2+}$ in EtOH—MeOH glass, recently obtained by *Bray, Ferguson* and *Hawkins* (*64*) as well as the fluorescence spectrum of the zinc complex, show a pronounced vibrational progression of ca. 1.4 *k*K which appears also in the spectrum of protonated bipy. *Schläfer* (*68*) treated the influence of the central metal ion as an electrostatic perturbation of the free ligand levels, calculating band shifts in reasonable agreement with experiment.

The UV spectral pattern of the spin-paired tris-bipy and -phen complexes of iron(II), ruthenium(II) and osmium(II) (see ref. (*23*)) is very similar to that of the complexes of other divalent metal ions. As compared with the latter complexes, the first internal ligand transition is shifted to larger wave numbers, the shift being greater for ruthenium(II) and osmium(II) than for iron(II). This shift was associated (*23, 69*) with the effect of back-donation which tends to push the filled and vacant π-levels of the ligands further apart.

The circular dichroism spectra of the optically active forms of the tris-bipy and phen complexes, display two CD bands of opposite sign in the region of the first ligand absorption band (*23, 70, 71*). Quite similar CD bands appear in the region of the first electron transfer band. According to *Mason* ((*23*) and further references therein), the UV CD bands reflect the splitting of the electronic transitions associated with the individual ligand molecules, resulting from electrostatic interactions among the ligands and, to a lesser degree, from interligand and metal-ligand π-bonding. Using the exciton splitting model (*72*), Mason has calculated the splitting of the long-axis polarized intraligand transitions (as three times the corrected exciton-splitting energy V), in fair agreement with the observed separation between the two CD peaks (1.5—4 *k*K). The theoretical prediction of the absolute configuration of (−)-$Fe(phen)_3^{2+}$, agrees with the direct determination by X-ray diffraction (*73*). The exciton-splitting model has been further applied to mixed bipy-phen complexes, such as $Fe(bipy)_2phen^{2+}$ (*74*).

Ferguson et al. (*64*) point out that there is no evidence in the absorption spectra of the bipy and phen complexes for splittings as large as those seen in the CD spectra. Low temperature spectra of $Ni(bipy)_3^{2+}$ and $Ni(bipy)(en)_2^{2+}$, show nearly identical vibrational splitting pattern. There is a very good mirror-image relationship between the vibrational patterns of the absorption and emission spectra of $Zn(bipy)_3^{2+}$. These and other spectral observations seem to put an upper limit of ca. 0.1 *k*K (of unknown sign) on V in the aforementioned tris-bipy complexes. The authors point out that the observed separation of the CD bands is not, necessarily, directly related with the value of V (vide supra). Using the vibronic coupling theory of *Perrin* and *Gouterman* (*75*), *Ferguson et al.* were able to reproduce correctly the pattern of the CD spectrum of $Ni(bipy)_3^{2+}$, using V = 0.07 *k*K.

c) Electron Transfer Transitions

The electron transfer[2]) character of the visible absorption spectra of iron complexes of conjugating imine ligands was first recognized by *Williams* (*77*). From the opposite effect of methyl substituents on λ_{max} of ferrous (or cuprous), and of ferric tris-phen complexes, it was concluded that the excitation involves a partial electron transfer from a filled metal *d*-orbital to a vacant ligand π^*-orbital (ferrous and cuprous), or from a filled ligand π-orbital to a hole in the *d*-shell (ferric). The concept of "inverted" $t_2 \rightarrow \pi^*$ electron transfer (ET) transitions in spin-paired iron(II) tris-diimine complexes, was further elaborated by *Jørgensen* (*78*).

[2]) Following *Jørgensen* (*76*) the term "electron transfer", rather than "charge transfer" spectra will be used throughout this review.

Earlier in this review the visible spectra of iron(II) complexes of di- and polymethine ligands were associated with the presence of a common structural element, viz. the "iron(II)-methine" chromophore III. If one admits that double-bonded structures such as IIIa and IIIb make a substantially larger contribution to the excited than to the ground state of the molecules, the chromophore picture becomes formally linked with the ET concept (*79*, *80*). This picture is, of course, not meant as a faithful representation of the π-electron distribution or to imply that the excited electron does not "see" beyond the N=C—C=N skeleton of the ligand[3]). It must, as well, be remembered that neither the intense color nor the low-spin behaviour of iron(II) complexes of unsaturated organic nitrogen ligands is, necessarily, connected with the presence of a butadiene-like conjugation in the ligand. This is exemplified by the strongly colored and spin-paired complexes of the terdentate ligands, tri-2-pyridylamine (*56*) and di-(2-pyridylmethyl)amine (*54*). The complexes of terdentate 8-quinolinalimine ligands (*82*, *83*) which contain a six-membered chelate ring, are further examples. In certain mixed-ligand complexes, even pyridine can give rise to an intense absorption in the visible (see later).

As proved by the intense color of complexes such as $Fe(phen)(CN)_4^{2-}$ (orange) (*84*, *85*), $Ru(bipy)Cl_4^{2-}$ (dark red) and $Ru(phen)(aca)_2$ (blue) (*86*), presence of a single iron(II)-dimethine (or ruthenium(II)-dimethine) chromophore is sufficient to produce strong absorption in the visible. It is apparent that the further ligands influence the spectral characteristics. In bis-dimethine, tris-dimethine and bis-trimethine complexes, several chromophoric units become linked through the common central metal ion. The different chemical architectures, or the different modes of "coupling" between the single chromophores, will be reflected in the spectral pattern of the complexes. Finally, in iron(II) complexes, the spin state of the metal will have to be considered. The spectral characteristics of complexes containing the iron(II)-methine or related chromophores will now be summarily examined.

The visible spectra of spin-paired iron(II) tris-diimine complexes display an intense ($\varepsilon_{max} \approx 10^4$) band with a maximum at 500—600 mμ (20—17 kK), a half width toward smaller wave numbers, δ-, (*58*) of 0.9 ± 0.1 kK and a characteristic shoulder toward larger wave numbers (see Figure 1)[4]). A second broad band or band system appears between 30 and 25 kK in the spectra of the complexes with ligands containing one or more heteroaromatic rings. In $Fe(phen)_3^{2+}$, this band seems to

[3]) Compare the recent criticism in reference (*87*).

[4]) The characteristic spectral pattern has helped to identify the first naturally occuring diimine ligand, 2-(2′-pyridyl)-Δ^1-pyrroline-5-carboxylic acid (*87*).

have moved into the first one. The spectra of $Ru(bipy)_3^{2+}$ and $Ru(phen)_3^{2+}$ (*88*) look rather similar, being shifted by some 3 *k*K toward larger wave numbers. The extended long wave length tail of the first ET band of $Ru(bipy)_3^{2+}$, has been associated with a single-triplet inverted ET transition (*60, 89*). In the spectrum of the corresponding osmium(II) complex (*4*), this tail raises to a broad, composite band of medium intensity which extends to ca. 14 *k*K.

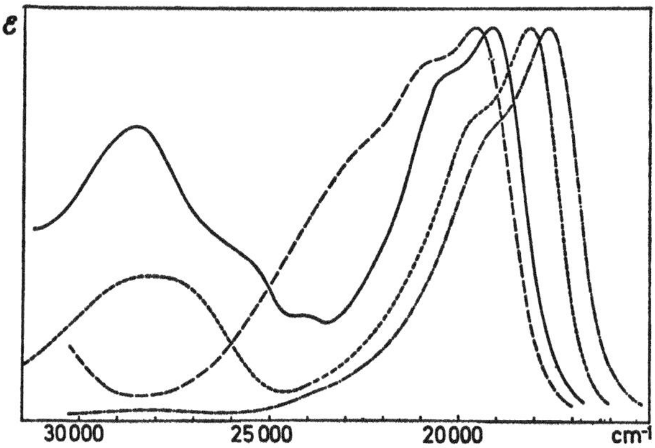

Fig. 1. Visible absorption spectra of iron(II) tris-diimine complexes. Ligands: 1,10-phenanthroline, — — —; 2,2′-bipyridine, ————; 2-pyridinalmethylimine, -----------; biacetylbismethylimine, —·—·—. Curves are drawn to equal height at the absorption maxima. (Reproduced by permission from: Theory and Structure of Complex Compounds. p. 218; *B. Jeżowska-Trzebiatowska*, edit. Oxford: Pergamon Press 1964.)

From a Gaussian analysis of the visible band of $Fe(GMI)_3^{2+}$, it has been concluded that the characteristic shoulder represents the first member of a vibrational progression with $\nu \approx 1.55$ *k*K (*80*). In the low temperature spectra of $Fe(bipy)_3^{2+}$ (*51*), $Fe(phen)_3^{2+}$ (*90*) and $Ru(bipy)_3^{2+}$ (*60, 91*), the double band structure is well resolved with a spacing of ca. 1.6, 1.5 and 1.6 *k*K, respectively. These wave numbers fall well within the range of strong ring vibrations (C—C + C—N) of the ligands. A pronounced vibrational progression with $\nu \approx 1.3$ *k*K has been observed in the emission spectra of $Ru(bipy)_3^{2+}$ and $Ru(phen)_3^{2+}$ (*92, 93*). The long dispute about the origin of this emission seems now to have been settled in terms of a very short-lived (ET) phosphorescence (*60, 89, 94*).

One notices in Fig. 1 that the first ET band shifts to higher energies as the number of aromatic rings in the ligand increases. *Busch* and *Bailar* (*16*) tried to explain this effect in terms of a "crossed resonance"; viz. fusing of another resonating (aromatic) ring on the iron(II)-diimine chromophore will tend to reduce the degree of resonance (conjugation) within the latter. It is curious to notice that a similar hypsochromic effect of the fusing of aromatic rings on a (weakly) conjugated five-membered ring is observed in fulvene (VI) (*95*) which, in a certain way, resembles the iron-diimine ring.

The spectra of iron(II) complexes of aromatic diimine ligands containing heterocyclic rings other than pyridine, e. g. 2-pyridyl-2′-thiazole (*96*), frequently show the characteristic spectral pattern of Fig. 1. When the π-system of the ligand becomes more extended, this pattern changes, the whole spectrum acquiring a more "individual" character (*96, 97*). The few published spectra of complexes of bidentate hydrazone and azine ligands (*98*) indicate a similar individual behaviour. Hydrazone complexes provide an example for the influence of the spin state of the metal ion on the characteristics of the ET spectra of iron(II) tris-dimethine complexes (*99*). For the complexes of 2-pyridinal-hydrazone, -methylhydrazone (both spin-paired) and -dimethylhydrazone (spin-free), $\lambda_{max}(\varepsilon_{max} \cdot 10^{-3})$ are 480 mμ (7.4), 460 mμ (6.4) and 375 mμ (4.6), respectively.

The visible spectra of iron(II) bis-terimine complexes have a characteristic pattern of their own. The main ET band is remarkably narrow (δ-, 0.55 $\pm$ 0.1 *k*K) with a low shoulder and a new peak at ca. 3 *k*K toward larger wave numbers. (In the parent terpy complex, this peak is fused into a single broad shoulder. A salient hump on the long wave length wing, is another "anomaly"). These specific spectral properties seem to result mainly from certain steric conditions imposed on the complex by fusing of a second chelate loop on the iron(II)-diimine ring as pictured in VII (*19*).

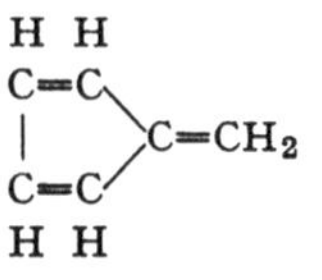

VI

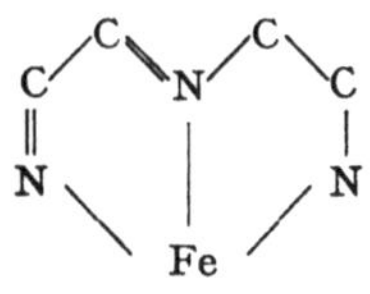

VII

This is substantiated by the spectra of the iron(II) complexes of 2-pyridinal-(2′-aminoethyl)imine and of the -(2′-dimethylaminoethyl) derivative. Steric hindrance prevents the latter from acting as a tridentate and its iron(II) complex displays the normal iron(II)-diimine pattern (*80*). The spectrum of the former complex is clearly of the iron(II)-terimine type. The nature of the bonds in the second chelate loop reflects primarily on the energy of the ET transition. Presence of a double bond conjugated with the N=C—C=N grouping, causes a substantial bathochromic shift. The narrowness of the main ET band can be connected with the greater rigidity of the fused chelate ring system VII which will tend to reduce vibrational amplitudes and to restrain the molecule from changing its equilibrium configuration when excited (*19, 100*). A similar effect can be noted in the spectra of iron(II)-bis-α-dioxime complexes (*101, 102*) where strong hydrogen-bonding between the ligands results in a rigid and planar macrocyclic ring system. No restraint of this kind is present in complexes of monodentate ligands. Accordingly ET bands in pyridine complexes (*78, 103*) are broad, with a half-width of ca. 1.5—2 *k*K. The appearance of an "extra" band in the bis-terimine complexes can be associated with the lower overall symmetry (D_{2d} or lower) of the complexes which may lead to an effective splitting of the cubic t_{2g} level of the metal ion.

We now turn to the effect of ligand substitution on the ET spectra of iron-methine complexes. *Smith* and his co-workers (*88, 104—107*) in the search for better analytical reagents, have extensively investigated the spectral properties of substituted phen, bipy and terpy complexes (see also (*26*)). From a naive interpretation of the ET concept, one would expect that electron donating (or withdrawing) substituents would produce a hypsochromic (or bathochromic) shift of the visible absorption bands. Inverted ET transitions involving π-orbitals of substituted pyridines in complexes of the type $Ru(NH_3)_5py^{2+}$ (*103*) and $Fe(dioxime)_2py_2$ (*101, 102*) show, in fact, the expected behaviour. Methyl groups produce a slight blue-shift, while electron-attracting, conjugating substituents such as COOR and CHO cause a large red-shift, especially when *para* to the nitrogen. In iron(II) tris-diimine complexes, substituent effects do not follow such a simple pattern. Methyl groups *para* to the nitrogen of bipy and phen produce red-shifts of ca. 0.1 and 0.05 *k*K (per substituent). With methyl in *meta* (5,5' in bipy), a blue-shift of ca. 0.2 and 0.3 *k*K is observed. Methyl groups in the 5,6-positions of phen cause a red shift of ca. 0.2 *k*K. For substituents other than alkyl (see Table 2), there is, apparently, no intelligible relation between spectral shifts and the electronic properties or the relative positions of the substituents on the ligand.

Table 3 lists the absorption maxima of various 2-pyridinalimine and 2-pyridyl ketoimine complexes. On passing from the complexes of the free imines (=NH), to the respective alkyl derivatives, one notices a strong hypsochromic shift. The bathochromic effect of replacing methyl

Table 2. *Absorption maxima of ligand-substituted iron(II) tris-2,2'-bipyridine complexes (from reference (107) and (26) (values in parenthesis))*

Substituted 2,2'-Bipyridine	λ_{max} (mμ)		Substituted 2,2'-Bipyridine	λ_{max} (mμ)	
none	349	522	4,4'-$(NH_2)_2$	379	569
4,4'-$(C_2H_5)_2$	355	529	4,4'-$(COOH)_2$	378	540 (547)
4,4'-Cl_2	355	532	4,4'-$(COO^-)_2$	—	(538)
4,4'-Br_2	357	534	4,4'-$(COOC_2H_5)_2$	384	541 (545)
4,4'-$(OCH_3)_2$	357	525	4,4'-$(CONH_2)_2$	384	540
4,4'-$(OC_6H_5)_2$	363	540	5,5'-$(COOC_2H_5)_2$	—	(570)
4,4'-$(C_6H_5)_2$	386	552	5,5'-$(COO^-)_2$	—	(562)
4,4'-$(NO_2)_2$	—	525			

Table 3. *Absorption maxima (λ_{max} in mμ) of $[Fe(C_5H_4N \cdot CR{=}NR')_3]^{2+}$ complexes in H_2O (from reference (108))*

R'/R	H	CH_3	C_6H_5	R'/R	H	CH_3	C_6H_5
H		572	590	$CH_2 \cdot C_6H_5$	562		
CH_3	551	558	565	$C_2H_4 \cdot C_6H_5$ [b]	564	568	
C_2H_5	556	565	572	C_6H_5	574	568	579
C_3H_7	560	566	573	$C_6H_4NH_2$(m)	574	568	579
C_4H_9	560	566	573	$C_6H_4NH_3^+$(m)		565	579
C_5H_{11}	562	567	574	$C_6H_4NH_2$(p)	583	578	590
C_6H_{11} [a]	565			$C_6H_4NH_3^+$(p)		565	578
$C_2H_4N(et)_2$	563	567		C_6H_4COOet(p)		563	

[a] cyclohexyl. [b] 2-phenylethyl.

by phenyl, decreases with increasing twisting angle between the plane of the chelate ring and that of the phenyl group, indicating that mesomeric effects are involved (*109*).

We finally consider the spectral effects of partially replacing diimine by other ligands. In spin-paired $Ru(bipy)_2L$ complexes, the main ET band shifts to smaller wavenumbers in the order:

$$L = bipy > (CN)_2 > py_2 > en > Cl_2 > oxalate$$

(at ca. 80 °K in rigid glasses). Quite similar shifts with a total spread of ca. 3 *k*K occur in the emission spectra (*91*). In solution (CH_2Cl_2, CH_3NO_2), band shifts seem to follow the order:

$$L = bipy > (NO_2)_2 > py_2 > (H_2O)_2 \geqslant (N_3)_2 \gg Br_2 > I_2 > Cl_2$$

(*110, 111, 112*). $Fe(bipy)_2(CN)_2$ and $Fe(phen)_2(CN)_2$ in $CHCl_3$ show a red shift of ca. 3 *k*K as compared with the parent tris-diimine complexes (*44*). The energy of the ET band is strongly affected by hydrogen-bonding solvents (*113*) and, quite dramatically, by protonation or by alkylation (*114*) on the CN nitrogens, which cause a blue-shift of 6—7 *k*K. Similar effects are observed in $Fe(phen)(CN)_4^{2-}$ and in $Fe(bipy)(CN)_4^{2-}$ (*115*).

Solid state reflection spectra of spin-free $Fe(phen)_2X_2$ complexes display a composite band of variable shape which is shifted by several *k*K to smaller wave numbers, as compared with $Fe(phen)_3^{2+}$ (*40*). In $Fe(phen)_2(NCS)_2$ and $Fe(bipy)_2(NCS)_2$, transition from the spin-free to the spin-paired form is accompanied by a blue-shift of ca. 1.5 *k*K (*33, 34*). Intensities for iron(II) bis-bipy halide complexes in CH_2Cl_2, are smaller by a factor of ca. 5 than for the tris-bipy complex (*110*). For iron(II) mono-bipy in aqueous solution, the intensity is ca. 25 times smaller and λ_{max} shifts down to 435 mμ (*116*).

Recently, deep blue $Fe(AP)_2$ $(NCS)_2$ (AP=2,2′-azopyridine) has been prepared (*117*). This is the first example of a colored iron(II) complex derived from a ligand containing the N=N−C=N grouping. Finally, orange-red $Fe(CO)_3N_4(CH_3)_2$ (*118*) is likely to contain the RN−N=N−NR grouping coordinated to the central metal.

A simple theoretical treatment of ET transitions in ferrous, cuprous and ferric phen complexes was given by *Day* and *Sanders* (*90, 119*). This treatment is based on *Murrell's* one-electron system approximation (*120*) to *Mulliken's* theory of charge transfer complexes (*121*). To a zero order, ground and excited states wave functions are represented by the orbitals of the "donor" and "acceptor", respectively. For $d \rightarrow \pi^*$, inverted ET, the transition energy is given by

$$E_{ET} = I_{Me} - A_L + C \quad (1)$$

where I_{Me} is the ionization potential of the metal ion, A_L the electron affinity of the ligand and C a Coulomb term approximated as the attraction between the electron transferred into the ligand π^*-orbital and the metal ion with its charge increased by one. The transition moment is zero unless ground and excited state wave functions are allowed to mix, producing (within the Hückel MO approximation)

$$\psi_0 = \varphi_{Me} + \lambda\varphi_L \qquad \psi_1 = \varphi_L - \lambda\varphi_{Me} \quad (2)$$

where the mixing coefficient λ is proportional to the resonance integral β_{Me-L}. The leading term in the resulting intensity expression is a "transfer" term, proportional to λ and thus to β_{Me-L}. Metal and ligand orbitals involved in the ET must thus belong to the same irreducible representation of the point group of the complex, in order to derive intensity from a "transfer" term. For a model 1:1 complex (point group C_{2v}), the respective transitions will be polarized from the metal to the centre of the ligand, illustrating the rule that "transfer" intensity must always be polarized in the direction of ET (*90*). Through β_{Me-L}, ET intensity and metal-ligand π-bond strength become intimately correlated.

In tris-diimine complexes (point group D_3), the metal *d*-orbitals of cubic t_{2g} parentage will transform as A_1 and E. The symmetry adapted combinations of ligand orbitals will transform as A_2 and E, or A_1 and E, depending whether the single π-orbitals are symmetrical or antisymmetrical with respect to a reflection in the symmetry plane perpendicular to the plane of the molecule. *Orgel* (*122*) and *Weber* (*123*) have given explicit expressions for the symmetry orbitals. Among the symmetry allowed transitions only those of the type E → E will derive intensity from a "transfer" term. As required, those transitions will have a component polarized perpendicularly to the threefold axis of the complex, i. e. in the direction of ET. *Palmer* and *Piper* (*51*), who have measured the polarized crystal spectra of $Fe(bipy)_3^{2+}$ and $Ru(bipy)_3^{2+}$ in a $Zn(bipy)_3SO_4 \cdot 7\,H_2O$ host, have found that more than 90% of the intensity is in the perpendicular spectrum.

Day and *Sanders* (*90, 119*) use the one-electron model to interpret the effect of methyl substituents on the ET energy of the tris-phen complexes. It is necessary to assume that the methyl groups perturb the molecule both by changing the energy of the ligand π-orbitals and the energy of the metal *d*-orbitals. The first effect is set proportional to the square of the Hückel MO coefficient at the substituent position and the second one to the change of the charge of the ligand nitrogens on substitution, accessed theoretically or approximated by the change in pK_a. The experimental results could be rationalized by postulating that ET in ferrous and ferric phen complexes involves the π_2^*- and π_1-orbitals respectively. The opposite direction of the shifts in ferrous and ferric complexes is correctly predicted. So far, no explanation of the effects of substituents other than alkyl (see Table 2) has been attempted.

One is tempted to explain the spectral effects of further ligands on the ET energy of mixed iron(II) and ruthenium(II) bis-diimine complexes by the influence of those ligands on the energy of the metal *d*-orbitals (see (*101, 102, 124, 125*)) for the similar case of mixed iron(II) bis-α-dioxime complexes). Increased σ- or π-donor strength of the further ligands will tend to increase the *d*-orbital energy and thus to cause a red shift of the ET bands. Increased π-acceptor strength should act in the opposite direction. However, in the series of $Ru(bipy)_2L$ complexes (see before), there is no clear cut relation between λ_{max} and the expected electronic properties of the ligands. The strong hypsochromic effect of protonating or alkylating the CN group in the mixed cyanide complexes, might be due to an increase of the π-acceptor strength of the CN ligand (*114*).

Recently, rather sophisticated configuration interaction calculations of the electronic states in $Fe(GMI)_3^{2+}$ (*65*), $Fe(bipy)_3^{2+}$ (*81*) and $Fe(phen)_3^{2+}$ (*126*), have been reported (the theoretical treatment is outlined in (*127*)). The ground, locally excited, and inverted ET configurations are considered. The energy of the latter is calculated from an expression analogous to 1, where I_{Me} is now the valence state ionization potential for the metal *d*-orbital in the field of the ligands, A_L the electron affinity of the ligand and C the negative of the Coulomb repulsion integral $J_{t_2,\pi}$. The experimental energy of the first ET transition is used for parametrization. For all three complexes the theory predicts an intense, perpendicularly polarized ET transition involving the lowest vacant π^*-orbital of the ligand (compare the different prediction of the simple theory for $Fe(phen)_3^{2+}$). Two other transitions involving the same π^*-level are expected to be of low intensity and to nearly coincide with the main transition. Calculations place the ET transition to the next higher π^*-level in $Fe(bipy)_3^{2+}$ (three, nearly coincident components) at ca. 13 *k*K (experimentally 6—10 *k*K) and in $Fe(phen)_3^{2+}$ at ca. 9 *k*K (possibly hidden in the broad shoulder at 6 *k*K or less) above the first transition. In $Fe(GMI)_3^{2+}$, the second ET band is expected to move into the far UV and to mix strongly with internal ligand transitions (one observes a strong band around 53 *k*K with a salient shoulder around 43 *k*K). The results of theoretical calculations (*81*) of the lowest ET energy of the three tris-diimine complexes is shown in Table 4.

Table 4. *The lowest ET energies for iron(II) tris-diimine complexes (from reference (81))*

	$Fe(GMI)_3^{2+}$	$Fe(bipy)_3^{2+}$	$Fe(phen)_3^{2+}$
I_{Me}	147.5	104.6	105.0
A_L	85.0	55.8	54.1
$J_{t_2,\pi}$	49.2	38.9	39.8
E_{ET}(calc)	13.3	9.9	11.1
E_{ET}(exp.)	18.0	19.1	19.6

(all values in *k*K)

One notices a surprisingly large difference between the I_{Me} value of the GMI complex and those for the bipy and phen complexes. In E_{ET}, this difference is largely smoothed out because of a cancellation among the single terms of Eq. 1.

From a naive HMO picture, one might expect that the sum of frequencies of the first ET band in the spectra of corresponding iron(II) and iron(III) complexes would fall in the frequency range of the first internal ligand transitions (*23*). At least for a limited number of ligands (bipy, phen, terpy, GMI and PMI (*108*)), the agreement is actually much better than expected, the sum of the wavenumbers of the ET bands, and the wavenumber of the first strong UV band in the iron(II) complex, differing by only 1.2 ± 0.5 *k*K.

4. Infrared Spectra

In a study of iron(II) complexes containing the iron(II)-methine (diimine) chromophore, *Busch* and *Bailar* (*16*) discovered that the C=N stretching frequencies of ligands such as biacetylbismethylimine or 2-pyridinalmethylimine, on complex formation with iron(II), either seemed to disappear completely or were shifted to considerably lower frequencies. This observation was taken as evidence for a strong conjugation within the chromophoric chelate ring with consequent reduction of the π-bond order of the aliphatic imine group. The subject was further pursued by *Busch* and co-workers (see references in Table 5) and extended, particularly, to hydrazone- and azine-type ligands. Some typical results are listed in Table 5.

In the series of imine, **1**—**5**, and azine, **6**, **7**, complexes one notices invariably a strong decrease of ν(C=N) in the iron(II) complexes, as compared with the free ligands and the cobalt(II) and nickel(II) compounds. The 1:2 iron(II) complex of 2-pyridinaldazine, **7**, has two C=N bands, one essentially the same as in the free ligand and one similar to that in the 2:3 iron(II) complex, **6**. In the 1:2 complex, the ligand acts as a terdentate and the two aliphatic C=N groups of each ligand belong to a 5-membered and 6-membered chelate ring, respectively. Apparently, conjugation within the latter is more limited and does not greatly affect the vibration of the C=N group (*20*). In the series of hydrazone complexes, **8**—**12**, spectral changes are much less dramatic. This may be due, at least in part, to the interplay between steric repulsion and resonance effects involving the amino nitrogen atom (*99*).

Using empirical frequency : bond-length : bond-order relations, *Bayer* (*129*) arrived at an estimate of the C=N π-bond order in compounds **1**, **3** and **4** (ca. 0.69, 0.66 and 0.72). This was taken as evidence for extensive conjugation and a "quasi-aromatic" character of the chromophoric chelate ring (*134*, *135*).

Table 5. *Infrared C=N stretching frequencies of methine ligands and their iron(II), cobalt(II) and nickel(II) complexes*

Ligand	Me_mL_n $m:n$	free	Fe^{2+}	ν(C=N) Co^{2+}	Ni^{2+}	Ref.
1. GMI	1:3	1650	1530		1620	(*128, 129*)
2. BMI	1:3		1540(w)	1608	1608	(*16, 128, 130*)
3. Bipip	1:3	1645 1660	1524			(*129*)
4. PMI	1:3	1634	1558	1652	1654	(*16, 129, 130*)
5. PDMI	1:2	1660	1532	1643	1643	(*130*)
6. PAA	2:3	1631	1544	1623	1625	(*20, 133*)
7. PAA	1:2	1631	1527 1628		1637	(*20*)
8. GdH	1:3	1583	1560(av)	1584	1583	(*48*)
9. BdH	1:3	1582	1595	1613	1613	(*48, 131, 132*)
10. PH	1:3	1566(sh)	1541	1558	1555	(*99, 131*)
11. PMH	1:3	1573	1538	1553	1553	(*99*)
12. PdMH[a])	1:3	1571	1530	1536	1538	(*99*)

Abreviations: Bipip = bis-Δ_1-piperideine; PDMI = 2,6-pyridinedialbismethylimine; PAA = 2-pyridinaldazine; PMK = 2-pyridylmethylketazine; PMH = 2-pyridinalmethylhydrazone; PdMH = 2-pyridinaldimethylhydrazone. For other abbreviations, see Table 1.

[a]) Spin free iron(II) complex.

From a normal coordinate treatment of the IR spectrum of $Fe(GMI)_3^{2+}$, *Nakamoto* (*128*) concluded that the concept of group frequencies, as used above, may not be very appropriate for the conjugated chelate chromophore because of a strong coupling between various vibrational modes. The assignment of (weak) bands at 745, 515 and 501 cm^{-1} to intermixed Fe—N and N—CH_3 stretches and the reported Fe—N force constant of 4.7×10^5 dyne/cm, seem surprisingly high.

The IR spectra of $Fe(bipy)_3^{2+}$ and $Fe(phen)_3^{2+}$ are, on the whole, similar to those of the corresponding complexes of other divalent metal ions (*136, 137*). The small increase of the frequency of the heterocyclic ring vibrations has been interpreted in terms of a cross-conjugation effect (*16*). *König et al.* (*33, 34, 138*) consider the strong decrease of the intensity of these bands, as one passes from high-spin $Fe(diimine)_2(NCS)_2$ to the spin-paired form (or to $Fe(diimine)_3^{2+}$), as indicative of an extensive delocalization of metal *d*-electrons towards the ligands. Similarly, the splitting of C—H bending modes which accompanies this transition is explained by the removal of the equivalency of the hydrogen atoms on the heterocyclic rings, caused by strong back donation. Finally, the higher ν(N—CS) of the mixed ligand complexes in their spin-paired form (*139*) is explained by a lesser availability of metal $d\pi$ electrons for back bonding to NCS. The splitting of the ν(N—CS) mode observed in both forms is indicative of a cis-arrangement of the two NCS groups. The same argument has been used earlier (*114, 140*) to infer the cis-configuration of $Fe(diimine)_2(CN)_2$.

5. Thermodynamic Properties

Complex formation between iron(II) and dimethine or trimethine ligands in aqueous solution is a stepwise process. However, while for most other metal ions successive stepwise formation constants with ligands such as bipy and phen decrease in magnitude, the trend for iron(II) is $K_1 > K_2 \ll K_3$ (*141—145*). Because of this anomalous trend, iron(II) tris-bipy and -phen complexes do not follow, the usual *Irving-Williams* order of stabilities (*146*), $..Fe^{2+} < Co^{2+} < Ni^{2+}..$, but place iron(II) between cobalt(II) and nickel(II). The unexpectedly high value of K_3 was associated with a change of the bond type on addition of the third ligand from "ionic" to ($d^2 sp^3$) "covalent" and the concomitant formation of metal-ligand double bonds (*143*). The process of providing extra bonding energy by the electronic rearrangement of the metal ion, was called by *Irving* and *Williams* (*147*) "orbital stabilization". As pointed out earlier in this review, such rearrangement is connected with a decrease of the metal-ligand distance and with a substantial increase of 10 Dq. The energy difference between the real spin-paired and the hypothetical spin-free form of an iron(II) tris-diimine complex may be considered as a measure of the "orbital stabilization energy".

The fundamental thermodynamic quantity related with the strength of a coordinate bond, is the enthalpy change of a gas phase reaction such as

$$Fe^{2+}_{gas} + nL_{gas} \rightleftharpoons FeL^{2+}_{n\,gas} \qquad \Delta H_{g(n)} \tag{3}$$

Because of the lack of pertinent gas phase data, the enthalpy change of the aqueous replacement reaction

$$Fe(H_2O)^{2+}_{6\,aqu} + nL_{aqu} \rightleftharpoons Fe(H_2O)_mL^{2+}_{n\,aqu} + (6\text{-}m)H_2O \qquad \Delta H_n \tag{4}$$

referred to a convenient standard state, is taken as measure of the bond strength increment as one passes from the aquo to the other complex. One has, however, to consider that ΔH_n differs from the enthalpy of the replacement reaction in the gas phase by hydrational terms (*148—150*), which may vary with the nature of the ligand and the spin-state of the metal-ion (vide infra). Because of the paucity and rather limited precision of ΔH_n data, free energy changes of the replacement reaction, ΔG_n, are frequently used as a relative measure of the strength of the coordinate bond. Unless the entropy changes, ΔS_n, are known to be small or to remain approximately constant within a series, this new source of uncertainty may well outweight the better precision of the free energy data. It is rather obvious that minor differences in the thermodynamic values of different complexes cannot be related unambiguously with changes of the intrinsic strength of the coordinate bond.

With these restrictions in mind, we shall consider the thermodynamic data for stepwise complex formation of iron(II) with several diimine ligands and with ethylenediamine (en), listed in Table 6 together with some data on the respective cobalt(II) and nickel(II) complexes. One

Table 6. *Rounded-off thermodynamic data for the formation of metal diimine and ethylenediamine complexes. ΔG values for 25⁰*

Ligand		FeL^{2+}	FeL_2^{2+}	FeL_3^{2+}	Ionic Strength	Ref.
en	$-\Delta G$	5.9	10.5	13.3	1	*(151)*
	$-\Delta H$	5	10.5	16	1	*(151)*
	ΔS	3	0	−9		
bipy	$-\Delta G$	5.9	11.0	23.9	0.1	*(145)*
	$-\Delta H$	6	—	28.5	0.33	*(116, 143)*
	ΔS	0	—	−15		
phen	$-\Delta G$	8.0	15.1	28.9	0.1	*(145)*
	$-\Delta H$	—	—	32	0.1	*(152, 153)*
	ΔS	—	—	−10		
5,6-$(CH_3)_2$-phen	$-\Delta G$	8.7	—	30.5	0—0.1	*(154, 155)*
	$-\Delta H$	—	—	31	0—0.1	*(155)*
	ΔS	—	—	−2		
2-CH_3-phen	$-\Delta G$	5.7	10.8	14.7	0.1	*(25)*
2,2′-PIM	$-\Delta G$	5.6	10.8	15.8	0.1	*(156)*
		CoL^{2+}	CoL_2^{2+}	CoL_3^{2+}		
bipy	$-\Delta G$	7.8	15.4	22.0	0.1	*(145)*
	$-\Delta H$	8	15	20.5	0.1—1	*(152, 157)*
	ΔS	0	0	5		
en	$-\Delta H$	7	14	22	1	*(151)*
		NiL^{2+}	NiL_2^{2+}	NiL_3^{2+}		
bipy	$-\Delta H$	9.5	18.5	27.5	0.1—1	*(152, 157)*

notices the singular behaviour of spin-paired $Fe(bipy)_3^{2+}$ and $Fe(phen)_3^{2+}$ ($\Delta G_3/\Delta G_1 > 3$), as contrasted with the normal behaviour ($\Delta G_3/\Delta G_1 < 3$) of the spin-free complexes of 2-methylphen and 2,2'-pyridylimidazole (PIM). One observes also that for $Fe(bipy)_3^{2+}$ the ratio $\Delta H_3/\Delta H_1$ is larger than $\Delta G_3/\Delta G_1$, because the negative value of ΔS_3 opposes the enthalpy effect. The negative entropy change for the tris-chelated iron(II) complex contrasts with the slightly positive value for the corresponding cobalt(II) (and other divalent metal ion) complex and may reflect the firmer immobilization of the ligands in, and of solvent molecules (*150*) around, the more compact spin-paired iron(II) complex. The greater stability of the 5,6-methyl substituted tris-phen complex as compared with the parent compound (see Table 6), may well be due to a more favourable entropy change.

The data in Table 6 allow one to infer a rather consistent value of the enthalpy of formation of the hypothetical spin-free iron(II) tris-bipy complex (see *152, 158*). One notices first of all that the ΔH values for cobalt(II) bipy and en complexes are very similar. The same holds for a number of other divalent metal ions. Accordingly $\Delta H(Fe(bipy)_3^{2+}$ spin-free) $\approx \Delta H(Fe(en)_3^{2+})$. Secondly, one may try to extrapolate the ΔH_3 values for nickel(II) and cobalt(II), to iron(II). Finally, one can apply the mean value of $\Delta H_3/\Delta H_1$ of bipy complexes, other than iron(II), to ΔH_1 of the iron(II) complex. The three procedures lead to an average value of 15 ± 2 kcal/mole. To obtain a value of the "orbital stabilization energy", one still has to evaluate the difference between the hydrational enthalpies of the spin-paired and the spin-free complex. This is, actually, impossible. However, the negative value of ΔS_3 for the spin-paired complex suggests that solvent molecules might be more firmly attached to this than to the spin-free complex (vide supra). The best one can do is to consider the difference of approximately 15 kcal/mole, between the enthalpies of formation of spin-paired and spin-free $Fe(bipy)_3^{2+}$ as an upper limit of the "orbital stabilization energy" and to hope that the hydrational terms will not be overimportant. For $Fe(phen)_3^{2+}$, the value should be substantially the same.

Stability constants for a number of tris-chelated iron(II)-(substituted phen) complexes have been measured and correlated with the basicity of the free ligands. For single CH_3, C_6H_5, Cl and NO_2 substitution in the 5-position, there is an approximately linear relation between log β_3 and the pK_a of the ligand (*159*). The 5,6-dimethyl derivative seems to fall on the same straight line, but the stability of the 4,7-dimethyl derivative is lesser than expected on the ground of its basicity (*154*). There is some evidence (*160*) that for the 4-cyano derivative the deviation might be in the opposite direction. As expected (*161*) ligands of a different structure, such as bipy and terpy, cannot be fitted into the same relation with the phen derivatives (*162*). *Williams et al.* (*163, 164*) have pointed out that no simple relation between the stability constants of metal complexes and the basicity of the ligands is to be expected, when metal-ligand π-bonding is an important factor. Changes in the π-acceptor or π-donor ability of the ligands produced by different substituents will affect the strength of the metal-ligand bond, but are unlikely to influence the pK of the ligands.

The most numerous thermodynamic data on iron, ruthenium and osmium diimine complexes, are the oxidation-reduction (redox) potentials of the MeL_3^{3+}/MeL_3^{2+} (and similar) couples[5]). The relation

$$F\,E^0_{compl} - F\,E^0_{aqu} = \Delta G_{MeL_3^{3+}} - \Delta G_{MeL_3^{2+}} \quad (5)$$

where E^0_{compl} and E^0_{aqu} are the standard potentials of the complex and aquo ion couples, respectively, provides a measure of the relative stabilities of the Me^{3+} and Me^{2+} complexes. For the $Fe(phen)_3^{3+}/Fe(phen)_3^{2+}$ couple, E^0_{compl} at 25° is 1.14 v (*166, 167*), as compared with $E^0_{aqu} = 0.77$ v. For the corresponding ruthenium couples (*168, 169*), the difference of the E^0 values is as large as 1.3 v. Thus, complex formation with phen stabilizes the spin-paired +2 states to a much larger extent than the +3 states. In a series of tris-chelated iron complexes of substituted phens (*159, 170–172*) or bipys (*165*), E^0_{compl} decreases, on the whole, with increasing ligand basicity. This indicates that the stability of the iron(III) complexes increases, with increasing σ-bond ability of the ligands, more rapidly than the stability of the iron(II) complexes (*165*). When looked upon more closely, the trend of E^0_{compl} with the pK_a of the ligands is less regular than that of log β_3. This was explained by *Williams et al.* (*164*) in terms of an opposite effect of the π-bonding properties of the ligands on the stabilities of the complexes of iron(II) (π-electron donor) and iron(III) (π-electron acceptor), respectively. Hydration and entropy effects may contribute to the observed irregularities. Iron(III) tris-bipy and -phen complexes are strongly stabilized by a very favourable entropy change ($\Delta S_3 \geqslant 30$ e. u.) (*153, 167*), which contrasts with the negative entropy of formation of the iron(II) complex. As a corollary, the difference between the partial molar entropies of the iron(III) and the iron(II) complex is slightly positive (5–8 e. u.), contrary to what one might expect from charge effects. However, the observation (*173*) that a single 5-methyl substituent in phen inverts the sign of this difference, cautions against any over-simplification of the problem.

[5]) Under ordinary conditions, the spin-paired iron(III) complexes obtained by oxidation of the iron(II) complexes, are metastable with respect to the spin-free mononuclear or polynuclear (hydroxo) species (*165*).

The redox potentials of the afore mentioned complex couples decrease strongly with increasing acid concentration (*13*). While activity and ion-pairing (*174*) effects may be important factors, the effect as a whole is not clearly understood. It is noteworthy that the redox potential of mixed cyano complexes, e.g. $Fe(bipy)_2(CN)_2$, increases with increasing acidity (*175*).

The redox potentials of a number of mixed-ligand complexes of ruthenium and osmium have been, rather thoroughly, analyzed in terms of the individual contribution of ligational and hydrational enthalpy, and of entropy terms (*176*, *177*). The importance of metal-ligand π-bonding on the relative stability of the $+3$ and $+2$ complexes, has again been emphasized. For further details the reader is referred to the original work.

In strong acid (10 M H_2SO_4) the tris-chelated iron(II) complexes of aliphatic diimines can be oxidized reversibly to the iron(III) state (*178*). At lower acidities, however, oxidative attack of the ligands sets in without rupture of the metal-ligand bond. Methyl or methylene groups attached to the methine carbon of the coordinated diimine ligand, are subject to attack by molecular oxygen (*179*). This explains the spectral changes which occur on prolonged standing of acidic solutions of $Fe(BMI)_3^{2+}$ and $Fe(Bipip)_3^{2+}$ (*11*). It is interesting to note that the methyl protons in iron(II) tris-biacetylbis-n-butylimine, are slowly replaced by deuterium on standing in CD_3OD (*52*).

6. Rate and Mechanism of Dissociation

Thermodynamic equilibrium data are not easily available if the free ligand is a labile species. This holds for all ligands containing aliphatic imine groupings and, in particular, for the fully aliphatic diimines, the lower members of which have not yet been obtained in the non-complexed form. Under those circumstances, kinetic parameters for the dissociations of the complex species may provide some information on the strength of the metal-ligand bonds. However, it is necessary to have some prior information on the mechanism of the dissociation process.

The dissociation of $Fe(bipy)_3^{2+}$ in aqueous solution follows a first order rate law and is acid catalized (*180*). Under conditions which minimize variations of activity coefficients (HCl—LiCl medium) the pseudo-first order rate constant follows closely the expression (*143*)

$$k_{\text{diss}} = \frac{k_0 k_x + k_\infty [\text{H}^+]}{k_x + [\text{H}^+]}, \tag{6}$$

where k_0 and k_∞ are the rate constants at zero and infinite acidity, respectively. The dissociation of $Fe(GMI)_3^{2+}$ follows the same rate expression (*178*). Mutatis mutandis, the same expression is valid for the formation of $Fe(bipy)_3^{2+}$ (*181*). The most likely mechanism compatible with equation 6 is the one proposed by *Basolo, Hayes* and *Neumann* (*182*) and which comprises as the first step of the dissociation process the rupture of one single metal-ligand bond (rate constant k_∞). The so formed "half-bonded" intermediate either recloses the chelate ring (rate constant k_c) or looses the first ligand completely (rate constant k_d). The loss of the further ligands is very fast and does not reflect on the rate law. For this mechanism, $k_0 = k_\infty k_d/(k_d + k_c)$. Acid catalysis consists in the protonation of the free nitrogen atom of the "half-bonded" intermediate which now looses the ligand at a faster rate. Another mechanism, compatible with the experimental rate law and which involves protonation of the original complex (*143*), was disproved by equilibrium measurements (*181*).

For the acid-catalized dissociation of iron(II) tris-(2-pyridinal-*n*-propylimine), k_{diss} does not converge, at high acidities, toward a constant value, as expected from Eq. 6 (*183*). It has been suggested that mono- and bi-protonated species might participate in the reaction. However, it is not necessary to invoke a different reaction mechanism if one considers that, for an unsymmetrical ligand such as a 2-pyridinalimine, there must exist two parallel dissociation paths differing by the sequence in which bonds between the metal and the two different ligand moieties are broken. Under these circumstances, k_{diss} becomes the sum of two terms like the one of Eq. 6, which very satisfactorily represents the experimental rate data (*184*).

The rate of dissociation of $Fe(phen)_3^{2+}$ is nearly independent of the acidity (*185*). The same holds for the dissociation of the tris-chelated iron(II) complexes of 2-pyridinal-isopropylimine (*183*) and of 2-pyridinal- and 2-pyridyl keto-N-phenylimines (*186, 187*)[6]. For the phen complex, this was explained on the basis that the rigid structure of the ligand would not easily permit the formation of a "half-bonded" intermediate capable of adding a proton (*183, 188*). Whether the dissociation still proceeds by a stepwise or else by a simultaneous rupture of the two metal-ligand bonds, is an open question. In the latter case, kinetic data for acid-catalized and acid-independent dissociation processes are not, necessarily, comparable quantities. The dissociation of $Fe(terpy)_2^{2+}$ is likely to involve the successive breaking of the three bonds between the metal and a single ligand which finds its expression in an exceptionally high

---

[6]) The apparently anomalous acid-depence found for some nitro-derivatives (*186*) may have been caused by the presence of impurities.

ratio $k_\infty/k_0 \approx 10^4$ (*189*). A "fascinatingly complex" dependence of the dissociation kinetics of bis-(2,4,6-tripyridyl-s-triazine) iron(II) on the *p*H has been reported (*190*).

Table 7 lists kinetic data for the dissociation of a number of iron(II) diimine complexes as compared with similar complexes of cobalt(II) and nickel(II). It is to be remembered that k_∞ (and the activation parameters derived therefrom) refer to the breaking of a single metal ligand bond, while k_0 refers, less unambigously, to the total loss of the first ligand. It is evident that both the energies and the entropies of activation are considerably greater for the dissociation of the spin-paired iron(II) complexes than for the spin-free complexes listed in Table 7. Since energy and entropy factors work in opposite directions, differences in the rate constants are less spectacular. Assuming that there is a retention of spin during activation, the anomalously high values of the activation energy of the spin-paired iron(II) complexes can be rationalized in terms of a loss of crystal field stabilization energy on passing from the octahedral

Table 7. *Rounded-off values of kinetic parameters for the dissociation of metal diimine complexes. Rate constants, sec*$^{-1}$*, at 25°*

Complex	–Log k_0	–Log k_∞	$E_0^\ddagger$ kcal	$E_\infty^\ddagger$ kcal	$\Delta S_0^\ddagger$ e. u.	$\Delta S_\infty^\ddagger$ e. u.	Ref.
$Fe(bipy)_3^{2+}$	3.9	2.9	29	27	19	17	(*143, 182, 191*)
$Fe(bipy)^{2+}$	(1)		(17)		(0)		(*192*)
$Fe(phen)_3^{2+}$	4.2		31		25		(*182, 193*)
$Fe(phen)^{2+}$	0		13		–16?		(*194*)
$Fe(GMI)_3^{2+}$	7.7	6.6	33.5	30.5	16	14	(*178*)
$Fe(PPI)_3^{2+}$	4.3		30	28 [a]	19	14 [a]	(*184*)
				26 [b]		10 [b]	
$Fe(terpy)_2^{2+}$	6.7	2.7	28.5		5		(*189, 191*)
$Fe(terpy)^{2+}$	2.2		18		–10		(*191*)
$Co(phen)_3^{2+}$	1		20.5		5		(*195*)
$Co(phen)^{2+}$	1.9		19.5		–4		(*195*)
$Co(terpy)_2^{2+}$	3.2		15		–25		(*191*)
$Ni(bipy)_3^{2+}$	2.5	1.7	22.5	22	3	4	(*182, 192*)
$Ni(bipy)^{2+}$	4.3		23.5		0		(*192, 196*)
$Ni(phen)_3^{2+}$	<5.0		25		2		(*182*)
$Ni(phen)^{2+}$	5.0		24.5		–1		(*196*)

[a] Iron(II)-aliphatic imine bond broken first.
[b] Iron(II)-pyridine bond broken first. PPI = 2-pyridinal-n-propylimine.

ground state to a square-pyramidal transition state (*188, 195*). However, one has to consider that somewhere during the loss of the first ligand a change of the spin-state must occur. As pointed out in the previous chapter, this is accompanied by the loss of the "orbital stabilization energy" of ca. 15 kcal/mole and by an entropy gain of some 10—20 e.u. These values correspond crudely to the expected difference between the activation parameters of low-spin and (hypothetical) high-spin $Fe(bipy)_3^{2+}$, $Fe(phen)_3^{2+}$ or $Fe(terpy)_2^{2+}$ (compare the data in Table 7). It is also apparent that E_∞ for $Fe(bipy)_3^{2+}$ is about twice as large as the "orbital stabilization energy". On these grounds one is lead to believe that the change of the spin state might actually preceed the breaking of the first metal-ligand bond or, at least, occur simultaneously with the latter. A change of the spin state during the activation process has been explicitly considered for the intramolecular racemization of iron(II) tris-diimine complexes (*188, 197*).

Change of the solvent exerts major effects on the dissociation of iron(II) tris-diimine complexes. In 100% methanol, the dissociation of $Fe(bipy)_3^{2+}$ is catalized by small anions, such as chloride, rather than by hydrogen ions (*198*). Thus, either the half-bonded mechanism is inoperative or else the proton catalized path depends critically on the presence of water. In water-methanol mixtures, the dissociation rate depends quite irregularly on the mole fraction. These and other apparent kinetic anomalies (*199*) can be rationalized, at least in part, in terms of a structural model presented by *Basolo et al.* (*198, 199*). In a complex such as $Fe(phen)_3^{2+}$, there exist three major pockets between the ligands which can accomodate a total of six water molecules. These water molecules can immediately occupy a coordination position made vacant by a departing ligand. Removal of these water molecules from their close proximity to the metal should thus decrease the rate of dissociation of the complex, unless another species takes over the role of water (see the effect of chloride in methanol). On the other hand, removal of the water molecules from the pockets might loosen the compact structure of the hexahydrate and give greater freedom of motion to the chelate rings, resulting in an rate accelerating effect.

While the rate of dissociation of $Fe(phen)_3^{2+}$ is not sensibly affected by hydrogen ions, hydroxyl (*200*) as well as cyanide and azide ions (*201*) greatly accelerate the rate of displacement of the diimine ligand. For the reaction with OH^-, the activation energy for the catalized path is lowered by ca. 6 kcal/mole (*202*). The preferred mechanism (*201*) consists in the replacement of water molecules in one of the pockets between the ligands, by the nucleophilic ion. Interaction between the π-orbitals of the latter and the antibonding metal *d*-orbitals is believed to destabilize the metal-diimine bonds.

Burgess (*193*, *203*) determined the activation parameters for the acid dissociation of a number of ligand-substituted iron(II) tris-phen complexes. There exists no simple relation between the activation energy on the one hand and the stability constants of the complexes or the basicity of the free ligands on the other. While 4,7-dimethyl phen is a stronger base than the 5,6-dimethyl derivative and their tris-chelated iron(II) complexes have the same stability, the activation energy for the dissociation of the former complex is ca. 1.5 kcal lower than that of the latter. This is explained in terms of a mesomeric (hyperconjugative) weakening of the metal-ligand π-bond by the methyl substituents *para* to the ligand nitrogens (4,7) which outweighs the increase of the σ-bond strength resulting from the greater basicity. It might be significant that in $Ni(phen)^{2+}$, where π-bonding effects can be expected to be of lesser importance, 4,7-dimethyl substitution of the ligand brings about the opposite kinetic effect (*196*). Still another effect considered is the influence of a substituent on the electron density around the metal atom which might determine the ease of approach of a water molecule to the metal in the transition state toward dissociation (*203*). The latter type of effect is considered as the dominant one in the dissociation promoted by nucleophylic reagents, such as hydroxyl (*202*).

In the series of iron(II) tris-(2-pyridinalimine) complexes, acid dissociation rates increase by a factor of ten as one goes from the N-methyl to the N-*n*-propyl derivative (*183*). The iso-propyl derivative reacts ca. 15 times faster than the *n*-propyl derivative and the rate becomes independent on the hydrogen ion concentration. This indicates that steric factors are involved. The acid-independent dissociation of the N-phenyl derivative is ca. 10^4 times faster than the dissociation of the N-methyl complex, the activation energy decreasing by some 5 kcal (*186*). The lower basicity of the phenyl substituted nitrogen is certainly a major factor. Methyl substituents on the phenyl ring lead to a further decrease of the activation energy, which has again been interpreted in terms of a mesomeric weakening of the metal-ligand-bond (*186*).

A dramatic decrease of the dissociation rate by a factor of ca. 10^5 is observed when the hydrogen atom on the methine carbon of a 2-pyridinal-N-phenylimine ligand is replaced by methyl or by phenyl (*109*). The corresponding increase of the activation energy amounts to at least 5 kcal (*187*). Mutual spatial interference of the substituents will force the phenyl groups out of the plane of the chelate ring and make them project into the "pockets" between the ligands. This will tend to retard the reaction rate both by restricting the freedom of motion of the ligands and by hindering the approach of water molecules toward the reaction centers.

Partial replacement of diimine ligands in tris-chelated iron(II) complexes, by cyanide, seems to exert a strong stabilizing effect on the remaining metal-diimine bonds. There is evidence that the rate determining step for the acid dissociation of $Fe(bipy)_2(CN)_2$ and $Fe(phen)_2(CN)_2$ involves a cyanide group rather than the diimine ligand (*204*). In both cases, the rate is independent on the *p*H up to *p*H ca. 1, and decreases on further increase of the acidity. The first order rate constants and activation energies (31 ± 1 kcal) have very similar values.

7. The Importance of Back Coordination

An important question in the understanding of the structure and properties of iron(II) diimine and related complexes is the extent of metal-ligand π-bonding. A partial double bond character of the coordinate bonds in $Fe(bipy)_3^{2+}$ was first postulated on the basis of the exceptional stability, the diamagnetism and the intense color of the complex (*143, 205*). Double bond formation was pictured in terms of *Pauling's* (*206*) valence bond representation, making use of metal *d*-orbitals with their associated electrons and nitrogen 2 *p* orbitals made available by a suitable structural rearrangement of the ligands (*11, 205*). *Williams* (*79, 207*) has discussed double bonding in iron(II) diimine complexes in terms of a MO picture in which electrons originally associated with the metal are placed into bonding molecular orbitals formed by use of metal $d\pi$ and ligand π^*-orbitals. This eliminates the requirement of the somewhat arbitrary rearrangement of the ligands needed to vacate the nitrogen 2 *p* orbitals. Both pictures express what is commonly called back coordination or back donation (*208, 209*). Configuration interaction between the "no-bond" ground configuration and excited ET configurations (*52, 210*) likewise implies the back donation concept. Some authors consider the iron(II) diimine ring as an aromatic or pseudo-aromatic system in which a metal $d\pi$ electron pair has completed the "aromatic sextet" (*134, 135, 211*). Most authors agree that π-bonding in spin-paired iron(II) diimine complexes is of major importance. We shall now try to examine the π-bonding hypothesis in the light of the experimental results presented earlier in this review, and with the aid of some additional information.

The most direct evidence for back coordination in iron(II) methine complexes comes from the observation of IR C=N stretching frequencies in complexes of ligands containing aliphatic or alicyclic imine or azine groupings (see Table 5). The substantial decrease of ν(C=N) (ca. 100 cm^{-1}) on complexation with iron(II) can, with some confidence, be taken as a qualitative indication of back coordination. However, at present it is difficult to assess the extent of π-electron delocalization in the iron(II) diimine ring from the observed IR shifts.

The very existence of intense ET transitions in iron(II) methine complexes can be taken as further evidence for back coordination. On the assumption that all of the intensity arises from a "transfer" term (see p. 152), *Sanders* and *Day* (*102, 210*) have obtained a rather consistent value of ca. $-5\,k$K for the iron-nitrogen ($d\pi - p\pi$) resonance integral β_{Me-N} in $Fe(GMI)_3^{2+}$, $Fe(phen)_3^{2+}$ and $Fe(niox)_2py_2$. Using the same procedure, *Mason* (*23*) has calculated $\beta_{Me-N} \approx -5\,k$K for $Ru(bipy)_3^{2+}$. Purely theoretical estimates from the Mulliken relation (*212*) between overlap and resonance integrals, lead to β_{Me-N} of ca. $-8\,k$K for $Fe(GMI)_3^{2+}$, $Fe(phen)_3^{2+}$ and $Fe(bipy)_3^{2+}$ (*52, 81, 126*). The net charge flow from the metal to a single ligand, obtained from configuration interaction calculations, amounts to ca. 0.1 electrons. This is close to the value obtained by *Sanders* and *Day* (*210*) from SCF calculations on a model 1:1 iron(II) phen complex. Finally, the theoretically calculated ground state stabilization through interaction with excited ET configurations is 32, 26 and 20 kcal/mol for $Fe(GMI)_3^{2+}$, $Fe(phen)_3^{2+}$ and $Fe(bipy)_3^{2+}$, respectively (*52, 81, 126*).

It is interesting to compare these numbers with the value of ca. 15 kcal/mol obtained earlier as the upper limit of the energy difference between the real spin-paired, and a hypothetical spin-free $Fe(bipy)_3^{2+}$ or $Fe(phen)_3^{2+}$ complex. It is unlikely that the whole of this difference is due to increased back donation, particularly because of the synergic interaction between metal-ligand σ- and π-bonding (*209*). On the other hand, one has to consider the possible existence of weak π-bonding in the spin-free complex as well as the increased interligand repulsion in the more compact spin-paired complex. Considering these factors, it still seems that purely theoretical calculations overestimate the energetic importance of back coordination and, in particular, the differences in π-bonding among the complexes of GMI, phen and bipy (vide supra).

Various thermodynamic and kinetic data provide qualitative evidence for back coordination in spin-paired iron(II) methine complexes. It was pointed out by *Craig et al.* (*213*) that, if only σ-bonds were formed, ethylene diamine ($pK_1 \approx 10$) should always be a stronger ligand than bipy ($pK_1 \approx 4.3$) and phen ($pK_1 \approx 5$). Since this is not the case (see Table 6), it was concluded that double bonding occurs between transition metals and the diimine ligands. The lack of a simple relation between the activation energy for dissociation of various ligand substituted iron(II) tris-phen complexes (*193, 203*), or the redox potentials of the respective iron(III)/iron(II) couples (*164*) on the one hand, and the basicity of the ligands on the other, likewise indicates that factors other than σ-bonding are operative.

NMR and Mössbauer spectral data have been used to infer the existence of back coordination in iron(II) methine complexes. The most salient feature of the NMR spectra of complexes as $Fe(bipy)_3^{2+}$ (*214, 215*), $Fe(phen)_3^{2+}$ and $Ru(phen)_3^{2+}$ (*216*), is a large high-field shift (1—1.3 ppm referenced against free ligand) of the 6,6′-bipy and 2,9-phen protons. This high-field shift is also observed in the complexes of zinc(II) and cobalt(III) and was found to increase with decreasing radius of the central metal. Two different explanations have been advanced: a) a direct metal-hydrogen interaction (*216*) or b) the influence of the magnetic anisotropy of a neighboring ligand (*217*). The variations in the resonance frequencies for the other protons in the zinc(II), iron(II) and ruthenium(II) tris-phen complexes have been tentatively explained by increasing back donation along the series (*216, 218*).

In $Fe(GMI)_3^{2+}$, the NMR signal of the C—H protons appears at a lower field than in glyoxalbiscyclohexylimine (*219*) or crotonal-n-butylimine (*52*) by 0.9 and 1.4 ppm, respectively. In $Fe(GdH)_3^{2+}$, a low field shift of 0.45 ppm, as compared with the free ligand, has been observed (*219*). These shifts have been attributed (*52*) to an increased deshielding of the protons by the magnetic anisotropy of the conjugated chelate ring. The possibility that the low-field shifts may occur also by the decrease of the electron density on the ligands caused by σ-bonding, has been acknowledged. The large high-field shift for one of the methylene protons in the complex of biacetylbis-*n*-butylimine, has been attributed to the effect of ring currents in a neighboring chelate loop.

Proton NMR contact shifts in complexes of paramagnetic metal ions can, in principle, provide more direct information on metal-ligand π-bonding (see e.g. (*220*)). *La Mar* and *Van Hecke* (*221, 222*) have investigated contact shifts in a series of ligand substituted chromium(II) tris-bipy and tris-phen complexes. These complexes ressemble the iron(II) compounds in having a spin-paired ($^3T_1$) ground state and displaying intense ET transitions (*223, 224*). For both phen and bipy, the 4-H and 4-CH_3 proton shifts in complexes of symmetrically disubstituted ligands (referenced against free ligand) have approximately equal magnitude and opposite signs, indicating that unpaired spin is delocalized in the ligand π-system. However, the more complex behaviour of the contact shifts in complexes of monosubstituted phen's suggests that in addition to a direct π-delocalization (back donation), an indirect delocalization mechanism might be operative (*222*). Since pseudo-contact shifts and σ-spin delocalization may also contribute to the observed isotropic shifts, it seems at present impossible to assess the extent of back coordination from the experimental NMR data (see (*225*)).

Extensive delocalization of magnetic electrons onto the ligand atoms in $Cr(bipy)_3^{2+}$ has been inferred from the spin-only value of the magnetic moment of this complex (*226*). However, an alternate interpretation (*227*) which postulates a strong trigonal distortion to quench the orbital contribution to the magnetic moment, requires only a modest delocalization of the unpaired spins.

Mössbauer spectral data have been obtained for a series of ligand substituted tris-bipy and tris-phen iron(II) complexes (*228, 229*). In the perchlorate salts, variations of the isomer shift are within ca. 0.05 mm/sec, i.e. smaller than those produced by a change of the anion and of the degree of hydration (*230*). Gradual substitution of the diimine ligand by cyanide, leads to a gradual decrease of the isomer shift (*231, 232*). Increased σ- and π-bonding, both tend to increase the electron density at the nucleus and, consequently, to decrease the isomer shift. Since the diimine ligands are likely to be better π-acceptors than cyanide (*233*), the observed trend of the isomer shifts seems rather to be caused by the better σ-bonding ability of the latter ligand. The large value of the quadrupole coupling constant for iron(II) bis-terimine complexes, e.g. $Fe(terpy)_2^{2+}$ (1.15 mm/sec) (*228*), indicates a significant distortion from cubic symmetry. This distortion may be the result of a cooperative effect of steric factors and of different π-bonding abilities along the axial and the equatorial directions in the complex (*19, 130*).

Summarizing the aforegoing discussion it can be concluded that, while the validity of single arguments for back coordination in spin-paired iron(II) methine complexes can certainly be questioned, the whole of the experimental evidence definitively substantiates its existence. It is much more difficult to assess the extent of back coordination from the available experimental data. The result of theoretical calculations, corresponding to a net charge flow of ca. 0.1 electrons to a single ligand, can probably be taken as an upper limit for the existing $d \rightarrow \pi^*$ delocalization.

Acknowledgment. The author wishes to thank Miss *Helena Li Chum* for her help in preparing the manuscript.

8. References

1. *Blau, F.:* Monatsh. Chem *10*, 367, 375 (1889); *19*, 647 (1898).
2. *Burstall, F. H.:* J. Chem. Soc. *1938*, 172.
3. *Dwyer, F. P., Humpoletz, J. E., Nyholm, R. S.:* J. Proc. Roy. Soc. N. S. Wales *80*, 212 (1946).
4. *Burstall, F. H., Dwyer, F. P., Gyarfas, E. C.:* J. Chem. Soc. *1950*, 953.
5. *König, E., Herzog, S.:* J. Inorg. Nucl. Chem. *32*, 585, 701, 613 (1970).
6. *Herzog, S., Praekel, H.:* Z. Chem. *5*, 469 (1965).
7. *Mahon, C., Reynolds, W. L.:* Inorg. Chem. *6*, 1927 (1967).
8. *Herzog, S., Weber, A.:* Z. Chem. *8*, 66 (1968).
9. *Jørgensen, C. K.:* Inorganic Complexes. London: Academic Press 1963.
10. *Sone, K.:* Bull. Chem. Soc. Japan *25*, 1 (1952).

11. *Krumholz, P.:* J. Am. Chem. Soc. *75*, 2163 (1953).
12. *Dwyer, F. P., Gill, N. S., Gyarfas, E. C., Lions, F.:* J. Am. Chem. Soc. *75*, 3834 (1953).
13. *Brandt, W. W., Dwyer, F. P., Gyarfas, E. D.:* Chem. Rev. *54*, 959 (1954).
14. *Bähr, G., Thämlitz, H.:* Z. Anorg. Allgem. Chem. *282*, 3 (1955).
15. *— Döge, H. G.:* Z. Anorg. Allgem. Chem. *292*, 119 (1957).
16. *Busch, D. H., Bailar, J. C.:* J. Am. Chem. Soc. *78*, 1137 (1956).
17. *Morgan, G. T., Burstall, F. H.:* J. Chem. Soc. *1932*, 20.
18. *Goodwin, H. A., Lions, F.:* J. Am. Chem. Soc. *82*, 5013 (1960).
19. *Krumholz, P.:* Inorg. Chem. *4*, 612 (1965).
20. *Stratton, W. J., Busch, D. H.:* J. Am. Chem. Soc. *82*, 4834 (1960).
21. *Lindoy, L. F., Livingstone, S. E.:* Coordin. Chem. Rev. *2*, 173 (1967).
22. *König, E.:* Coordin. Chem. Rev. *3*, 409 (1968).
23. *Mason, S. F.:* Inorg. Chim. Acta Rev. *2*, 89 (1968).
24. *Mc Whinnie, W. R., Miller, J. D.:* Advan. Inorg. Chem. Radiochem. *12*, 135 (1969).
25. *Irving, H., Cabell, M. J., Mellor, D. H.:* J. Chem. Soc. *1953*, 3417.
26. *James, B. R., Parris, M., Williams, R. J. P.:* J. Chem. Soc. *1961*, 4630.
27. *Goodwin, H. A., Sylva, R. N.:* Australian J. Chem. *21*, 83 (1968).
28. *Dosser, R. J., Eilbeck, W. J., Underhill, A. E., Edwards, P. R., Johnson, C. E.:* J. Chem. Soc. A *1969*, 810.
29. *Goodgame, D. M. L., Machado, A. A. S. C.:* Inorg. Chem. *8*, 2031 (1969).
30. — — Chem. Commun. *1969*, 1420.
31. *Sylva, R. N., Goodwin, H. A.:* Australian J. Chem. *20*, 479 (1967).
32. *Baker, W. A., Jr., Bobonich, H. M.:* Inorg. Chem. *3*, 1184 (1965).
33. *König, E., Madeja, K.:* Inorg. Chem. *6*, 48 (1967).
34. *— — Watson, K. J.:* J. Am. Chem. Soc. *90*, 1146 (1968).
35. *— Huffner, S., Steichele, E., Madeja, K.:* Z. Naturforsch. *22a*, 1543 (1967); *23a*, 632 (1968).
36. *Casey, A. T., Isaac, F.:* Australian J. Chem. *20*, 2765 (1967).
37. *Jesson, J. P., Trofimenko, S., Eaton, D. R.:* J. Am. Chem. Soc. *89*, 3158 (1967).
38. *— Weiher, F., Trofimenko, S.:* J. Chem. Phys. *48*, 2058 (1968).
39. *Basolo, F., Dwyer, F. P.:* J. Am. Chem. Soc. *76*, 1454 (1954).
40. *Madeja, K., König, E.:* J. Inorg. Nucl. Chem. *25*, 377 (1963).
41. — Chem. Zvesti *19*, 186 (1965).
42. *— Wilke, W., Schmidt, S.:* Z. Anorg. Allgem. Chem. *346*, 306 (1966).
43. *König, E., Chakravarty, A. S., Madeja, K.:* Theoret. Chim. Acta *9*, 171 (1967).
44. *Schilt, A. A.:* J. Am. Chem. Soc. *82*, 3000 (1960).
45. *Beck, W., Schuierer, E.:* Chem. Ber. *95*, 3048 (1962).
46. *König, E., Madeja, K.:* Inorg. Chem. *7*, 1848 (1968).
47. *Baker, W. A., Jr., Garland, R. I.:* In: Progress in Coordination Chemistry, p. 599; *M. Cais*, editor. Amsterdam: Elsevier Publishing Company 1968.
48. *Fisher, H. M., Stoufer, R. C.:* Inorg. Chem. *5*, 1172 (1966).
49. *Spacu, P., Teodorescu, M., Ciomartan, D.:* Rev. Roumaine Chim. *12*, 145 (1967).
50. *König, E., Schläfer, H. L.:* Z. Physik. Chem. (Frankfurt) *34*, 355 (1962).
51. *Palmer, R. A., Piper, T. S.:* Inorg. Chem. *5*, 864 (1966).
52. *Ito, T., Tanaka, N.:* J. Inorg. Nucl. Chem. *32*, 155 (1970).
53. *Robinson, M. A., Curry, J. D., Busch, D. H.:* Inorg. Chem. *2*, 1178 (1963).
54. *Nelson, S. M., Rodgers, J.:* J. Chem. Soc. A *1968*, 272.
55. *Curry, J. D., Robinson, M. A., Busch, D. H.:* Inorg. Chem. *6*, 1570 (1967).
56. *Mc Whinnie, W. R., Poller, R. C., Thevarasa, M.:* J. Chem. Soc. *1967*, 1671.

57. *Orgel, L. E.*: Reports to the Tenth Solvey Council, Quelques Problèmes de Chimie Minérale, p. 289, Bruxelles 1956.
58. *Jørgensen, C. K.*: Absorption Spectra and Chemical Bonding, p. 128. Oxford: Pergamon Press 1962.
59. *Van Santen, J. H., Van Wieringen, J. S.*: Rec. Trav. Chim. *71*, 420 (1952).
60. *Lytle, F. E., Hercules, D. M.*: J. Am. Chem. Soc. *91*, 253 (1969).
61. *Yamasaki, K.*: Bull. Chem. Soc. Japan *12*, 390 (1937); *13*, 538 (1938); *15*, 130, 461 (1940).
62. *Gondo, Y.*: J. Chem. Phys. *41*, 3928 (1964).
63. *Clar, E.*: Chem. Ber. *69*, 607 (1936).
64. *Bray, R. G., Ferguson, J., Hawkins, C. J.*: Australian J. Chem. *22*, 2091 (1969).
65. *Ito, T., Tanaka, N., Hanazaki, I., Nagakura, S.*: Bull. Chem. Soc. Japan *41*, 365 (1968).
66. *Krumholz, P.*: J. Am. Chem. Soc. *73*, 3487 (1951).
67. *Sone, K., Krumholz, P., Stammreich, H.*: J. Am. Chem. Soc. *77*, 777 (1955).
68. *Schläfer, H. L.*: Z. Physik. Chem. (Frankfurt) *8*, 373 (1956).
69. *Gil, L., Moraga, E., Bunel, S.*: Mol. Phys. *12*, 333 (1967).
70. *McCaffery, A. J., Mason, S. F.*: Proc. Chem. Soc. *1963*, 211.
71. *Hidaka, J., Douglas, B.*: Inorg. Chem. *3*, 1180 (1964).
72. *Moffitt, M.*: J. Chem. Phys. *25*, 467 (1956).
73. *Templeton, D. H., Zalkin, A., Ueki, T.*: Acta Cryst. *21*, A154 (1966).
74. *Bosnich, B.*: Inorg. Chem. *7*, 178, 2379 (1968).
75. *Perrin, M. H., Gouterman, M.*: J. Chem. Phys. *46*, 1019 (1967).
76. *Jørgensen, C. K.*: Oxidation States and Oxidation Numbers. New York: Springer 1969.
77. *Williams, R. J. P.*: J. Chem. Soc. *1955*, 137.
78. *Jørgensen, C. K.*: Acta Chem. Scand. *11*, 166 (1957).
79. *Williams, R. J. P.*: Chem. Rev. *56*, 299 (1956).
80. *Krumholz, P.*: In: Theory and Structure of Complex Compounds, p. 217. *B. J. Trzebiatowska*, editor. Oxford: Pergamon Press 1964.
81. *Hanazaki, I., Nagakura, S.*: Inorg. Chem. *8*, 648 (1969).
82. *Lions, F., Martin, K. V.*: J. Am. Chem. Soc. *79*, 2733 (1957).
83. *Dekkers, J., Goodwin, H. A.*: Australian J. Chem. *20*, 69 (1969).
84. *Barbieri, G. A.*: Atti Accad. Nazl. Lincei *20*, 273 (1934).
85. *Schilt, A. A.*: J. Am. Chem. Soc. *79*, 5421 (1957).
86. *Dwyer, F. P., Goodwin, H. A., Gyarfas, E. C.*: Australian J. Chem. *16*, 42 (1963).
87. *Shiman, R., Neilands, J. B.*: Biochemistry *4*, 2233 (1965).
88. *Brandt, W. W., Smith, G. F.*: Anal. Chem. *21*, 1313 (1949).
89. *Paris, J. P.*: Ph. D. Thesis, Purdue University, Lafayette, Ind. 1960.
90. *Day, P., Sanders, N.*: J. Chem. Soc. *1967*, 1536.
91. *Klassen, D. M., Crosby, G. A.*: J. Chem. Phys. *48*, 1853 (1968).
92. *Paris, J. P., Brandt, W. W.*: J. Am. Chem. Soc. *81*, 5001 (1959).
93. *Crosby, G. A., Perkins, W. G., Klassen, D. M.*: J. Chem. Phys. *43*, 1498 (1965).
94. *Demas, J. N., Crosby, G. A.*: J. Mol. Spectry. *26*, 72 (1968).
95. *Pullman, B., Pullman, A.*: Les Théories Electroniques de la Chimie Organique, pp. 495—497. Paris: Masson et Cie. 1952.
96. *Knott, R. F., Breckenridge, J. G.*: Can. J. Chem. *32*, 512 (1954).
97. *Crowne, F. R., Breckenridge, J. G.*: Can. J. Chem. *32*, 641 (1954).
98. *Stratton, W. J.*: Inorg. Chem. *9*, 517 (1970).
99. *Robinson, M. A., Busch, D. H.*: Inorg. Chem. *2*, 1171 (1963).
100. *Orgel, L. E.*: J. Chem. Phys. *23*, 1824 (1955).
101. *Jillot, B. A., Williams, R. J. P.*: J. Chem. Soc. *1958*, 462.

102. *Sanders, N., Day, P.:* J. Chem. Soc. *1969*, 2303.
103. *Ford, P., Rudd, De F. P., Gaunder, R., Taube, H.:* J. Am. Chem. Soc. *90*, 1187 (1968).
104. *McCurdy, W. H., Smith, G. F.:* Analyst *77*, 846 (1952).
105. *Smith, G. F.:* Anal. Chem. *26*, 1534 (1954).
106. *Wilkins, D. H., Schilt, A. A., Smith, G. F.:* Anal. Chem. *27*, 1574 (1955).
107. *Smith, G. F., Banick, W. M., Jr.:* Analyst *83*, 661 (1958).
108. *Krumholz, P.:* Proc. 8th. Inter. Conf. Coordination Chem., p. 89; Vienna 1964.
109. — Inorg. Chem. *4*, 609 (1965).
110. *Ferguson, J. E., Harris, G. M.:* J. Chem. Soc. A *1966*, 1293.
111. *Bosnich, B., Dwyer, F. P.:* Australian J. Chem. *19*, 2235 (1966).
112. *Davies, N. R., Mullins, T. L.:* Australian J. Chem. *20*, 657 (1967).
113. *Bjerrum, J., Adamson, A. W., Bostrup, O.:* Acta Chem. Scand. *10*, 329 (1956).
114. *Hamer, N. R., Orgel, L. E.:* Nature *190*, 439 (1961).
115. *George, P., Hanania, G. I. H., Irvine, D. H.:* J. Chem. Soc. *1959*, 779.
116. *Krumholz, P.:* J. Am. Chem. Soc. *71*, 3654 (1949).
117. *Baldwin, D. A., Lever, A. B. P., Parish, R. V.:* Inorg. Chem. *8*, 107 (1969).
118. *Dekker, M., Knox, G. R.:* Chem. Commun. *1967*, 1243.
119. *Day,* P., *Sanders, N.:* J. Chem. Soc. *1967*, 1530.
120. *Murrel, J. N.:* Quart. Rev. *15*, 191 (1961).
121. *Mulliken, J.:* J. Am. Chem. Soc. *72*, 600 (1950); *74*, 811 (1952).
122. *Orgel, L. E.:* J. Chem. Soc. *1961*, *3683*.
123. *Weber, G.:* Z. Physik. Chem. (Leipzig) *218*, 204 (1961).
124. *De Alti, G., Galasso, V., Bigotto, A.:* Inorg. Chim. Acta *3*, 527 (1969).
125. ———, *Costa, G.:* Inorg. Chim. Acta *3*, 533 (1969).
126. *Ito, T., Tanaka, N., Nagakura, S.:* Bull. Chem. Soc. Japan *42*, 702 (1969).
127. *Hanazaki, I., Hanazaki, F., Nagakura, S.:* J. Chem. Phys. *50*, 265 (1969).
128. *Nakamoto, K.:* Advances in the Chemistry of Coordination Compounds, p. 437; *S. Kirschner*, editor. New York: MacMillan Company 1961.
129. *Bayer, E., Häfelinger, G.:* Chem. Ber. *99*, 1689 (1966).
130. *Figgins, P. E., Busch, D. H.:* J. Phys. Chem. *65*, 2236 (1961).
131. *Stoufer, R. C., Busch, D. H.:* J. Am. Chem. Soc. *78*, 6016 (1956).
132. —— J. Am. Chem. Soc. *82*, 3491 (1960).
133. *Stratton, W. J., Busch, D. H.:* J. Am. Chem. Soc. *80*, 1286 (1958).
134. *Bayer, E.:* Angew. Chem. *73*, 533 (1961).
135. — Angew. Chem. *76*, 76 (1964).
136. *Schilt, A. A., Taylor, R. C.:* J. Inorg. Nucl. Chem. *9*, 211 (1959).
137. *Inskeep, R. G.:* J. Inorg. Nucl. Chem. *24*, 763 (1963).
138. *König, E., Madeja, K.:* Spectrochim. Acta *23* A, 45 (1967).
139. *Baker, W. A., Jr., Jones, G. J.:* Chem. Commun. *1965*, 368.
140. *Schilt, A. A.:* Inorg. Chem. *3*, 1323 (1964).
141. *Baxendale, J. H., George, P.:* Nature *163*, 725 (1949); Trans. Faraday Soc. *46*, 55 (1950).
142. *Krumholz, P.:* Nature *163*, 724 (1949).
143. — Anais Acad. Brasil. Cienc. *22*, 263 (1950).
144. *Kolthoff, I. M., Leussing, D. L., Lee, T. S.:* J. Am. Chem. Soc. *72*, 2173 (1950).
145. *Irving, H., Mellor, D. H.:* J. Chem. Soc. *1962*, 5222.
146. —, *Williams, R. J. P.:* Nature *162*, 746 (1948).
147. —— J. Chem. Soc. *1953*, 3192.
148. *Grinberg, A. A., Yatsimirskii, K. B.:* Bull. Acad. Sci. USSR, Div. Chem. Sci. (English Transl.) *1952*, 239.
149. *George, P., McClure, D. S.:* Progr. Inorg. Chem. *1*, 380 (1959).

150. *Rossotti, F. J. C.:* In: Modern Coordination Chemistry; *J. Lewis* and *R. G. Wilkins*, editor. New York: Interscience Publ. 1960.
151. *Ciampolini, M., Paoletti, P., Sacconi, L.:* In: Advances in the Chemistry of Coordination Compounds, p. 303; *S. Kirschner*, editor. New York: McMillan Company 1961; J. Chem. Soc. *1960*, 4553.
152. *Anderegg, G.:* Helv. Chim. Acta *46*, 2397 (1962).
153. *Lahiri, S. C., Aditya, S.:* Z. Physik. Chem. (Frankfurt) *41*, 173 (1964).
154. *Brisbin, D. A., McBryde, W. A. E.:* Can. J. Chem. *41*, 1135 (1963).
155. *Lahiri, S. C., Aditya, S.:* J. Inorg. Nucl. Chem. *30*, 2487 (1968).
156. *Eilbeck, W. J., Holmes, F.:* J. Chem. Soc. A *1967*, 1777.
157. *Davies, R. L., Dunning, K. W.:* J. Chem. Soc. *1965*, 4168.
158. *Albrecht, G., Tschirnich, J., Madeja, K.:* Z. Chem. *5*, 312 (1965).
159. *Brandt, W. W., Gullstrom, D. K.:* J. Am. Chem. Soc. *74*, 3532 (1952).
160. *Hawkins, C. J., Duewell, H., Pickering, W. F.:* Anal. Chim. Acta *25*, 257 (1961).
161. *Irving, H., Rossotti, H.:* Acta. Chem. Scand. *10*, 72 (1956).
162. *Brandt, W. W., Wright, J. P.:* J. Am. Chem. Soc. *76*, 3082 (1954).
163. *Jones, J. G., Poole, J. B., Tomkinson, J. C., Williams, R. J. P.:* J. Chem. Soc. *1958*, 2001.
164. *Tomkinson, J. C., Williams, R. J. P.:* J. Chem. Soc. *1958*, 2010.
165. *James, B. R., Lions, J. R., Williams, R. J. P.:* Biochemistry *1*, 379 (1962).
166. *Hume, D. N., Kolthoff, I. M.:* J. Am. Chem. Soc. *65*, 1895 (1943).
167. *George, P., Hanania, G. I. H., Irvine, D. H.:* J. Chem. Soc. *1959*, 2548.
168. *Dwyer, F. P.:* J. Proc. Roy. Soc. N. S. Wales *83*, 138 (1949).
169. *Mercier, E. E., Buckley, R. R.:* Inorg. Chem. *4*, 1692 (1965).
170. *Smith, G. F., Richter, F. P.:* Ind. Eng. Chem. Anal. Ed. *16*, 580 (1944).
171. *Brandt, W. W., Smith, G. F.:* Anal. Chem. *21*, 1313 (1949).
172. *Smith, G. F., Banick, W. M., Jr.:* Talanta *2*, 348 (1959).
173. *Kratochwil, B., Knoeck, J.:* J. Phys. Chem. *70*, 944 (1966).
174. *Dickens, J. E., Basolo, F., Neumann, H. M.:* J. Am. Chem. Soc. *79*, 1286 (1957).
175. *Schilt, A. A.:* Anal. Chem. *35*, 1599 (1963).
176. *Buckingham, D. A., Sargeson, A. M.:* In: Chelating Agents and Metal Chelates; *F. P. Dwyer* and *D. P. Mellor*, editor. New York: Academic Press 1964.
177. —, *Dwyer, F. P., Sargeson, A. M.:* Inorg. Chem. *5*, 1243 (1966).
178. *Krumholz, P.:* Proc. 7th. Inter. Conf. Coordination Chem., p.280, Stockholm 1962.
179. —, *Li Chum, H.:* Unpublished results.
180. *Baxendale, J. H., George, P.:* Nature *162*, 777 (1948); Trans. Faraday Soc. *46*, 736 (1950).
181. *Krumholz, P.:* J. Phys. Chem. *60*, 87 (1956).
182. *Basolo, F., Hayes, J. C., Neumann, H. M.:* J. Am. Chem. Soc. *76*, 3807 (1954).
183. *Murmann, R. K., Healy, E. A.:* J. Am. Chem. Soc. *83*, 2092 (1961).
184. *Krumholz, P., Vichi, E. J. S.:* Proc. 3rd. Symposium on Coordination Chemistry, p. 357, Debrecen, Hungary, 1970.
185. *Lee, T. S., Kolthoff, I. M., Leussing, D. L.:* J. Am. Chem. Soc. *70*, 3596 (1948).
186. *Burgess, J., Prince, R. H.:* J. Chem. Soc. *1967*, 434.
187. — J. Chem. Soc. A *1968*, 497.
188. *Basolo, F., Pearson, R. G.:* Mechanisms of Inorganic Reactions. New York: John Wiley 1958, 1967.
189. *Farina, R., Hogg, R., Wilkins, R. G.:* Inorg. Chem. *7*, 170 (1968).
190. *Pagenkopf, G. K., Margerum, D. W.:* Inorg. Chem. *7*, 2514 (1968).
191. *Hogg, R., Wilkins, R. G.:* J. Chem. Soc. *1962*, 341.
192. *Holyer, R. H., Hubbard, C. D., Kettle, S. F. A., Wilkins, R. G.:* Inorg. Chem. *4*, 929 (1965).

193. *Burgess, J., Prince, R. H.:* J. Chem. Soc. *1963*, 5752.
194. *Bell, R. S., Sutin, N.:* Inorg. Chem. *1*, 359 (1962).
195. *Ellis, P., Wilkins, R. G.:* J. Chem. Soc. *1959*, 299.
196. *—, Hogg, R., Wilkins, R. G.:* J. Chem. Soc. *1959*, 3308.
197. *Davies, N. R., Dwyer, F. P.:* Trans. Faraday Soc. *48*, 244 (1952).
198. *Seiden, L., Basolo, F., Neumann, H. M.:* J. Am. Chem. Soc. *81*, 3809 (1959).
199. *Jensen, A., Basolo, F., Neumann, H. M.:* J. Am. Chem. Soc. *80*, 2354 (1958).
200. *Margerum, D. W.:* J. Am. Chem. Soc. *79*, 2728 (1957).
201. *—, Morgenthaler, L. P.:* J. Am. Chem. Soc. *84*, 706 (1962).
202. *Burgess, J., Prince, R. H.:* J. Chem. Soc. *1965*, 4697, 6061.
203. — J. Chem. Soc. *1967*, 431.
204. *Balzani, V., Carassiti, V., Moggi, L.:* Inorg. Chem. *3*, 1252 (1964).
205. *Burstall, F. H., Nyholm, R. S.:* J. Chem. Soc. *1952*, 3570.
206. *Pauling, L.:* The Nature of the Chemical Bond. Ithaca, N.Y.: Cornell University Press 1939, 1960.
207. *Williams, R. J. P.:* J. Chem. Soc. *1956*, 8.
208. *Chatt, J.:* Chemistry of the Coordination Compounds, p. 515. New York: Pergamon Press 1958.
209. *Orgel, L. E.:* An Introduction to Transition-Metal Chemistry: Ligand Field Theory. London: Methuen & Co. Ltd. 1960.
210. *Sanders, N., Day, P.:* J. Chem. Soc. A *1970*, 1190.
211. *Figgins, P. F., Busch, D. H.:* J. Am. Chem. Soc. *82*, 820 (1960).
212. *Mulliken, R. S.:* J. Phys. Chem. *56*, 295 (1952).
213. *Craig, D. P., Maccoll, A., Nyholm, R. S., Orgel, L. E., Suttin, L. E.:* J. Chem. Soc. *1954*, 332.
214. *Castellano, S., Gunther, H., Ebersole, S.:* J. Phys. Chem. *69*, 4166 (1965).
215. *Kanaker, C. K., Khetrapal, C. L., Nipankar, S. V.:* J. Phys. Chem. *73*, 276 (1969).
216. *Miller, J. D., Prince, R. H.:* J. Chem. Soc. *1965*, 3185, 4706.
217. *Dove, M. F. A., Hallett, J. G.:* J. Chem. Soc. A *1969*, 1204.
218. *Rosenberger, H., Pettig, M., Madeja, K.:* Z. Chem. *6*, 30 (1966).
219. *Bayer, E., Breitmaier, E., Schurig, V.:* Chem. Ber. *101*, 1594 (1968).
220. *Eaton, D. R., Phillips, W. D.:* Advan. Magnetic Resonance *1*, 103 (1965).
221. *La Mar, G. N., Van Hecke, G. R.:* J. Am. Chem. Soc. *91*, 3442 (1969); *92*, 3021 (1970).
222. — — Inorg. Chem. *9*, 1546 (1970).
223. *Zwickel, A. M., Taube, H.:* Discussions Faraday Soc. *29*, 42 (1960).
224. *Murray, R., Waind, G. M.:* Proc. 7th. Intern. Conf. Coordination Chem., p. 309; Stockholm 1962.
225. *De Simone, R. E., Drago, R. S.:* J. Am. Chem. Soc. *92*, 2343 (1970).
226. *Terezakis, E. G., Carlin, R. L.:* Inorg. Chem. 6, 2125 (1967).
227. *Udachin, Y. M., Dyatkina, M. E.:* J. Struct. Chem. (USSR) *8*, 325 (1967).
228. *Epstein, L. M.:* J. Chem. Phys. *40*, 435 (1964).
229. *Collins, R. L., Pettit, R., Baker, W. A., Jr.:* J. Inorg. Nucl. Chem. *28*, 1001 (1966).
230. *Duncan, J. F., Mok, K. F.:* J. Chem. Soc. A *1966*, 1493.
231. *Berrett, R. R., Fitzsimmons, B. W., Owusu, A. A.:* J. Chem. Soc. A 1968, 1575.
232. *Bancroft, G. M., Mays, M. J., Prater, B. E.:* J. Chem. Soc. A *1970*, 956.
233. *Jørgensen, C. K.:* J. Inorg. Nucl. Chem. *24*, 1571 (1962).

Received August 31, 1970

The Nephelauxetic Effect

Calculation and Accuracy of the Interelectronic Repulsion Parameters I. Cubic High-Spin d^2, d^3, d^7, and d^8 Systems*

Dr. E. König

Institut für Physikalische Chemie II, Universität Erlangen-Nürnberg, Erlangen, Germany

Table of Contents

---

* This study has been sponsored in part by research grants of the Deutsche Forschungsgemeinschaft, the Stiftung Volkswagenwerk, and the Fonds der Chemischen Industrie whose support is gratefully acknowledged.

I. Introduction

In a comprehensive theoretical investigation, *Tanabe* and *Sugano* (*60, 61*) calculated, in 1954, the splittings of all energy levels arising from the electronic configurations d^1 to d^9 under the action of a ligand field of octahedral symmetry. Within the last few years, these results are being used by an increasing number of inorganic chemists to interpret the electronic spectra of transition metal compounds. Most frequently employed is a semi-empirical form of the theory. In this case, the octahedral ligand field splitting parameter $10\,Dq$ ($= \Delta$) and the interelectronic repulsion parameters (Racah parameters) B and C (frequently approximated as $C = 4\,B$) are directly determined from the spectra. If the number of observed electronic absorption bands exceeds the number of parameters determined therefrom, the numerical values of these parameters are used to calculate the transition energies corresponding to the extra electronic bands.

The numerical values of the interelectronic repulsion parameters, B and C, as determined from the spectrum of an actual transition metal compound (or at least the value of B[1]) are, in general, lower than the values in a free (gaseous) transition metal ion. This observation which was actually pointed out by several workers in the field (*42, 43, 53, 60, 61*), is called the *nephelauxetic effect* (*19, 53*). In a free $3d$ metal ion, the interelectronic repulsion integrals are approximately proportional to the average reciprocal radius $\langle r^{-1}\rangle_{3d}$ of the partly filled $3d$ shell. The decrease in B (and, possibly, in C) was therefore attributed to an expanded radial distribution of d electrons in the complex, i.e.

$$\langle r^{-1}\rangle_{3d}^{\text{complex}} \quad < \quad \langle r^{-1}\rangle_{3d}^{\text{free ion}} \qquad (1)$$

So far no information about details of the expansion of $3d$ orbitals has been obtained from the optical spectra. Valuable data on this problem may be extracted from ligand hyperfine splittings in the EPR (*26*) where the related relativistic nephelauxetic effect (*18*) is operative.

As far as optical spectra are concerned, *Jørgensen* (*19, 20, 22*) has rationalized a large body of experimental data in terms of the parameters $10\,Dq$ and B. Two different origins of the nephelauxetic effect have been advanced (*19*): (i) the *central-field covalency* is due to screening the nuclear charge of the central ion by electrons of the ligands; (ii) the *symmetry-restricted covalency* is caused by delocalization of metal d electrons onto the ligands (the delocalization is symmetry dependent: e_g electrons for

---

[1]) There seems to be evidence (*63*) that, in fact, C is *not* decreasing with decreasing B, or at least not decreasing by the same amount.

σ, t_{2g} for π bonding). However, no quantum mechanical theory has yet been developed which would provide a quantitative basis for the nephelauxetic effect. It is believed that such attempts will not be successful until criteria on the accuracy and reliability of the parameter values determined from the spectra will become available. To this end, the present study has been initiated.

The best starting point for the investigation is immediately apparent. Certain transition metal ions (electronic configurations d^2, d^3, d^7, and d^8) are characterized, within the approximation of Russell-Saunders coupling, by a ^{3}F or ^{4}F ground term followed, at higher energy, by a ^{3}P or ^{4}P term, respectively. If we limit our discussion to the spin-allowed (and parity-forbidden) transitions, these ions exhibit, in an octahedral ligand field, three electronic absorption bands. These bands arise as a consequence of the splitting of the ^{3}F or ^{4}F ground term by the ligand field. Thus in the d^3 configuration, e.g., the ^{4}F splits into $^4A_{2g}$, $^4T_{2g}$, and $^4T_{1g}$(F) in order of increasing energy, whereas ^{4}P changes over into $^4T_{1g}$(P). The $^4A_{2g}$ level being lowest, the transitions $^4A_{2g} \rightarrow {}^4T_{2g}$, $^4A_{2g} \rightarrow {}^4T_{1g}$(F), and $^4A_{2g} \rightarrow {}^4T_{1g}$(P) are expected. The energy of these transitions depends on the values of 10 Dq and B only[2]). If all three transitions are observed experimentally and if the afore-mentioned semi-empirical procedure is to be used, there are several possible ways how to determine 10 Dq and B. In practice, however, the $^4A_{2g} \rightarrow {}^4T_{1g}$(P) transition is rarely located due to the much more intense ligand and/or charge-transfer bands in the uv region. Thus most of the numerical values of 10 Dq and B compiled in the literature have been extracted from spectra where only two bands were observed. This holds true particularly in the octahedral configurations d^2, d^3, and d^7, and similar arguments apply to tetrahedral d^7 and d^8 systems. The usefulness of these data is greatly diminished as long as no critical investigation on the *complete* systems of three electronic transitions becomes available.

Another complication is frequently encountered, if the position of a particular ligand in the nephelauxetic series (*19*, *53*) ist to be determined: the parameter B is obtained[3]) from the electronic spectra of the complex species under study using one of the numerical procedures available. This

---

[2]) In this case, the question whether the nephelauxetic effect operates alone on B or on both B and C is eliminated.

[3]) Actually, different B values were found from interconfigurational versus intraconfigurational transitions. To account for this, the notations B_{35}, B_{33}, and B_{55} have been introduced (*19*), the subscripts being related to *Bethe's* notation of d orbitals in cubic symmetry, $\gamma_3(e)$ and $\gamma_5(t_2)$. The transitions $^4A_{2g}(t_{2g}^3) \rightarrow {}^2E_g(t_{2g}^3)$ of a d^3 electron and $^3A_{2g}(t_{2g}^6\, e_g^2) \rightarrow {}^1E_g(t_{2g}^6\, e_g^2)$ of a d^8 electron are intraconfigurational giving rise to B_{55} and B_{33}, respectively. On the other hand, all spin-allowed transitions are interconfigurational transitions and thus the parameter B_{35} only will be considered here.

value is subsequently compared with B values of some other complex ions taken from the literature. However, if these B values have their origin in a different type of numerical analysis, this practice is, in general, not justified. That this is so will be shown below.

Without doubt, there are numerous ways how the values of the interelectronic repulsion parameters may be influenced, e.g., mixing with excited states of different multiplicity, lower than octahedral symmetry, varying effective charge on the metal ion, differential expansion of t_{2g} and e_g orbitals etc. For the sake of argument, however, we will assume in what follows that the three parameter (10 Dq, B, C) ligand field theory is valid *exactly*. In section II, the ligand field theory of d^2, d^3, d^7, and d^8 configurations is reviewed and the full expressions for the transition energies are derived. Equations for the evaluation of 10 Dq and B resulting therefrom under the assumption of the various numerical fitting procedures are listed in section III. These expressions are subsequently applied (in section IV) to those spectra of transition metal complexes where *all three spin-allowed bands* are observed. In section V, the results are discussed with regard to accuracy and dependability of the parameter values involved. Possible reasons for some of the discrepancies are outlined.

II. Ligand Field Theory of Spin-Allowed Transitions in d^n Configurations, $n=2, 3, 7, 8$

The general aspects of the theory of ligand fields in octahedral and tetrahedral symmetries are well understood and adequately covered in various textbooks (*1, 8, 11, 55*). It will be assumed, therefore, that the reader is familiar with the basic principles of the method. The electron systems considered here have various features in common. For the sake of clarity, however, the configurations d^n, $n=2, 3, 7, 8$ in octahedral environment will be treated separately. Results for tetrahedral symmetry are not listed, since octahedral and tetrahedral systems are simply related by an inversion in the order of terms. Only terms having the same multiplicity as the ground term will be considered.

A. Configuration d^2

Two triplet terms, ^3F and ^3P, and three singlet terms, ^1D, ^1G, and ^1S, derive from the spherical d^2 configuration. Of these, ^3F is the ground term with ^3P being 15 B higher in energy. In an octahedral ligand field, ^3F splits into the $^3T_{1g}$, $^3T_{2g}$, and $^3A_{2g}$ states and ^3P changes over into

$^3T_{1g}$. Within the weak-field coupling scheme, the relevant matrix elements are

$$\begin{aligned}
\langle {}^3T_{2g}(F)\,|\,V_{oct}\,|{}^3T_{2g}(F)\rangle &= E\,({}^3T_{2g}(F)) = 2\,Dq\\
\langle {}^3A_{2g}(F)\,|\,V_{oct}\,|{}^3A_{2g}(F)\rangle &= E\,({}^3A_{2g}(F)) = 12\,Dq\\
\langle {}^3T_{1g}(F)\,|\,V_{oct}\,|{}^3T_{1g}(F)\rangle &= -6\,Dq\\
\langle {}^3T_{1g}(F)\,|\,V_{oct}\,|{}^3T_{1g}(P)\rangle &= 4\,Dq
\end{aligned} \tag{2}$$

The first two matrix elements determine the energies of two of the triplet states relative to the energy of the free ^3F ground state. The two $^3T_{1g}$ states are mixed by configuration interaction and their energies have thus to be determined from the secular determinant

$$\begin{array}{c} \\ {}^3T_{1g}(F) \\ {}^3T_{1g}(P) \end{array}
\begin{array}{c} \begin{array}{cc} {}^3T_{1g}(F) & \quad {}^3T_{1g}(P) \end{array} \\ \begin{vmatrix} -6\,Dq-E & 4\,Dq \\ 4\,Dq & 15\,B-E \end{vmatrix} \end{array} = 0 \tag{3}$$

which results in

$$E_{1,2}({}^3T_{1g}) = 7.5\,B - 3\,Dq \pm \tfrac{1}{2}(225\,B^2 + 100\,Dq^2 + 180\,Dq\cdot B)^{1/2} \tag{4}$$

Alternatively, the strong-field coupling scheme may be employed. In this case, the lowest energy configuration, t_{2g}^2, of energy $-8\,Dq$, gives rise to a $^3T_{1g}$ state as well as to the singlet states ^1E_g, $^1T_{2g}$, and $^1A_{1g}$ which will not be considered here. The excited configurations $t_{2g}e_g$ and e_g^2 having the respective energies of $+2\,Dq$ and $+12\,Dq$ produce the $^3T_{2g}$ and $^3T_{1g}$ states and the $^3A_{2g}$ state, respectively. The matrix elements of the electronic interaction being

$$\begin{aligned}
&\langle {}^3T_{2g}(t_{2g}e_g)\,|\,1/r_{ij}\,|{}^3T_{2g}(t_{2g}e_g)\rangle\\
&\qquad = \langle {}^3A_{2g}(e_g^2)\,|\,1/r_{ij}\,|{}^3A_{2g}(e_g^2)\rangle = A - 8\,B\\
&\langle {}^3T_{1g}(t_{2g}e_g)\,|\,1/r_{ij}\,|{}^3T_{1g}(t_{2g}e_g)\rangle = A + 4\,B\\
&\langle {}^3T_{1g}(t_{2g}^2)\,|\,1/r_{ij}\,|{}^3T_{1g}(t_{2g}^2)\rangle = A - 5\,B\\
&\langle {}^3T_{1g}(t_{2g}^2)\,|\,1/r_{ij}\,|{}^3T_{1g}(t_{2g}e_g)\rangle = -6\,B
\end{aligned}$$

the energies of these states result from

$$\begin{aligned} E({}^3T_{2g}(t_{2g}e_g)) &= A - 8\,B + 2\,Dq \\ E({}^3A_{2g}(e_g^2)) &= A - 8\,B + 12\,Dq \end{aligned} \tag{5}$$

$$\begin{array}{c|cc|} & {}^3T_{1g}(t_{2g}^2) & {}^3T_{1g}(t_{2g}\,e_g) \\ {}^3T_{1g}(t_{2g}^2) & A - 5\,B - 8\,Dq - E & -6\,B \\ {}^3T_{1g}(t_{2g}\,e_g) & -6\,B & A + 4\,B + 2\,Dq - E \end{array} = 0 \tag{6}$$

where in Eq. (6) the configuration interaction between the two ${}^3T_{1g}$ states again is taken into account. Substracting the energy of the free ion ground state ${}^3F$, $E({}^3F) = A - 8\,B$, energy expressions identical to those of Eq. (2) and Eq. (4) are obtained.

The ground term of the d^2 configuration in an octahedral environment invariably is ${}^3T_{1g}$. In order of increasing energy, three spin-allowed transitions are expected, viz. $a^3T_{1g} \rightarrow {}^3T_{2g}$, $a^3T_{1g} \rightarrow b^3T_{1g}$, and $a^3T_{1g} \rightarrow {}^3A_{2g}$. If these transitions are labeled ν_1, ν_2, and ν_3, respectively, the transition energies may be obtained from Eq. (2) to Eq. (4) as

$$\begin{aligned} \nu_1 &= \tfrac{1}{2}(10\,Dq - 15\,B) + \tfrac{1}{2}[(10\,Dq + 15\,B)^2 - 12\,B \cdot 10\,Dq]^{1/2} \\ \nu_2 &= [(10\,Dq + 15\,B)^2 - 12\,B \cdot 10\,Dq]^{1/2} \\ \nu_3 &= \tfrac{1}{2}(30\,Dq - 15\,B) + \tfrac{1}{2}[(10\,Dq + 15\,B)^2 - 12\,B \cdot 10\,Dq]^{1/2} \end{aligned} \tag{7}$$

B. Configuration d^3

The configuration d^3 gives rise to two quartet terms, ${}^4F$ and ${}^4P$, with ${}^4F$ the ground term and ${}^4P$ again being 15 B higher in energy. In addition, there are six doublet terms, viz. ${}^2G$, ${}^2H$, ${}^2P$, a^2D, ${}^2F$, and b^2D. The action of an octahedral ligand field on the two terms we are concerned with splits the ${}^4F$ into ${}^4A_{2g}$, ${}^4T_{2g}$, and ${}^4T_{1g}$ states, whereas ${}^4P$ goes over into a ${}^4T_{1g}$ state. If the weak-field coupling scheme is applied, the energies of these states follow from

$$\begin{aligned} E({}^4A_{2g}(F)) &= -12\,Dq \\ E({}^4T_{2g}(F)) &= -\;\,2\,Dq \end{aligned} \tag{8}$$

$$\begin{array}{c|cc|} & {}^4T_{1g}(F) & {}^4T_{1g}(P) \\ {}^4T_{1g}(F) & 6\,Dq - E & 4\,Dq \\ {}^4T_{1g}(P) & 4\,Dq & 15\,B - E \end{array} = 0 \tag{9}$$

where Eq. (9) is equivalent to

$$E_{1,2}(^4T_{1g}) = 7.5\,B + 3\,Dq \pm \tfrac{1}{2}(225\,B^2 + 100\,Dq^2 - 180\,Dq \cdot B)^{1/2}. \quad (10)$$

If, on the other hand, the strong-field coupling scheme is assumed, the $^4A_{2g}$ state arises from the configuration of lowest energy, t_{2g}^3, followed by $^4T_{2g}$ and a $^4T_{1g}$ state originating in $t_{2g}^2\,e_g$, whereas the other $^4T_{1g}$ state results from $t_{2g}\,e_g^2$. In this case, the energies derive from

$$\begin{aligned} E(^4A_{2g}(t_{2g}^3)) &= 3\,A - 15\,B - 12\,Dq \\ E(^4T_{2g}(t_{2g}^2\,e_g)) &= 3\,A - 15\,B - \ \ 2\,Dq \end{aligned} \quad (11)$$

$$\begin{array}{c|cc|l} & {}^4T_{1g}(t_{2g}^2\,e_g) & {}^4T_{1g}(t_{2g}\,e_g^2) & \\ {}^4T_{1g}(t_{2g}^2\,e_g) & 3\,A - 3\,B - 2\,Dq - E & 6\,B & \\ {}^4T_{1g}(t_{2g}\,e_g^2) & 6\,B & 3\,A - 12\,B + 8\,Dq - E & = 0. \end{array} \quad (12)$$

These expressions become identical with Eq. (8) and Eq. (10) if one substracts the energy of the free ion ground state, $E(^4F) = 3\,A - 15\,B$.

In the d^3 configuration in a field of octahedral symmetry, the $^4A_{2g}$ state is always the ground state, irrespective of Dq. Three spin-allowed transitions again are expected, viz. $^4A_{2g} \rightarrow {}^4T_{2g}$, $^4A_{2g} \rightarrow a^4T_{1g}$, and $^4A_{2g} \rightarrow b^4T_{1g}$. If these transitions are labeled, in order of increasing energy, ν_1, ν_2, and ν_3, the transition energies follow from Eq. (8) to Eq. (10):

$$\begin{aligned} \nu_1 &= 10\,Dq \\ \nu_2 &= \tfrac{1}{2}(15\,B + 30\,Dq) - \tfrac{1}{2}[(15\,B - 10\,Dq)^2 + 12\,B \cdot 10\,Dq]^{1/2} \\ \nu_3 &= \tfrac{1}{2}(15\,B + 30\,Dq) + \tfrac{1}{2}[(15\,B - 10\,Dq)^2 + 12\,B \cdot 10\,Dq]^{1/2}. \end{aligned} \quad (13)$$

C. Configuration High-spin d^7

From the equivalence of d^n and d^{10-n} electron systems it follows that the configurations d^3 and d^7 give rise to the same type and number of terms. Thus, in an octahedral ligand field, we will have again to deal with the $^4A_{2g}$, $^4T_{2g}$, a^4T_{1g}, and b^4T_{1g} states. In addition, d^n and d^{5+n} ligand field systems are interrelated on account of the spherically symmetrical half-filled d^5 configuration. The consequence is that, apart from multiplicity, the energy *expressions* of d^2 and d^7 systems are identical. Thus Eq. (2) to (4) apply to the quartet states of d^7.

The ground state of octahedral d^7 systems is a T_{1g} state similar to the d^2 case as long as $10\,Dq < 4\,B + 4\,C$ (high-spin ground state). However, the experimentally observed spin-allowed transitions are, in order of increasing energy, $a^4T_{1g} \rightarrow {}^4T_{2g}$, $a^4T_{1g} \rightarrow {}^4A_{2g}$, and $a^4T_{1g} \rightarrow b^4T_{1g}$. If these transitions are denoted ν_1, ν_2 ,and ν_3, the transition energies result from Eq. (2) to Eq. (4):

$$\nu_1 = \tfrac{1}{2}(10\,Dq - 15\,B) + \tfrac{1}{2}[(10\,Dq + 15\,B)^2 - 12\,B \cdot 10\,Dq]^{1/2}$$

$$\nu_2 = \tfrac{1}{2}(30\,Dq - 15\,B) + \tfrac{1}{2}[(10\,Dq + 15\,B)^2 - 12\,B \cdot 10\,Dq]^{1/2} \tag{14}$$

$$\nu_3 = [(10\,Dq + 15\,B)^2 - 12\,B \cdot 10\,Dq]^{1/2}.$$

D. Configuration d^8

The equivalence of d^2 and d^8 electron systems again produces the same type and number of terms. Thus the a^3T_{1g}, $^3T_{2g}$, $^3A_{2g}$, and b^3T_{1g} states will have to be considered in an octahedral ligand field. If one ignores multiplicity, the energy expressions of d^3 and d^8 systems become identical due to the relationship between d^n and d^{5+n} ligand field systems. Therefore, Eq. (8) to Eq. (10) are valid for the triplet states of d^8.

Additional simplifications are possible. The ground state of octahedral d^8 as well as d^3 systems is an A_{2g} state. Moreover, the order of excited states in both configurations is the same. If the spin-allowed transitions in octahedral d^8 systems, viz. $^3A_{2g} \rightarrow {}^3T_{2g}$, $^3A_{2g} \rightarrow a^3T_{1g}$, and $^3A_{2g} \rightarrow b^3T_{1g}$, are labeled ν_1, ν_2, and ν_3, respectively, the transition energies are directly determined by Eq. (13).

III. Numerical Procedures for Determination of the Parameters of Interelectronic Repulsion and of the Ligand Field, B and 10 *Dq*

The energies of spin-allowed transitions in octahedral and tetrahedral d^2, d^3, d^7, and d^8 systems, viz. Eq. (7), Eq. (13), and Eq. (14), depend on the parameters $10\,Dq$ and B only. In particular, these expressions are independent of the interelectronic repulsion parameter C which, however, may be obtained from the spin-forbidden transitions. In order to determine specific values of the parameters $10\,Dq$ and B from experimental band positions, numerical fitting procedures have to be employed. These will be discussed below for the configurations d^2, d^3, d^7, and d^8 separately.

A. Configuration d^2, Octahedral

The transition energies corresponding to the three $d \rightarrow d$ bands of octahedral d^2 systems are listed in Eq. (7). Since the ground term is a^3T_{1g}, the ligand field splitting parameter, $10\,Dq$, is not determined directly by the energy of the first band, cf. d^3 and d^8 systems.

From Eq. (7), however, it is easy to verify the simple relation

$$10\,Dq = \nu_3 - \nu_1. \tag{15}$$

This relation is, of course, applicable only if both the first and the third band have been observed.

In general, the following methods may be followed depending on the experimental data available.

a) Fitting the First and Second Band. From the expressions of ν_1 and ν_2 in Eq. (7), it may be demonstrated that

$$\begin{aligned} 10\,Dq &= 2\,\nu_1 - \nu_2 + 15\,B \\ B &= \tfrac{1}{30}[-(2\,\nu_1 - \nu_2) \pm \{-\nu_1^2 + \nu_2^2 + \nu_1\,\nu_2\}^{1/2}], \end{aligned} \tag{16}$$

where in the expression of B only the positive sign gives physically meaningful results. Instead of Eq. (16), the equivalent relations

$$\begin{aligned} B &= (\nu_2 - 2\,\nu_1 + 10\,Dq)/15 \\ 10\,Dq &= \tfrac{1}{2}[(2\,\nu_1 - \nu_2) \pm \{-\nu_1^2 + \nu_2^2 + \nu_1\,\nu_2\}^{1/2}] \end{aligned} \tag{17}$$

may be employed.

b) Fitting the First and Third Band. The expressions of ν_1 and ν_3 in Eq. (7) produce

$$\begin{aligned} 10\,Dq &= \nu_3 - \nu_1 \\ B &= (2\,\nu_1^2 - \nu_1\,\nu_3)/(12\,\nu_3 - 27\,\nu_1). \end{aligned} \tag{18}$$

c) Fitting the Second and Third Band. Employing the expressions of ν_2 and ν_3 in Eq. (7) one obtains

$$\begin{aligned} 10\,Dq &= \tfrac{1}{3}(2\,\nu_3 - \nu_2) + 5\,B \\ B &= \tfrac{1}{510}[7\,(\nu_2 - 2\,\nu_3) \pm 3\{81\,\nu_2^2 - 16\,\nu_3\,(\nu_3 - \nu_2)\}^{1/2}], \end{aligned} \tag{19}$$

where in the expression of B only the positive sign gives physically meaningful results. Likewise, the equivalent equations

$$\begin{aligned} B &= (\nu_2 - 2\,\nu_3 + 30\,Dq)/15 \\ 10\,Dq &= \tfrac{1}{34}[9(2\,\nu_3 - \nu_2) \pm \{81\,\nu_2^2 - 16\,\nu_3(\nu_3 - \nu_2)\}^{1/2}] \end{aligned} \tag{20}$$

may be used.

d) Fitting the Difference Between the First and Third Band. The specific form of Eq. (7) characterized in that ν_2 occurs as the second part in both the expressions of ν_1 and ν_3, gives rise to a very simple relation between the band energies in d^2. Thus the equations

$$\begin{aligned} 10\,Dq &= \nu_3 - \nu_1 \\ B &= (\nu_2 + \nu_3 - 3\,\nu_1)/15 \end{aligned} \tag{21}$$

may be used, if all three bands are observed.

B. Configuration d^3 and d^8 Octahedral, d^7 Tetrahedral

The energies of the three spin-allowed d–d bands of d^3 systems in a field of octahedral symmetry are determined by Eq. (13). Since $10\,Dq$ is given exactly by the energy of the first band, the fitting procedure will take advantage of this relation, provided the first band is observed. In actual fact, one of the following methods may be applied depending on the band energies available.

Since, in d^3 and d^8 systems, the expressions of transition energies are identical (cf. section II. D), the methods listed below apply to octahedral d^8 ions as well. In addition, due to the inversion of terms between octahedral and tetrahedral systems, the expressions are applicable to tetrahedral d^7 ions.

a) Fitting the Second Band. Solving the expression of ν_2 in Eq. (13) for B, one may base the fitting on the relations

$$\begin{aligned} 10\,Dq &= \nu_1 \\ B &= (2\,\nu_1^2 + \nu_2^2 - 3\,\nu_1\,\nu_2)/(15\,\nu_2 - 27\,\nu_1). \end{aligned} \tag{22}$$

b) Fitting the Third Band. Likewise, using ν_3 instead of ν_2, one obtains

$$\begin{aligned} 10\,Dq &= \nu_1 \\ B &= (2\,\nu_1^2 + \nu_3^2 - 3\,\nu_1\,\nu_3)/(15\,\nu_3 - 27\,\nu_1). \end{aligned} \tag{23}$$

c) Fitting the Sum of the Second and Third Band. The specific form of Eq. (13) has the consequence that an extremely simple relation may be obtained from the sum of ν_2 and ν_3. Thus the equations

$$\begin{aligned} 10\,Dq &= \nu_1 \\ B &= (\nu_2 + \nu_3 - 3\,\nu_1)/15 \end{aligned} \tag{24}$$

may be used to advantage, if all three *d–d* bands are observed.

d) Fitting the Difference Between the Second and Third Band. A considerable simplification is also achieved if the difference between ν_2 and ν_3 is taken as a basis for a calculation of B. In this case, the relevant expressions to be used are

$$\begin{aligned} 10\,Dq &= \nu_1 \\ B &= \tfrac{1}{75}[3\,\nu_1 \pm \{25\,(\nu_3 - \nu_2)^2 - 16\,\nu_1^2\}^{1/2}], \end{aligned} \tag{25}$$

where again only the positive sign in B has a physical significance.

e) Fitting the Second and Third Band Alone. In a few octahedral d^8 and tetrahedral d^7 complex ions, only the second and third band were observed. In these cases, determination of $10\,Dq$ and B from ν_2 and ν_3 is of importance. Since these systems are equivalent to octahedral d^3 systems, they will be treated here. The equations resulting from ν_2 and ν_3 of Eq. (13) are

$$\begin{aligned} 10\,Dq &= \tfrac{1}{34}[9\,(\nu_2 + \nu_3) \pm \{81\,(\nu_2^2 + \nu_3^2) - 178\,\nu_2\,\nu_3\}^{1/2}] \\ B &= (\nu_2 + \nu_3 - 30\,Dq)/15. \end{aligned} \tag{26}$$

Instead of Eq. (26), the equivalent relations

$$\begin{aligned} 10\,Dq &= \tfrac{1}{3}(\nu_2 + \nu_3) - 5\,B \\ B &= \tfrac{1}{510}[7\,(\nu_2 + \nu_3) \pm \{49\,(\nu_2 + \nu_3)^2 + 680\,(\nu_2 - \nu_3)^2\}^{1/2}] \end{aligned} \tag{27}$$

may be used.

C. Configuration High-spin d^7 Octahedral, d^8 Tetrahedral

The situation is closely similar to that in the d^2 configuration. The ground term being a^4T_{1g}, the parameter $10\,Dq$ cannot be determined directly from the energy of one band alone. In addition, in the experimental data which are of interest to us here, viz. high-spin cobalt(II), the order of the excited terms is different from d^2 and, therefore, the transition energies are determined by Eq. (14). From these equations it follows that

$$10\,Dq = \nu_2 - \nu_1, \tag{28}$$

which simple relation may always be used if the first and second band are observed.

In general, the methods listed below may be followed to determine 10 Dq and B. These methods apply, in addition, to tetrahedral d^8 ions as well.

a) Fitting the First and Second Band. The expressions of ν_1 and ν_2 in Eq. (14) produce

$$\begin{aligned} 10\,Dq &= \nu_2 - \nu_1 \\ B &= (2\,\nu_1^2 - \nu_1\,\nu_2)\,/\,(12\,\nu_2 - 27\,\nu_1). \end{aligned} \tag{29}$$

b) Fitting the First and Third Band. From the expressions of ν_1 and ν_3 in Eq. (14) one obtains

$$\begin{aligned} 10\,Dq &= 2\,\nu_1 - \nu_3 + 15\,B \\ B &= \tfrac{1}{30}[-(2\,\nu_1 - \nu_3) \pm \{-\nu_1^2 + \nu_3^2 + \nu_1\,\nu_3\}^{1/2}] \end{aligned} \tag{30}$$

where, in B, the positive sign has a physical meaning only. Instead of Eq. (30), the equivalent relations

$$\begin{aligned} B &= (\nu_3 - 2\,\nu_1 + 10\,Dq)\,/\,15 \\ 10\,Dq &= \tfrac{1}{2}[(2\,\nu_1 - \nu_3) \pm \{-\nu_1^2 + \nu_3^2 + \nu_1\,\nu_3\}^{1/2}] \end{aligned} \tag{31}$$

may be employed.

c) Fitting the Second and Third Band. Employing the expressions of ν_2 and ν_3 in Eq. (14) one obtains

$$\begin{aligned} 10\,Dq &= \tfrac{1}{3}(2\,\nu_2 - \nu_3) + 5\,B \\ B &= \tfrac{1}{510}[7\,(\nu_3 - 2\,\nu_2) \pm 3\,\{81\,\nu_3^2 - 16\,\nu_2(\nu_2 - \nu_3)\}^{1/2}], \end{aligned} \tag{32}$$

where again only the positive sign has a physical meaning. Instead of Eq. (32), the equivalent equations

$$\begin{aligned} B &= (\nu_3 - 2\,\nu_2 + 30\,Dq)\,/\,15 \\ 10\,Dq &= \tfrac{1}{34}[9\,(2\,\nu_2 - \nu_3) \pm \{81\,\nu_3^2 - 16\,\nu_2(\nu_2 - \nu_3)\}^{1/2}] \end{aligned} \tag{33}$$

may be used.

d) Fitting the Difference Between the First and Second Band. The specific form of Eq. (14) where ν_3 occurs in the second part of the expressions due to ν_1 and ν_2, enables one to set up a very simple relation between the band energies in d^7 systems

$$\begin{aligned} 10\,Dq &= \nu_2 - \nu_1 \\ B &= (\nu_2 + \nu_3 - 3\,\nu_1)/15, \end{aligned} \tag{34}$$

which may be employed, if all three *d—d* bands are observed.

IV. Application to the Electronic Spectra of Transition Metal Compounds

In Tables 2 to 8, experimental absorption band energies observed in the spectra of relevant transition metal compounds are compared to transition energies calculated on the basis of the various numerical procedures outlined in section III. Experimental results are listed, for each single compound, in line 1. Subsequent lines contain calculated transition energies, their deviation from the corresponding experimental value, $\delta = \nu_{calc} - \nu_{obs}$ (last column), and values of the parameters $10\,Dq$, B_{35}, and β_{35}. Each line applies to a different fitting method marked accordingly in column 2 (column 1 in Table 1). It is $\beta_{35} = B_{35,\,complex}/B_{free\ ion}$ (*19*). With regard to the notation B_{35} refer to the footnote on p. 177. The free ion B values employed in this study are listed in Table 1.

In the formulae of the complex ions studied here, the following abbreviations of ligands have been used throughout.

Table 1. *Values of the electronic repulsion parameter B in the free (gaseous) ions employed in the calculation of* β_{35} (*17, 30*)

Configuration	Ion	$B_{free\ ion}$, cm^{-1}
$3\,d^2$	V^{3+}	861
$3\,d^3$	V^{2+}	766
	Cr^{3+}	918
$3\,d^7$	Co^{2+}	971
$3\,d^8$	Ni^{2+}	1041

Ligand Abbreviations

den	Diethylenetriamine
dimen-NO	N,N-Dimethylethylenediamine-N-oxide
dmso	Dimethylsulphoxide
en	Ethylenediamine
gly⁻	Glycinate = Aminoacetate
$N(C_2H_5)_3$	Triethylamine
nma	N-Methylacetamide
pyNO	Pyridine-N-oxide
urea	Urea

A. Octahedral Vanadium(III), d^2

There seems to be only one example of a spectrum of the vanadium(III) ion in a ligand field of approximately octahedral symmetry where all three spin-allowed *d—d* bands were observed, viz. V^{3+} in Al_2O_3 (*33, 49*). The results of a numerical fit of the spectrum are listed in Table 2. The fitting procedure used is marked in column 1 by a letter refering to the corresponding paragraph of section III. A. Procedure (b) did not produce any meaningful values of *B*. The reason is probably the large weight given to experimental uncertainties if the expression Eq. (18) of *B* is applied. Equivalent expressions (viz. Eq. (16) and Eq. (17) in method (a), Eq. (19) and Eq. (20) in method (c)) yield identical numerical results.

Table 2. *Observed and calculated transition energies (in* cm^{-1}*) within the* d^2 *electronic configuration,* Al_2O_3*:* V^{3+} (*33, 49*)

	ν_1 $a^3T_{1g} \to {}^3T_{2g}$	ν_2 $a^3T_{1g} \to b^3T_{1g}$	ν_3 $a^3T_{1g} \to {}^3A_{2g}$	10 *Dq*	B_{35}	β_{35}	δ
exp	17,400	25,200	34,500	—	—		
a	fitted	fitted	36,081	18,681	605.4	0.70	+1,581
c	16,577	fitted	fitted	17,923	664.6	0.77	— 823
d	16,009	fitted	33,109	17,100	500.0	0.58	—1,390
exp[a])	17,190	24,610	34,500	—	—		
a	fitted	fitted	35,614	18,424	576.9	0.67	+1,114
c	16,609	fitted	fitted	17,894	619.0	0.72	— 581
d	16,211	fitted	33,521	17,310	502.7	0.58	— 979

[a]) Corrected for trigonal splitting (*39*).

B. Octahedral Vanadium(II) and Chromium(III), d^3

The results of numerical fits to the spectra of octahedrally coordinated vanadium(II) and chromium(III) compounds are compiled in Tables 3 and 4, respectively. The fitting method employed is indicated, with reference to section III. B, in column 2. Procedure (d) does not always yield a real B value and it did not, in fact, produce a reasonable value where it is not listed. Thus, e. g., $B = 892.9\ cm^{-1}$ and $B = 1002.6\ cm^{-1}$ are obtained by method (d) in $Al_2O_3:Cr^{3+}$ and in emerald:Cr^{3+}, respectively. Procedure (e) has not been considered at all, since it is not useful in d^3 compounds. Here, the first and second absorption band is, in general, easily accessible, whereas the third band is often masked by strong uv absorption.

C. Octahedral and Tetrahedral Cobalt(II), d^7

The experimental data and the results of numerical fits to the spectra of octahedrally and tetrahedrally coordinated cobalt(II) ions are presented in Tables 5 and 6, respectively. Again, the formulae used in the calculations are indicated in column 2. In octahedral cobalt(II), these letters refer to section III. C. In tetrahedral cobalt(II), the methods are identical to those of octahedral d^3 and d^8 ions and are referenced accordingly. Procedure (a) of octahedral d^7 compounds did not produce meaningful values of B and the results of its application are listed, therefore, for $KCoF_3$ only. In the other compounds, much too small or even negative values of B are obtained. Likewise, method (a) yields unsatisfactory results in the examples available in tetrahedral cobalt(II). Equivalent expressions (viz. Eq. (30) and Eq. (31) in method (b), Eq. (32) and Eq. (33) in method (c)) yield identical numerical results.

D. Octahedral and Tetrahedral Nickel(II), d^8

The experimental data and the results of numerical fits to these data, with respect to octahedrally and tetrahedrally coordinated nickel(II) compounds, are listed in Tables 7 and 8, respectively. The numerical procedures employed in octahedral nickel(II) are the same as those in octahedral chromium(III), cf. section III. B, and are referenced correspondingly. The numerical methods applied in tetrahedral nickel(II) are identical to those for octahedral cobalt(II), cf. section III. C. In octahedral nickel(II), the procedures (a) through (e) have been applied, although only the results of methods (a) through (d) are listed extensively in Table 7. The results of the application of method (e) are included in the discussion. In tetrahedral nickel(II), no reasonable values of B could be obtained if procedure (a) was applied.

Table 3. *Observed and calculated transition energies (in cm^{-1}) within the d^3 electronic configuration: Octahedral vanadium(II)*

Compound (Reference)		ν_1 $^4A_{2g} \rightarrow {}^4T_{2g}$	ν_2 $^4A_{2g} \rightarrow a^4T_{1g}$	ν_3 $^4A_{2g} \rightarrow b^4T_{1g}$	B_{35}	β_{35}	δ
VCl_6^{4-} [a]	exp	7,200	12,000	19,000	—		
(*10*)	a	10 *Dq*	fitted	21,600	800.0	1.04	+2,600
	b	10 *Dq*	11,587	fitted	599.1	0.78	− 413
	c	10 *Dq*	11,660	19,340	626.7	0.82	∓ 340
	d	10 *Dq*	11,449	18,449	553.2	0.72	− 551
VCl_2 (298 °K)	exp	8,980	13,920	21,840	—		
(*24*)	a	10 *Dq*	fitted	21,914	592.9	0.77	+ 74
	b	10 *Dq*	13,891	fitted	586.1	0.77	− 29
	c	10 *Dq*	13,899	21,861	588.0	0.77	∓ 21
	d	10 *Dq*	13,871	21,791	581.5	0.76	− 49
$VCl_2/AlCl_3$	exp	8,000	13,100	20,300	—		
(*44*)	a	10 *Dq*	fitted	22,277	758.5	0.99	+1,977
	b	10 *Dq*	12,664	fitted	597.6	0.78	− 436
	c	10 *Dq*	12,759	20,641	626.7	0.82	∓ 341
	d	10 *Dq*	12,449	19,649	539.9	0.71	− 651
$V(H_2O)_6^{2+}$	exp	12,300	18,500	27,900	—		
(*20*)	a	10 *Dq*	fitted	28,790	692.7	0.90	+ 890
	b	10 *Dq*	17,938	fitted	595.8	0.78	− 562
	c	10 *Dq*	18,167	28,233	633.3	0.83	∓ 333

$VBr_2 \cdot 6\,H_2O$[b, c]	exp	11,700	17,600	~30,000	—		
(*28*)	a	10 *Dq*	fitted	27,390	659.3	0.86	−2,610
	b	10 *Dq*	18,610	fitted	900.7	1.18	+1,010
	c	10 *Dq*	18,377	29,223	833.3	1.09	± 777
	d	10 *Dq*	18,927	31,327	1,010.2	1.32	+1,327
$VI_2 \cdot 6\,H_2O$[b, c]	exp	12,000	18,200	~30,000	—		
(*28*)	a	10 *Dq*	fitted	28,376	705.1	0.92	−1,624
	b	10 *Dq*	18,857	fitted	857.1	1.12	+ 657
	c	10 *Dq*	18,688	29,512	813.3	1.06	± 488
	d	10 *Dq*	19,131	30,931	937.4	1.22	+ 931
V $en_3Cl_2 \cdot H_2O$[c]	exp	15,600	21,500	~31,700	—		
(*28*)	a	10 *Dq*	fitted	33,998	579.8	0.76	+2,298
	b	10 *Dq*	17,324	fitted	148.3	0.19	−4,158
	c	10 *Dq*	20,187	33,013	426.7	0.56	∓1,313
$[V(en)_3]Br_2$[c]	exp	15,500	21,400	~32,000	—		
(*28*)	a	10 *Dq*	fitted	33,814	580.9	0.76	+1,814
	b	10 *Dq*	18,524	fitted	268.3	0.35	−2,876
	c	10 *Dq*	20,385	33,015	460.0	0.60	∓1,015
$[V(en)_3]I_2$[c]	exp	15,600	21,400	31,600	—		
(*28*)	a	10 *Dq*	fitted	33,909	567.3	0.74	+2,309
	b	10 *Dq*	17,018	fitted	121.2	0.16	−4,382
	c	10 *Dq*	20,063	32,937	413.3	0.54	∓1,337

[a] In LiCl/KCl eutectic at 400°.
[b] $[V(H_2O)_6]X_2$ structures are assumed.
[c] Diffuse reflectance spectra.

Table 4. *Observed and calculated transition energies (in cm^{-1}) within the d^3 electronic configuration: Octahedral chromium(III)*

Compound (Reference)		ν_1 $^4A_{2g} \rightarrow {}^4T_{2g}$	ν_2 $^4A_{2g} \rightarrow a^4T_{1g}$	ν_3 $^4A_{2g} \rightarrow b^4T_{1g}$	B_{35}	β_{35}	δ
CrF_3	exp	14,900	22,700	34,400	—		
(*20*)	a	10 *Dq*	fitted	35,442	896.1	0.98	+1,042
	b	10 *Dq*	22,134	fitted	788.9	0.86	− 566
	c	10 *Dq*	22,344	34,756	826.7	0.90	∓ 356
K_2NaCrF_6	exp	16,050	23,260	35,460	—		
(*64*)	a	10 *Dq*	fitted	36,211	754.7	0.82	+ 751
	b	10 *Dq*	22,617	fitted	661.8	0.72	− 643
	c	10 *Dq*	22,923	35,797	704.7	0.77	∓ 337
$Cr(H_2O)_6^{3+}$	exp	17,400	24,600	37,800	—		
(*20*)	a	10 *Dq*	fitted	38,528	728.6	0.79	+ 728
	b	10 *Dq*	23,844	fitted	629.6	0.69	− 756
	c	10 *Dq*	24,239	38,161	680.0	0.74	∓ 361
Al_2O_3:Cr^{3+}	exp	18,100	24,400	~39,100	—		
(*22*)	a	10 *Dq*	fitted	38,988	605.9	0.66	− 112
	b	10 *Dq*	24,540	fitted	622.7	0.68	+ 140
	c	10 *Dq*	24,463	39,037	613.3	0.67	± 63
emerald:Cr^{3+}	exp	16,260	23,700	37,740	—		
(*64*)	a	10 *Dq*	fitted	36,865	785.7	0.86	− 875
	b	10 *Dq*	24,275	fitted	882.3	0.96	+ 575
	c	10 *Dq*	24,056	37,384	844.0	0.92	± 356
$Cr(urea)_6^{3+}$	exp	16,100	22,600	34,400	—		
(*20, 22*)	a	10 *Dq*	fitted	35,481	652.0	0.71	+1,081
	b	10 *Dq*	21,328	fitted	495.2	0.54	−1,272
	c	10 *Dq*	22,042	34,958	580.0	0.63	∓ 558

Table 5. *Observed and calculated transition energies (in cm^{-1}) within the d^7 electronic configuration: Octahedral cobalt(II)*

Compound (Reference)		ν_1 $a^4T_{1g} \rightarrow {}^4T_{2g}$	ν_2 $a^4T_{1g} \rightarrow {}^4A_{2g}$	ν_3 $a^4T_{1g} \rightarrow b^4T_{1g}$	10 *Dq*	B_{35}	β_{35}	δ
$KCoF_3$	exp	7,150	15,200	19,200	—	—		
(6)	a	fitted	fitted	15,319	8,050	604.7	0.62	−3,881
	b	fitted	15,363	fitted	8,213	874.2	0.90	+ 163
	c	7,071	fitted	fitted	8,129	879.2	0.91	− 79
	d	7,005	15,055	fitted	8,050	863.3	0.89	− 145
$CoCl_2$	exp	6,600	13,300	17,150	—	—		
(6)	b	fitted	14,161	fitted	7,561	767.4	0.79	+ 861
	c	6,182	fitted	fitted	7,119	793.7	0.82	− 418
	d	5,834	12,534	fitted	6,700	710.0	0.73	− 766
$CoBr_2$	exp	5,700	11,800	16,000	—	—		
(36)	b	fitted	12,270	fitted	6,570	744.7	0.77	+ 470
	c	5,473	fitted	fitted	6,327	758.8	0.78	− 227
	d	5,283	11,383	fitted	6,100	713.3	0.74	− 417
$Co(H_2O)_6^{2+}$	exp	8,100	~16,000	19,400	—	—		
(20)	b	fitted	17,314	fitted	9,215	827.7	0.85	+1,314
	c	7,456	fitted	fitted	8,543	868.7	0.90	− 644
	d	6,927	14,827	fitted	7,900	740.0	0.76	− 1173
$MgO:Co^{2+}$	exp	8,500	17,200	19,600	—	—		
(35, 46)	b	fitted	18,137	fitted	9,637	815.8	0.84	+ 937
	c	8,040	fitted	fitted	9,161	845.5	0.87	− 460
	d	7,663	16,363	fitted	8,700	753.3	0.78	− 837
$Co(dmso)_6^{2+}$	exp	7,407	14,600	18,700	—	—		
(54)	b	fitted	15,872	fitted	8,465	823.4	0.85	+1,272
	c	6,787	fitted	fitted	7,812	862.5	0.89	− 620
	d	6,274	13,467	fitted	7,193	738.6	0.76	−1,133

Table 6. *Observed and calculated transition energies (in cm^{-1}) within the d^7 electronic configuration: Tetrahedral cobalt(II)*

Compound (Reference)		ν_1 $^4A_2 \rightarrow {}^4T_2$	ν_2 $^4A_2 \rightarrow a^4T_1$	ν_3 $^4A_2 \rightarrow b^4T_1$	B_{35}	β_{35}	δ
$ZnO:Co^{2+}$	exp	4,140	6,750	15,500	—		
(*46*)	a	10 *Dq*	fitted	11,358	379.2	0.39	−4,142
	b	10 *Dq*	7,111	fitted	679.4	0.70	+ 361
	c	10 *Dq*	7,096	15,154	655.3	0.68	± 346
	d	10 *Dq*	7,126	15,876	705.5	0.73	+ 376
$ZnAl_2O_4:Co^{2+}$	exp	4,138	7,396	16,410	—		
(7)	b	10 *Dq*	7,142	fitted	742.6	0.77	− 254
	c	10 *Dq*	7,150	16,656	759.5	0.78	∓ 246
	d	10 *Dq*	7,133	16,147	724.5	0.75	− 263

Table 7. *Observed and calculated transition energies (in cm^{-1}) within the d^8 electronic configuration: Octahedral nickel(II)*

Compound (Reference)		ν_1 $^3A_{2g} \rightarrow {}^3T_{2g}$	ν_2 $^3A_{2g} \rightarrow a^3T_{1g}$	ν_3 $^3A_{2g} \rightarrow b^3T_{1g}$	B_{35}	β_{35}	δ
$KNiF_3$	exp	7,250	12,530	23,810	—		
(*56*)	a	10 *Dq*	fitted	29,223	1,333.7	1.28	+5,413
	b	10 *Dq*	12,269	fitted	955.2	0.92	− 261
	c	10 *Dq*	12,286	24,054	972.7	0.93	∓ 244
	d	10 *Dq*	12,247	23,527	935.0	0.90	− 283
$NiCl_2$	exp	7,200	11,500	21,900	—		
(*36*)	a	10 *Dq*	fitted	18,641	569.4	0.55	−3,259
	b	10 *Dq*	12,032	fitted	822.1	0.79	+ 532
	c	10 *Dq*	11,980	21,420	786.7	0.76	± 480
	d	10 *Dq*	12,089	22,489	865.3	0.83	+ 589
$NiBr_2$	exp	6,800	11,800	20,600	—		
(*36*)	a	10 *Dq*	fitted	29,054	1,363.6	1.31	+8,454
	b	10 *Dq*	11,355	fitted	770.3	0.74	− 445
	c	10 *Dq*	11,396	21,004	800.0	0.77	∓ 404
	d	10 *Dq*	11,299	20,099	733.2	0.70	− 501
$Ni(H_2O)_6^{2+}$	exp	8,500	13,800	25,300	—		
(*20*)	a	10 *Dq*	fitted	23,007	753.8	0.72	−2,293
	b	10 *Dq*	14,144	fitted	929.6	0.89	+ 344
	c	10 *Dq*	14,108	24,993	906.7	0.87	± 307
	d	10 *Dq*	14,187	25,687	958.3	0.92	+ 387
$MgO:Ni^{2+}$	exp	8,600	13,500	24,600	—		
(*20, 47*)	a	10 *Dq*	fitted	21,456	610.4	0.59	−3,144
	b	10 *Dq*	14,183	fitted	865.5	0.83	+ 683
	c	10 *Dq*	14,092	24,008	820.0	0.79	± 592
	d	10 *Dq*	14,285	25,385	924.7	0.89	+ 785

Table 7 (continued)

Compound (Reference)		ν_1 $^3A_{2g} \rightarrow {}^3T_{2g}$	ν_2 $^3A_{2g} \rightarrow a^3T_{1g}$	ν_3 $^3A_{2g} \rightarrow b^3T_{1g}$	B_{35}	β_{35}	δ
NiO	exp	8,800	13,900	23,900	—		
(*2, 38*)	a	10 *Dq*	fitted	22,227	648.5	0.62	−1,673
	b	10 *Dq*	14,303	fitted	786.8	0.76	+ 403
	c	10 *Dq*	14,236	23,564	760.0	0.73	± 336
	d	10 *Dq*	14,391	24,391	825.5	0.79	+ 491
$Ni(CH_3OH)_6^{2+}$	exp	8,430	14,226	25,000	—		
(*15*)	a	10 *Dq*	fitted	27,168	1,073.6	1.03	+2,168
	b	10 *Dq*	14,017	fitted	915.1	0.88	− 209
	c	10 *Dq*	14,039	25,187	929.1	0.89	∓ 187
	d	10 *Dq*	13,988	24,762	897.3	0.86	− 238
$Ni(dmso)_6^{2+}$	exp	7,728	12,970	24,038	—		
(*37*)	a	10 *Dq*	fitted	24,072	923.8	0.89	+ 34
	b	10 *Dq*	12,967	fitted	921.4	0.89	− 3
	c	10 *Dq*	12,967	24,041	921.6	0.89	∓ 3
	d	10 *Dq*	12,966	24,034	921.1	0.89	− 4
$Ni(pyNO)_6^{2+}$	exp	8,403	14,124	25,840	—		
(*37*)	a	10 *Dq*	fitted	26,407	1,021.5	0.98	+ 567
	b	10 *Dq*	14,071	fitted	980.1	0.94	− 53
	c	10 *Dq*	14,076	25,888	983.7	0.95	∓ 48
	d	10 *Dq*	14,065	25,781	975.8	0.94	− 59
$Ni(dimen\text{-}NO)_3^{2+}$	exp	9,800	16,400	27,000	—		
(*59*)	a	10 *Dq*	fitted	30,032	1,135.5	1.09	+3,032
	b	10 *Dq*	15,998	fitted	906.5	0.87	− 402
	c	10 *Dq*	16,057	27,343	933.3	0.90	∓ 343
	d	10 *Dq*	15,907	26,507	867.6	0.83	− 493

$Ni(NH_3)_6^{2+}$ (20)	exp	10,750	17,500	28,200	—		
	a	10 *Dq*	fitted	29,345	973.0	0.94	1,146
	b	10 *Dq*	17,261	fitted	880.7	0.85	− 239
	c	10 *Dq*	17,306	28,394	896.7	0.86	∓ 194
	d	10 *Dq*	17,183	27,883	854.4	0.82	− 317
$Ni(en)_3^{2+}$ (20)	exp	11,200	18,350	29,000	—		
	a	10 *Dq*	fitted	31,249	1,066.6	1.03	+2,249
	b	10 *Dq*	17,890	fitted	886.0	0.85	− 460
	c	10 *Dq*	17,981	29,369	916.7	0.88	∓ 369
	d	10 *Dq*	17,713	28,363	831.8	0.80	− 637
$Ni(gly)_3^-$ (20)	exp	10,100	16,600	27,600	—		
	a	10 *Dq*	fitted	28,510	987.3	0.95	+ 910
	b	10 *Dq*	16,447	fitted	916.5	0.88	− 153
	c	10 *Dq*	16,471	27,729	926.7	0.89	∓ 129
	d	10 *Dq*	16,412	27,412	901.6	0.87	− 188
$Ni(CH_3NH_2)_6^{2+}$ (4)	exp	10,000	16,779	27,322	—		
	a	10 *Dq*	fitted	31,104	1,192.2	1.15	+3,782
	b	10 *Dq*	16,283	fitted	907.0	0.87	− 496
	c	10 *Dq*	16,359	27,743	940.1	0.90	∓ 421
	d	10 *Dq*	16,161	26,705	857.8	0.82	− 617
$Ni(iPr \cdot NH_2)_6^{2+}$ (4)	exp	9,480	15,267	26,042	—		
	a	10 *Dq*	fitted	25,066	792.9	0.76	− 976
	b	10 *Dq*	15,463	fitted	871.0	0.84	+ 195
	c	10 *Dq*	15,433	25,877	857.9	0.82	± 165
	d	10 *Dq*	15,503	26,278	889.5	0.85	+ 236
$Ni(N\{C_2H_5\}_3)_6^{2+}$ (25)	exp	10,500	17,100	27,500	—		
	a	10 *Dq*	fitted	28,700	953.3	0.92	+1,200
	b	10 *Dq*	16,849	fitted	856.6	0.82	− 251
	c	10 *Dq*	16,896	27,704	873.3	0.84	∓ 204
	d	10 *Dq*	16,766	27,166	828.8	0.80	− 334

Table 7 (continued)

Compound (Reference)		ν_1 $^3A_{2g} \rightarrow {}^3T_{2g}$	ν_2 $^3A_{2g} \rightarrow a^3T_{1g}$	ν_3 $^3A_{2g} \rightarrow b^3T_{1g}$	B_{35}	β_{35}	δ
$Ni(CH_3CN)_6^{2+}$	exp	10,700	17,400	27,810	—		
(*13*)	a	10 *Dq*	fitted	29,109	960.6	0.92	+1,299
	b	10 *Dq*	17,118	fitted	855.2	0.82	− 282
	c	10 *Dq*	17,173	28,037	874.0	0.84	∓ 227
	d	10 *Dq*	17,017	27,427	822.9	0.79	− 383
$Ni(acetamide)_6^{2+}$	exp	8,240	13,370	24,510	—		
(*3*)	a	10 *Dq*	fitted	22,263	727.5	0.70	−2,247
	b	10 *Dq*	13,709	fitted	900.0	0.87	+ 339
	c	10 *Dq*	13,673	24,207	877.3	0.84	± 303
	d	10 *Dq*	13,752	24,892	928.3	0.89	+ 382
$Ni(nma)_6^{2+}$	exp	7,520	12,660	23,700	—		
(*3*)	a	10 *Dq*	fitted	23,865	931.0	0.89	+ 165
	b	10 *Dq*	12,646	fitted	919.1	0.88	− 14
	c	10 *Dq*	12,647	23,713	920.0	0.88	∓ 13
	d	10 *Dq*	12,644	23,684	917.9	0.88	− 16
$Ni(den)_2^{2+}$	exp	11,500	18,700	29,100	—		
(*20*)	a	10 *Dq*	fitted	31,280	1,032.0	0.99	+2,180
	b	10 *Dq*	18,181	fitted	852.1	0.82	− 519
	c	10 *Dq*	18,296	29,504	886.7	0.85	∓ 404
	d	10 *Dq*	17,925	28,325	783.3	0.75	− 775

Table 8. *Observed and calculated transition energies (in cm^{-1}) within the d^8 electronic configuration: Tetrahedral nickel(II)*

Compound (Reference)		ν_1 $a^3T_1 \rightarrow {}^3T_2$	ν_2 $a^3T_1 \rightarrow {}^3A_2$	ν_3 $a^3T_1 \rightarrow b^3T_1$	10 *Dq*	B_{35}	β_{35}	δ
$ZnO:Ni^{2+}$	exp	4,400	8,600	15,500	—	—		
(*47*)	b	fitted	9,551	fitted	5,151	790.0	0.76	+ 951
	c	3,948	fitted	fitted	4,652	817.0	0.79	− 452
	d	3,567	7,767	fitted	4,200	726.7	0.70	− 833
$CdS:Ni^{2+}$	exp	4,130	8,150	12,750	—	—		
(*45*)	b	fitted	8,924	fitted	4,794	618.9	0.60	+ 774
	c	3,760	fitted	fitted	4,391	641.5	0.62	− 370
	d	3,448	7,468	fitted	4,020	567.3	0.55	− 682

V. Discussion

A. Methods Adopted Previously to the Calculation of B and $10\,Dq$

Most previous authors engaged in a systematic investigation of the nephelauxetic effect applied, as a matter of convenience, an *approximate* form of the relevant energy expressions. Thus *Jørgensen* (*19, 22, 23*), in his extensive compilation of interelectronic repulsion parameters on octahedral d^3 and tetrahedral d^7 systems, used the first of the two implicit equations

$$\begin{aligned} \nu_2 - \nu_1 &= 12\,B - x \\ \nu_3 - 2\,\nu_1 &= 3\,B + x, \end{aligned} \tag{35}$$

where

$$x = 36\,B^2/(10\,Dq - 9\,B + x).$$

These equations may be deduced easily (*19*) if Eq. (11) and Eq. (12) are rewritten in terms of transition energies ν_n, $n = 1, 2$, or 3, viz.

$$\begin{vmatrix} 10\,Dq - \nu_n & \cdot & \cdot \\ \cdot & 12\,B + 10\,Dq - \nu_n & 6\,B \\ \cdot & 6\,B & 3\,B + 20\,Dq - \nu_n \end{vmatrix} = 0 \tag{36}$$

Eq. (35) determines the correction, x, for the intermixing of the subshell configurations $t_{2g}^2\,e_g$ and $t_{2g}\,e_g^2$ in Eq. (36) and is obtained by a second-order perturbation calculation (*17*). Another method proposed by *Jørgensen et al.* (*21*) is based on a tabulation of $(\nu_2 - \nu_1) = \varkappa\,B$ as a function of the variable (ν_2/ν_1). On the other hand, relation (24) was used by *Jørgensen* (*22, 23*) and by others on octahedral d^3 and d^8 systems if all three bands are observed, cf. nickel(II) compounds. As an alternative to the derivation given in section III.B, this simple formula may be obtained by the diagonal sum rule (*23*).

Graphical methods have likewise been used extensively, the simplest being the direct location of a Dq/B value on the Tanabe-Sugano diagram where the ratio of two transition energies agrees with the experimental ratio. From Dq/B and the intercept on the vertical axis, E/B, both Dq and B can be evaluated. This method which is described in detail by *Figgis* (*8*) is not very accurate, however, given the rather small size of published Tanabe-Sugano plots (*60, 61*). Therefore, a modified approach has been suggested recently by *Lever* (*29, 30*). In this method, the transition energy ratios ν_3/ν_1, ν_3/ν_2, and ν_2/ν_1 are directly plotted against Dq/B and E/B. In actual practice, the application suffers from the same shortcomings as the direct method: the diagrams available are small, the results therefrom are of very limited accuracy. Larger scale diagrams, although announced (*29*), have not been published.

If the actual aim is the assessment of *accuracy* of the parameters 10 Dq and B, then *exact* solutions of the secular equations are required. Therefore, the complete expressions applicable to the various situations of interest have been detailed in section III above. It seems to us that an exhaustive listing and a critical appreciation of these equations has never been attempted, although some of the expressions were used previously and a few are applied throughout. Thus *Poole* (*48*) proposed the equation

$$B = (2\,\nu_1 - \nu_2)\,(\nu_2 - \nu_1)/(27\,\nu_1 - 15\,\nu_2)$$

for the calculation of B from the first and the second band in the spectra of d^3 systems. Of course, this expression is identical to Eq. (22). In order to determine 10 Dq and B from ν_2 and ν_3 in d^7 and d^8 systems, *Underhill and Billing* (*62*) deduced the quadratic equations

$$340\,(Dq)^2 + 18\,(\nu_3 - 2\,\nu_2)\,Dq + \nu_2^2 - \nu_2\,\nu_3 = 0$$

and

$$340\,(Dq)^2 - 18\,(\nu_2 + \nu_3)\,Dq + \nu_2\,\nu_3 = 0.$$

These expressions are equivalent to Eq. (33) and Eq. (26), respectively. Recently, *Perumareddi* (*16*) applied Eq. (22) to Eq. (25) to the spectra of four well known nickel(II) complexes. Unfortunately, the paucity of analyzed numerical data precluded any conclusions of general applicability to be drawn.

B. Results of the Present Theoretical Analysis of Spectral Data

1. Octahedral Vanadium(III). It has been pointed out by *Jørgensen* (*19*) that the vanadium(III) ion may not be the most suitable example on which to study the reduction of interelectronic repulsion parameters. In principle, both the vanadium(III) and the cobalt(II) ions to be discussed below are configurationally instable owing to the influence of the Jahn-Teller effect (*16*) on the orbitally degenerate ground states $^3T_{1g}$ and $^4T_{1g}$, respectively. However, in contrast to E_g states, it may be expected that the distortion should be relatively small in electronic T_{1g} and T_{2g} states. Since any indication of a dissymmetry is lacking in the electronic spectra studied here, the influence of the Jahn-Teller effect will not be considered further.

An additional complication arises since, in the d^2 system, the only example of a reasonably complete *d—d* spectrum is furnished by a V^{3+} ion incorporated into the Al_2O_3 lattice (*33, 49*). In this host crystal, the site symmetry of the V^{3+} ion is close to C_{3v} (*14*). Experimentally, the $^3T_{2g}$ excited state exhibits, at 77 °K, a trigonal splitting of 90 cm^{-1}, whereas the splitting of the b^3T_{1g} state amounts to 380 cm^{-1} (0.5% and 1.5% of the transition energy, respectively) (*39*). Although the splitting of the a^3T_{1g} ground state may be calculated from these data, the somewhat higher value of 960 cm^{-1} seems to be in better agreement with spectroscopic and magnetic evidence (*14*). This value corresponds to a splitting parameter $K = -320$ cm^{-1}, and it was determined by *McClure* (*39*) from the temperature variation in the intensity of the *d—d* absorption peaks. For more details the recent review of *Hush* and *Hobbs* (*14*) should be consulted.

If, for the present purpose, we disregard the trigonal splitting, the numerical fit of the spectrum produces the data set out in the upper part of Table 2. The best duplication of experimental band positions is obtained using procedure (a), since the deviation between the calculated and observed ν_3 energy, i. e. $\delta\nu_3 = 1580$ cm^{-1}, is not more than 4.4% relative to ν_3. At the same time, 10 *Dq* is in error by as much as 8.5%. It is a consequence of Eq. (15) that, in d^2 systems, 10 *Dq* can be obtained accurately only if ν_1 and ν_3 are used in the fit. In this case, the somewhat low value $\beta_{35} = 0.58$ and a mean absolute deviation of 6.5% in the calculated energies have to be accepted.

The results may be improved considerably if corrections due to the effect of trigonal splitting are applied. We have used the experimental data in parallel polarization which apply to the $^3A_1$ and $^3A_2$ components of $^3T_{2g}$ and b^3T_{1g}. The correction amounts to -2 K for the a^3T_{1g} ground state and to $-$ K for the $^3T_{2g}$ and b^3T_{1g} excited states. The results of the fit are presented in the lower part of Table 2. It is easily recognized that the deviation in ν_3 calculated according to (a) is reduced to 3.1% and the mean absolute deviation in ν_1 and ν_3 calculated according to (d) is 4.5% only, although again $\beta_{35} = 0.58$.

2. Octahedral Vanadium(II) and Chromium(III). The only single crystal spectrum of octahedrally coordinated vanadium(II) is that of VCl_2 (*24*). In this crystal, each V^{2+} ion is surrounded by an almost regular octahedron of Cl^- ions in a CdI_2-type lattice. Since, at 298 °K, no structure has been observed in the first two spin-allowed *d—d* bands, only the energy of the band maxima could be used in the fitting process. The agreement is, in fact, unexpectedly accurate and the mean absolute deviation amounts to e.g. about 0.1% if method (c) is employed. The reproduction of the corresponding 22 °K spectrum of the VCl_2 crystal is even more accurate and will be commented upon elsewhere (*27*). All other data originate in the absorption spectra of solutions or melts or the diffuse reflectance spectra of powder samples.

In the d^3 configuration, 10 *Dq* is fixed by the energy of the ${}^4A_{2g} \rightarrow {}^4T_{2g}$ transition, ν_1. Of the numerical procedures applicable, the best fit is achieved, in general, according to procedure (c). Thus, if we consider the deviation, δ, between the values calculated according to procedures (a), (b), (c), and (d), respectively, and the corresponding observed transition energies, the following values are obtained (in percent). VCl_6^{4-}: 12.0, 3.56, 2.33, 3.80; $V(H_2O)_6^{2+}$: 3.09, 3.13, 1.50, --; $Cr(urea)_6^{3+}$: 3.04, 5.97, 2.06, --. The first two values apply to ν_3 and ν_2, respectively, the last two are mean average deviations of both ν_2 and ν_3. It is obvious that, in the examples quoted, the deviation in the value of ν_3 calculated according to method (a) is always larger and, in fact, up to five times larger (viz. VCl_6^{4-}) than if procedure (c) is employed. Although the example of the VCl_6^{4-} ion is at the extreme, the observation casts considerable doubt on the accuracy of the many values of the parameters 10 *Dq* and *B* calculated following procedure (a), i.e. evaluated from ν_1 and ν_2 alone (*9, 19, 22, 30, 50*).

It has been pointed out (*23, 52*) that, if a third band (${}^4A_{2g} \rightarrow b{}^4T_{1g}$) is observed, its energy, ν_3, would *not* follow Eq. (24). On the basis of our present investigation, this statement is not born out, however. In all compounds where three spin-allowed *d–d* bands are observed, the band energies ν_1, ν_2, and ν_3 conform to Eq. (24) within the experimental uncertainties. This is exemplified by the values of δ included in Tables 3 and 4. It is difficult to understand why this statement is repeated even in very recent reviews of chromium(III) spectra (*9, 23, 50*). An often quoted example (*9, 38*) is that of $MgO:Cr^{3+}$ where bands at 16,200, 22,700, and 29,700 cm^{-1} were reported (*34*). However, this spectrum seems to be more involved than usually appreciated since Cr^{3+} substitutes as a trivalent ion for a considerably larger divalent ion. In his original paper, *Low* (*34*) gave no definite assignment of the bands and, in fact, the reported spectrum is atypical of a Cr^{3+} ion within an octahedral field. The problem certainly deserves an accurate reinvestigation and the possibility of additional impurities which might influence the optical spectrum should not be overlooked. In case that any "real" discrepancies between theory and experiment are encountered, it is rather due to the large uncertainty in the "vertical" transition energies (cf. section V.C below) if Eq. (24) is not accurately followed.

It appears to be profitable to study, therefore, the effect of experimental uncertainty in band energies, particularly with respect to ν_3, on the calculated values of the transition energies ν_1 and ν_2 and on *B*. Relevant examples included in Table 3 are provided by the spectra of the compounds $V\,en_3Cl_2 \cdot H_2O$, $[V(en)_3]Br_2$, and $[V(en)_3]\,I_2$ (*28*). In the reflectance spectra, the $^4A_{2g} \rightarrow b^4T_{1g}$ band was observed as a shoulder, the corresponding energy thus not being known with any precision. Large differences in the computed values are recognized in Table 3. Thus, e.g., errors up to 25% in ν_2 are encountered if procedure (b) is applied.

The spectra of several single crystals containing Cr^{3+} ions have been studied at room temperature or at 77 °K. Thus in both CrF_3 and K_2NaCrF_6, the Cr^{3+} ion is at the center of a regular octahedron of F^- ions. The resulting differences in the parameter values of 10 *Dq* and *B* are somewhat unexpected and have been attributed (*5, 64*) to the influence on the ligand field and on the electron interaction of the next-nearest neighbors in the lattice.

The isotropic spectrum of the Cr^{3+} ion in Al_2O_3 (ruby), cf. Table 4, permits an excellent numerical fit by methods (a), (b), and (c) producing deviations in the calculated transition energies of 0.31, 0.46, and 0.21%, respectively. This demonstrates again the superiority of method (c), although the complete truth certainly cannot be presented by a succesful numerical fit. In fact, the site symmetry of a Cr^{3+} ion in Al_2O_3 is, ideally, C_{3v} and *McClure* (*39*) has shown that, in polarized light, a trigonal splitting of the excited states $^4T_{2g}$, a^4T_{1g}, and b^4T_{1g} is observed. The experimental transition energies of the individual trigonal states and the splittings resulting therefrom are listed in Table 9. From these data, the energy of the unsplit terms $^4T_{2g}$, a^4T_{1g}, and b^4T_{1g} is obtained as 18,150, 24,670, and 39,130 cm^{-1}, respectively. If the numerical fit is based on these energies, values of 10 *Dq* and *B* practically unchanged from those of Table 4 are deduced.

Table 9. *Energy levels from the polarized single crystal spectra of* $Al_2O_3:Cr^{3+}$ (*39*)

Excited State		Energy	Splitting
Cubic	Trigonal		
$^4T_{2g}$	^4E	18,000 ⊥	450
	$^4A_1$	18,450 ‖	
a^4T_{1g}	^4E	24,400 ⊥	800
	$^4A_2$	25,200 ‖	
b^4T_{1g}	^4E	39,000 ⊥	400
	$^4A_2$	39,400 ‖	

The rather large value of B_{35} for emerald (Cr^{3+} in beryl, $Be_3Al_2Si_6O_{18}$) which was found previously by *Wood et al.* (*64*) was commented upon recently by *Jørgensen* (*23*). Most likely it is caused, in conjunction with the low value of 10 Dq, by unusually large Cr—O distances.

3. Octahedral High-spin Cobalt(II) and Tetrahedral Nickel(II). The results on octahedral cobalt(II) and tetrahedral nickel(II) compounds will be considered jointly, since identical expressions apply to the calculation of the transition energies, 10 Dq, and B. The Jahn-Teller effect (*16*) in the ground state, a^4T_{1g}, of octahedral d^7 ions is expected to be weak and may be disregarded for the present purpose. On the other hand, a strong Jahn-Teller effect is predicted for tetrahedral d^8 ion ground states (*57*).

First let us focus our attention on octahedral cobalt(II), viz. Table 5. Single crystal spectra taken at low temperatures (4° or 20 °K) are available for $KCoF_3$, $CoCl_2$, $CoBr_2$, and $MgO:Co^{2+}$ (*6, 46*). However, in none of these spectra has a pronounced vibrational structure been detected in *all* of the quartet bands which could be used to determine accurately the energy of the "vertical" transitions. In $KMgF_3:Co^{2+}$ where a structure in the $a^4T_{1g} \rightarrow {}^4T_{2g}$ band is apparent, it has not been analyzed and, in addition, the $a^4T_{1g} \rightarrow {}^4A_{2g}$ band has not been studied (*6*). Similarly, the $a^4T_{1g} \rightarrow {}^4A_{2g}$ band has not been recorded in the spectra of $MgO:Co^{2+}$ measured at 4 °K (*46*) and, therefore, the room temperature data of *Low* (*35*) will be employed here. In $CoBr_2$, data from a room temperature reflectance spectrum (*36*) will be used. Under these circumstances, only the band maxima can be fitted to the theory. Some very recent low temperature data which provide an improved agreement between theory and experiment will be commented upon elsewhere (*27*).

Of the numerical procedures outlined in section III, method (a) has been practiced on $KCoF_3$ only; no reasonable values of B could be obtained with this method on the remaining compounds and, even in $KCoF_3$, the result of the calculated value of ν_3 is not encouraging. Looking at the transition energies calculated by methods (b), (c), and (d), the smallest percentage deviation (varying between 1.1 and 8.0%) is achieved using method (b). Here the only exception is the $Co(H_2O)_6^{2+}$ ion where the best result is produced by method (c) closely followed by that of method (d) (percentage error 0.86 and 1.25%, respectively). It is typical of the spectra of octahedral cobalt(II) ions that the accurate location of the $a^4T_{1g} \rightarrow {}^4A_{2g}$ band is often made difficult by its low intensity. Consequently, the experimental accuracy of ν_2 is expected to be sometimes lower than that of ν_1 and ν_3. Since ν_2 is not used in (b), the superior performance of this method is not surprising. However, the octahedral d^7 configuration has the peculiar property that the splitting parameter is determined exactly according to 10 $Dq = \nu_2 - \nu_1$. Since, in most cases,

Eq. (29) does not produce a reasonable result, method (d) should be the best applicable procedure in this configuration. In fact, this is verified easily, if both the differences in calculated values of the transition energies *and* 10 Dq are considered.

A few specific comments on the site symmetry should be added with respect to the single crystal data included in Table 5. $KCoF_3$ has the cubic perovskite structure, each Co^{2+} ion being surrounded by a perfect octahedron of F^- ions. $CoCl_2$ and $CoBr_2$ have layered structures, in which each Co^{2+} ion is at the center of a slightly trigonally distorted octahedron of anions, the site symmetry being D_{3d}. Since, in the latter two crystals, band splittings which could be attributed to low symmetry ligand fields were not observed (*6, 36*), the data are analyzed on the basis of octahedral symmetry.

In order to take account of both the excited quartet and doublet states, the combined ligand-field and spin-orbit theory has been applied to some of the spectra. Thus the parameter values obtained for $KCoF_3$ may be compared to $10\,Dq = 7700$ cm^{-1} and $B = 880$ cm^{-1} ($C = 4.40\,B$, $\zeta = 500$ cm^{-1}) resulting from the four-parameter ($10\,Dq$, B, C, ζ) theory. Likewise, in $CoCl_2$, $10\,Dq = 6900$ cm^{-1} and $B = 780$ cm^{-1} ($C = 4.40\,B$, $\zeta = 420$ cm^{-1}) were reported to provide a reasonable fit to the spectra.

Turning our attention to tetrahedral nickel(II), spectral data on two relevant crystals are available, viz. Ni^{2+} in ZnO (*47*) and Ni^{2+} in CdS (*45*). Both crystals have been studied down to 4.2 °K. However, a complete analysis of the splittings observed in the low-temperature spectra has not been performed, and, in fact, seems to be complicated considerably by the Ham effect (*57*). Thus only the centers of the bands will be used here for comparison with theory. The wurzite ZnO structure is hexagonal, the point symmetry at the cation site being C_{3v}. CdS also has the wurzite structure, although the Cd^{2+} ion is in almost exactly tetrahedral coordination. In both crystals, the approximate symmetry T_d is predominantly reflected in the absorption spectra. Since, in addition, reliable information about the size of energy level splittings caused by the Jahn-Teller effect is either lacking or complicated by the Ham effect (*12, 57*), the data were fitted as set out in Table 8. The general conclusions are similar to those infered from octahedral cobalt(II). If transition energies are considered on an isolated basis, method (b) yields the best results. However, since again $10\,Dq = \nu_2 - \nu_1$ is valid exactly, method (d) should be prefered always, if both the energies and 10 Dq have to be calculated.

4. Octahedral Nickel(II) and Tetrahedral Cobalt(II). The results on octahedral nickel(II) compounds are numerous, viz. Table 7. Single crystal spectra at 4.2 °K were studied in $KNiF_3$ and in $MgO{:}Ni^{2+}$ (*47, 56*). Although considerable fine structure has been observed in $KNiF_3$, details of the spectrum were not listed (*56*). The situation is somewhat complicated since $KNiF_3$ becomes antiferromagnetic below 275 °K and, therefore, only the room temperature spectrum is considered here. In $MgO{:}Ni^{2+}$, fine structure was detected in all three triplet-triplet bands. However, a vibronic analysis of these data is lacking and thus again only the room temperature spectra are employed.

The analysis of the spectral data is straightforward. Similar to the d^3 configuration, 10 Dq is determined directly by the energy, ν_1, of the first spin-allowed transition, $^3A_{2g} \rightarrow {}^3T_{2g}$. In all compounds studied, the best fit is obtained according to method (c). As an example, we may consider the deviation of transition energies calculated according to the methods (a), (b), (c), and (d) (in percent). Thus one obtains from Table 7: NiF_6^{4-} in $KNiF_3$: 18.5, 2.12, 1.50, 1.75; $Ni(H_2O)_6^{2+}$: 9.96, 2.42, 1.70, 2.11; $Ni(NH_3)_6^{2+}$: 4.04, 1.40, 0.90, 1.49. The first two values apply to ν_3 and ν_2, respectively, the last two are the mean absolute deviations relative to both ν_2 and ν_3. It is obvious that transition energies calculated according to method (a) show particularly large differences to the observed band energies. Checking through the results of Table 7, it is found that the deviation in the value of ν_3 as obtained by method (a) is at least four times as large (viz. $Ni(CH_3CN)_6^{2+}$) and at most twelve times as large (viz. $KNiF_3$) as if method (c) is used. This observation is striking, although it is not of as high a practical importance as in the d^3 configuration since, in most six-coordinated nickel(II) compounds, the third band ($^3A_{2g} \rightarrow b^3T_{1g}$) is observed easily.

The site symmetry in some of the single crystals employed in Table 7 may deserve additional comments. $KNiF_3$ has the cubic perovskite structure (*56*), the Ni^{2+} ion being located at the center of a NiF_6^{4-} regular octahedron. $NiCl_2$ and $NiBr_2$ have layered structures (*36*), the Ni^{2+} ion being surrounded by a slightly trigonally distorted octahedron of Cl^- and Br^- ions, respectively. However, band splittings due to lower than octahedral symmetry were not observed. In analogy to the corresponding compounds of cobalt(II), the spectra are therefore analyzed assuming octahedral microsymmetry. In MgO, any substitution of Mg^{2+} by a divalent cation such as Ni^{2+} will be in a site of octahedral symmetry.

Finally, $Ni(dimen\text{-}NO)_3^{2+}$, $Ni(en)_3^{2+}$, $Ni(gly)_3^-$, and $Ni(den)_2^{2+}$ were included in Table 7 and fitted on the basis of octahedral microsymmetry, since no band splittings attributable to lower than octahedral fields were apparent.

There seem to be only two tetrahedral cobalt(II) ions amenable to the present study, viz. $ZnO{:}Co^{2+}$ and $ZnAl_2O_4{:}Co^{2+}$ (*46*, *7*). Both crystals have been investigated down to 4.2 °K. However, a vibronic analysis of the complicated fine structure in $ZnO{:}Co^{2+}$ is unavailable and thus only the reported band centers will be fitted here. In

$ZnAl_2O_4$:Co^{2+}, the spin-orbit components of the three quartet-quartet transitions have been recently assigned (7) and the centers of gravity used in this study were calculated therefrom. The results of a fit to these data are set out in Table 6. In $ZnAl_2O_4$:Co^{2+}, no reasonable results are obtained if method (a) is applied. The general conclusions, however, are similar to those obtained in octahedral nickel(II) compounds. Again method (c) provides the best agreement between experimental and calculated transition energies.

C. Limitations of the Ligand Field Theoretical Approach

The arguments presented in this study have been confined completely to the semi-empirical form of ligand field theory. It is not argued here that this treatment is correctly reflecting all interactions involved. In fact, it is only on the basis of M.O. theory that 10 Dq has been calculated to the correct order of magnitude (*58, 40, 41*). On the other hand, the three-parameter (10 Dq, B, C) theory has been used successfully on a wide variety of transition metal compounds, although it does not account for finer details of the spectra.

There have been attempts (*29, 32, 50*) to attribute apparent deviations of calculated transition energies to the inadaquacy of ligand field theory as compared to the general M.O. theory. Such differences might be indeed expected. Nevertheless, the numerical analysis of spectral data does not lend itself to the support of this assumption. Two different trends may be distinguished in the numerical results quoted above: (i) The *accuracy* of 10 Dq and B clearly depends on the method adopted to their calculation. In some of the equations, inaccuracies of experimental data propagate to a larger extent into the resulting parameter values, whereas in other equations only minor influence subsists. Because of this, consequences emerge as to which methods may be used successfully; these have been discussed in section V.B. above and, in addition, will be briefly collected in section VI below. (ii) The values of transition energies calculated by a specific method are sometimes larger and sometimes smaller than the observed band energy. It is the opinion of the author that such an *unsystematic* variation in the results is caused by insufficient accuracy of some of the experimental data employed.

The second point above deserves particular attention. It is obvious that calculated transition energies always correspond to differences between the potential curves of the excited state and the ground state taken at a fixed molecular configuration. The spin-allowed and parity-forbidden transitions within the partly filled d^n shell of transition metal ions considered here are associated with changes in electron configuration, e.g., ${}^4A_{2g}(t_{2g}^3) \rightarrow {}^4T_{2g}(t_{2g}^2\, e_g)$. Consequently, ground and excited state potential minima do not coincide. The forbiddenness of such a transition may be relaxed by the absorption of one quantum of an odd vibration. According to the Franck-Condon principle in conjunction with overlap considerations, the most intense "vertical" transition at 0 °K will be of the type $0'' \rightarrow \nu'_{odd} + n\nu'_{even}$ with less intense transitions for $n+1$ and $n-1$; $n+2$ and $n-2$; and so on. The result is a progression in the totally symmetric mode of the excited state superimposed upon one quantum of the odd vibration. With increasing temperature the vibrational structure of the transition will be progressively blurred and the maximum shifted to lower energies due to population of the higher vibrational levels of the electronic ground state.

It is obvious that the band maxima observed in room temperature spectra cannot, in general, be directly compared to calculated transition energies. Since details of the vibrational structure will be different from substance to substance, an unsystematic variation between the results of theory and experiment is indeed expected if room temperature data are used. In addition, even if spectra at cryogenic temperatures are available, a direct comparison with theory may not be possible. This situation is encountered if the vibrational structure is complicated due to lower than cubic effective symmetry, spin-orbit interaction (*31*, *32*), or the influence of the Jahn-Teller (*41*, *57*), Renner, and Ham (*12*, *57*) effects.

It is difficult to asses the actual magnitude of the inaccuracy introduced by the usual procedure, but it may be of the order of several hundred cm^{-1}. Apart from this there is an experimental uncertainty in the location of maxima in broad bands of again several 100 cm^{-1}. There seems to be only one way out of these difficulties: if single crystal spectra at cryogenic temperatures become available on a considerable number of transition metal compounds of cubic microsymmetry. Such results are available only on a very limited number of substances since low temperature studies on dichroic crystals have always been prefered in the past. If these spectra are carefully analyzed, the energy of the highest intensity "vertical" transition originating in the zero-point vibrational level of the ground state may be determined or calculated with reasonable accuracy. The application of the theory to these results should then yield data of sufficient precision for further theoretical speculations and practically oriented applications. The extremely small deviations of some of the single crystal data included in this report are believed to illustrate the point.

Of course, there are additional limitations which are due to the approximate nature of the three-parameter theory. These include first order effect of large magnitude which cannot be accounted for by suitable corrections to the simple theory. These limitations, however, may be overcome by extending the theory in the required direction.

VI. Summary and Conclusions

The present paper reviews and critically tests methods which have been used in the past to determine the parameters of the ligand field, 10 *Dq*, and of interelectronic repulsion, *B* and *C*, from absorption spectra. For convenience, the discussion is limited to the spin-allowed *d*–*d* bands in high-spin d^2, d^3, d^7, and d^8 complexes of octahedral and tetrahedral microsymmetry. Emphasis is placed on the values of the parameter *B*, since 10 *Dq* may be fixed by a suitable choice of the calculation method.

The complete expression relevant to the different methods which provide a means for the calculation of 10 *Dq* and *B* from the experimental band energies (ν_1, ν_2, and ν_3) are deduced on the basis of ligand field theory. These equations are applied to numerous systems where all three spin-allowed *d*–*d* bands are observed. It is demonstrated that different methods of analyzing the same spectrum may result in considerably different values of 10 *Dq* and *B*. The reliability of a specific method is judged from the percentage error in band energies calculated using parameter values determined therefrom.

Thus in octahedrally coordinated d^3 and d^8 ions as well as in tetrahedral d^7 ions, method (c) which results from fitting the sum of the second and third band (viz. Eq. (24)) should be prefered if all three bands are observed. In the octahedral d^2 configuration, the only example available indicates the superiority of method (d), i.e. fitting the difference of the first and the third band (viz. Eq. (21)). Finally, in the octahedral d^7 and tetrahedral d^8 configuration, the best result is again produced by method (d), i.e. by fitting the difference of the first and the second band (viz. Eq. (34)). If, due to experimental reasons, another choice has to be made, the results of this study may be employed to estimate the error thus introduced in the determination of 10 *Dq* and *B*.

The observed irregular variation in the results obtained by different numerical methods is attributed to the lack of accuracy in some of the experimental data. It is suggested that low-temperature single crystal spectra are needed to obtain 10 *Dq* and *B* values of sufficient precision.

Obviously the present study will have to be extended to include the energy of spin-forbidden transitions as well as low-spin configurations of transition metal compounds. The analysis of these spectra is somewhat more complicated since it involves the Racah parameter *C* and the possible influence of spin-orbit interactions. There is limited evidence (*63*) available which casts some doubt on the common practice to assume $C = 4B$ and to distinguish the quantities B_{35} and B_{55}, at least within the d^3 configuration. In any case, ligand field theory should be applied with caution, and conclusions from apparent deviations between theory and experiment should be rigorously tested.

VII. References

1. *Ballhausen, C. J.*: Introduction to Ligand Field Theory. New York: McGraw-Hill 1962.
2. *Doyle, W. P., Lonergan, G. A.*: Discussions Faraday Soc. *26*, 27 (1958).
3. *Drago, R. S., Meek, D. W., Joesten, M. D., LaRoche, L.*: Inorg. Chem. *2*, 124 (1963).
4. — — *Longhi, R., Joesten, M. D.*: Inorg. Chem. *2*, 1056 (1963).
5. *Ferguson, J., Knox, K., Wood, D. L.*: J. Chem. Phys. *35*, 2236 (1961); Erratum, ibid. *37*, 193 (1962).
6. — *Wood, D. L., Knox, K.*: J. Chem. Phys. *39*, 881 (1963).
7. — — *Van Uitert, L. G.*: J. Chem. Phys. *51*, 2904 (1969).
8. *Figgis, B. N.*: Introduction to Ligand Fields. New York: Interscience 1966.
9. *Forster, L. S.*: Transition Metal Chem. *5*, 1 (1969).
10. *Gruen, D. M., McBeth, R.*: Proc. 7th Internat. Conf. Coordin. Chem., p. 23; Stockholm 1962.
11. *Griffith, J. S.*: The Theory of Transition Metal Ions. Cambridge: University Press 1961.
12. *Ham, F. S.*: Phys. Rev. *138*, A 1727 (1965).
13. *Hathaway, B. J., Holah, D. G.*: J. Chem. Soc. 2400 (1964).
14. *Hush, N. S., Hobbs, R. J. M.*: Progr. Inorg. Chem. *10*, 259 (1968).
15. *Imhof, V., Drago, R. S.*: Inorg. Chem. *4*, 427 (1965).
16. *Jahn, H. A., Teller, E.*: Proc. Roy. Soc. (London) *A 161*, 220 (1937).
17. *Jørgensen, C. K.*: Absorption Spectra and Chemical Bonding in Complexes. London: Pergamon Press 1962.
18. — Orbitals in Atoms and Molecules. London: Academic Press 1962.
19. — Progr. Inorg. Chem. *4*, 73 (1962).
20. — Advan. Chem. Phys. *5*, 33 (1963).
21. — *Pappalardo, R., Rittershaus, E.*: Z. Naturforsch. *20a*, 54 (1965).
22. — Struct. Bonding *1*, 3 (1966).
23. — Oxidation Numbers and Oxidation States. Berlin–Heidelberg–New York: Springer 1969.
24. *Kim, S. S., Reed, S. A., Stout, J. W.*: Inorg. Chem. *9*, 1584 (1970).
25. *Kiser, R. W., Lapp, T. W.*: Inorg. Chem. *1*, 401 (1962).
26. *König, E.*: Electron Paramagnetic Resonance. In: *Hill, H. A. O., Day, P.*, Ed.; Physical Methods in Advanced Inorganic Chemistry. London: Interscience 1968.
27. — Struct. Bonding, to be published.
28. *Larkworthy, L. F., Patel, K. C., Phillips, D. J.*: Proc. 12th Internat. Conf. Coordin. Chem., p. 122; Sydney 1969.
29. *Lever, A. B. P.*: Werner Centennial. Advan. Chem. Ser. *62*, 430 (1966).
30. — Inorganic Electronic Spectroscopy. Amsterdam: Elsevier 1968.
31. *Liehr, A. D., Ballhausen, C. J.*: Ann. Phys. *6*, 134 (1959).
32. — J. Phys. Chem. *67*, 1314 (1963).
33. *Low, W.*: Z. Physik. Chem. (Frankfurt) *13*, 107 (1957).
34. — Phys. Rev. *105*, 801 (1957).
35. — Phys. Rev. *109*, 247 (1958).
36. *Ludi, A., Feitknecht, W.*: Helv. Chim. Acta *46*, 2226 (1963).
37. *Meek, D. W., Drago, R. S., Piper, T. S.*: Inorg. Chem. *1*, 285 (1962).
38. *McClure, D. S.*: Solid State Phys. *9*, 399 (1959).
39. — J. Chem. Phys. *36*, 2757 (1962); Erratum, ibid. *37*, 1571 (1962).
40. *Offenhartz, P. O'D.*: J. Chem. Phys. *47*, 2951 (1967).
41. — J. Am. Chem. Soc. *91*, 5699 (1969).

42. *Orgel, L. E.:* J. Chem. Phys. *23*, 1004 (1955).
43. *Owen, J.:* Proc. Roy. Soc. (London) *A 227*, 183 (1955).
44. *Øye, H. A., Gruen, D. M.:* Inorg. Chem. *4*, 836 (1965).
45. *Pappalardo, R., Dietz, R. E.:* Phys. Rev. *123*, 1188 (1961).
46. — *Wood, D. L., Linares, R. C.:* J. Chem. Phys. *35*, 2041 (1961).
47. — — — J. Chem. Phys. *35*, 1460 (1961).
48. *Poole, C. P.:* J. Phys. Chem. Solids *25*, 1169 (1964).
49. *Pryce, M. H. L., Runciman, W. A.:* Discussions Faraday Soc. *26*, 34 (1958).
50. *Reinen, D.:* Struct. Bonding *6*, 30 (1969).
51. *Sartori, G., Cervone, E., Cancellieri, P.:* Atti Accad. Nazl. Lincei *35*, 226 (1963).
52. *Schäffer, C. E.:* J. Inorg. Nucl. Chem. *8*, 149 (1958).
53. — *Jørgensen, C. K.:* J. Inorg. Nucl. Chem. *8*, 143 (1958).
54. *Schläfer, H. L., Opitz, H. P.:* Z. Elektrochem. *65*, 372 (1961).
55. — *Gliemann, G.:* Einführung in die Ligandenfeldtheorie. Frankfurt am Main: Akademische Verlagsgesellschaft 1967.
56. *Shulman, R. G., Sugano, S., Knox, K.:* Phys. Rev. *130*, 512 (1963).
57. *Sturge, M. D.:* Solid State Phys. *20*, 91 (1967).
58. *Sugano, S., Shulman, R. G.:* Phys. Rev. *130*, 517 (1963).
59. *Summers, J. T., Quagliano, J. V.:* Inorg. Chem. *3*, 1767 (1964).
60. *Tanabe, Y., Sugano, S.:* J. Phys. Soc. Japan *9*, 753 (1954).
61. — — J. Phys. Soc. Japan *9*, 766 (1954).
62. *Underhill, A. E., Billing, D. E.:* Nature *210*, 834 (1966).
63. *Witzke, H.:* Theor. Chim. Acta, in press.
64. *Wood, D. L., Ferguson, J., Knox, K., Dillon, J. F., Jr.:* J. Chem. Phys. *39*, 890 (1963).

Received September 2, 1970

The Ab-Initio Calculation of Molecular Vibrational Frequencies and Force Constants

Prof. Dr. C. J. H. Schutte

Department of Chemistry, University of South Africa, Pretoria, South Africa

Table of Contents

1. Introduction

The infra-red band spectra of molecules were first "explained" in terms of whole-molecule rotation and intramolecular atomic vibrations by *Bjerrum* (*1*) in 1912. This explanation was consolidated by *Kratzer* (*2*) in 1920 who used the classical formulation (Bohr-formalism), and confirmed by the calculations of the quantum mechanics of *Schrödinger* and *Heisenberg* in the rigid-rotator, harmonic-oscillator, rigid-rotator-*cum*-harmonic oscillator, and non-rigid-rotator-*cum*-anharmonic oscillator approximations (*3*).

In theory, all computations of infra-red spectroscopic frequencies of molecules start from the *complete Schrödinger equation* for the molecule, viz. from

$$H\Psi = E\Psi, \tag{1.1}$$

The non-relativistic Hamiltonian operator H includes *all* the nuclear-nuclear, nuclear-electronic and electron-electron repulsions/attractions V operating within the molecule as well as the kinetic energies T of all the particles:

$$H = T_e + T_N + V_{ee} + V_{eN} + V_{NN}\,, \tag{1.2}$$

where e refers to electrons and N to nuclei. V is taken to be Coulombic in origin, viz.

$$V_{ij} = \frac{q_i q_j}{r}\,, \tag{1.3}$$

where the signs of the charged particles i and j are to be taken into account.

In practice, however, these calculations are rather intractible for many-electron molecules, so that, using the Born-Oppenheimer approximation (see below), it is assumed that the electronic motion can be factored off from Eq. (1.1), yielding an effective potential-energy field in which the nuclear motions (vibrations) take place and which couple with the rotatory motion of the molecule. Normally perturbation theory is used in the first and second approximations to give the vibrational and rotational eigenvalues of H. Instead of a theoretically-obtained potential-energy function (see below), it is normally assumed that the vibratory motion follows Hooke's law (harmonic oscillator) with

$$V_{\text{harmonic}} = \tfrac{1}{2}\,k\,(r - r_e)^2 \tag{1.4}$$

with k the *force constant* (in dynes cm^{-1}), or an anharmonic potential, e. g. *Morse's* function (*4*); these approximations are adequately described by *Nielsen* (*5*).

The force constant k_{ij} between two atoms i and j in a molecule is a measure of the *restoring forces* which act upon the atoms in question to bring them back to their equilibrium configuration when the molecule has been distorted. These very important force constants can only be determined in an *a priori* way for very few molecules; this means that the k's of Eq. (1.4) can be determined in an arbitrary way. The usual procedure is to express the kinetic and potential energies in terms of some appropriate velocities $\dot{q}_i$ and displacement co-ordinates q_i:

$$2\,T = \sum a_{ij}\,\dot{q}_i\,\dot{q}_j \tag{1.5}$$

$$2\,V = \sum b_{ij}\,q_i\,q_j\,. \tag{1.6}$$

The cross products in T and V may be eliminated by a linear transformation to a set of new (normal) co-ordinates Q_k in which each nucleus executes a simple harmonic motion about its equilibrium position with frequency ν_k and in phase with all other nuclei. ν_k is determined from:

$$2\pi\nu_k = \sqrt{\lambda_k}\,. \tag{1.7}$$

The lambdas are solutions of the secular equation

$$|\, b_{ij} - a_{ij}\,\lambda \,| = 0, \tag{1.8}$$

where b_{ij} are the relevant force constants. The experimental ν_k's are used *via* Eq. (1.8) to calculate the b_{ij}'s. Very elegant methods were described to execute this calculation, e. g. the FG-matrix method; these calculations are adequately described by *Mizushima* (*6*), *Wilson, Decius* and *Cross* (*7*), and *Nakamoto* (*8*). *The force constants calculated in this way can only be compared with force constants of other molecules which were calculated using the same type of potential-energy function*, e. g. the Urey-Bradley function with transferable force constants:

$$V = \sum \tfrac{1}{2}\,K\,(\Delta r_i)^2 + \sum \tfrac{1}{2}\,H_{ij}\,r_i^0\,r_j^0\,(\Delta\alpha_{ij})^2 + \sum \tfrac{1}{2}\,F_{ij}\,(\Delta q_{ij})^2\,, \tag{1.9}$$

where K, H and F are stretching, bending and repulsive[1]) force constants, respectively; r_i and r_j bond lengths with equilibrium values r_i^0 and r_j^0; α_{ij} interbond angles and q_{ij} non-bonded distances. K varies between 1.49×10^5 dynes cm^{-1} for C—Br to 6.22 for C—O; H between 0.028×10^{-5} dynes cm^{-1} for Br—Sn—Br to 0.40 for H—C—H and F between 0.11×10^5 dynes cm^{-1} for Br———(Sn)—————Br to 1.24 for F—————(C)—————F (See Ref. (*6*), p. 187). The spectra of complex molecules can be unravelled by means of these transferable force constants and they contribute towards an insight in the *bonding situation* in any particular molecule, as well as to the non-bonded interactions which play such a big rôle in *conformational analysis*.

It is the purpose of this paper to examine the progress which has been made since 1926 in the *ab-initio* calculation of both the *vibrational frequencies* of molecules as well as the *forces* acting upon individual atoms when the equilibrium of the molecule has been disturbed. In a way, this type of calculation is the "inverse" of the FG-method discussed above, because no experimental data are utilized in these calculations — therefore the name *ab-initio* is used. The main problems in this type of calculation are the calculation of the exact electronic wave functions and the construction of an energy hypersurface with reference to the

---

[1]) Between non-bonded atoms.

various possible nuclear positions. The assumption inherent to the argument to be presented below, is that the *adiabatic principle* holds, i. e. that the electronic motion is sufficiently rapid so that they instantaneously readjust themselves to the motion of the nuclei (their motions with respect to the nuclei are thus without "backlash").

2. The Born-Oppenheimer Approximation

One of the central problems in the wave mechanics of vibrating nuclei in a molecule is the determination of the potential energy function V(Q) for the nuclear motion. This potential energy can be calculated using the Born-Oppenheimer approximation (9—13) which depends upon the ratio of the nuclear masses M_a to the electron mass m_0 and the absence of near-degenerate vibronic states (see, e. g. Ref. (*13*)).

In what follows, the nuclear co-ordinates will be numbered Q_a, the α referring to a particular nucleus of the ν nuclei (note that each α occurs *thrice*, i. e. for the three Q-directions X, Y, Z at each nucleus). The electronic co-ordinates x_i, y_i, z_i are designated q_i, the i referring to a particular electron of the n electrons. It must be noted that Q and q are *not* displacement co-ordinates!

The Schrödinger equation for a molecular system with α nuclei and n electrons is

$$\left[-\sum_a^{\nu} \frac{h^2}{8\pi^2 M_a} \nabla_a^2 - \sum_i^{n} \frac{h^2}{8\pi^2 m_0} \nabla_i^2 + V(Q_a, q_i)\right] \Psi(Q_a, q_i) = E\,\Psi(Q_a, q_i)\,. \tag{2.1}$$

The Laplacian operators ∇^2 are given by

$$\nabla_a^2 = \sum_a^{\nu} \frac{\partial^2}{\partial Q_a^2} \quad \text{and} \quad \nabla_i^2 = \sum_i^{n} \frac{\partial^2}{\partial q_i^2}\,, \tag{2.2}$$

$$\text{and} \quad \frac{\partial}{\partial Q_a} = \frac{\partial}{\partial X_a} + \frac{\partial}{\partial Y_a} + \frac{\partial}{\partial Z_a}\,, \tag{1.2a}$$

with a similar expression for $\partial/\partial q_i$.

The potential-energy term $V(Q_a, q_i)$ is derived from Eq. (1—2):

$$\begin{aligned} V(Q_a, q_i) &= V_{NN} + V_{Ne} + V_{ee} \\ &= \sum_{a>\beta}^{\nu}\sum^{\nu} \frac{Z_a Z_\beta e^2}{r_{a\beta}} - \sum_a^{\nu}\sum_i^{n} \frac{Z_a e^2}{r_{ai}} + \sum_{i>j}^{n}\sum^{n} \frac{e^2}{r_{ij}}\,. \end{aligned} \tag{2.3}$$

It is clear that for a certain *fixed internuclear geometry* (configuration) Q′ the first term in Eq. (2.3) is a constant, i. e. the potential energy reduces to

$$V(Q_a', q_i) = V_{NN} + V(Q_a', q_i) \tag{2.4}$$

The first term in Eq. (2.1) disappears for a fixed internuclear configuration Q′ because $\nabla_a \Psi(Q_a', q_i)$ becomes zero in terms of *molecule-fixed axes x, y, z.* If, in addition, the constant term $V_{NN}\,\Psi(Q_a', q_i)$ is transposed to the right and $(E - V_{NN})$ is called $E(Q')$ *to emphasize that it is only valid for the particular internuclear geometry* Q_a' *chosen for the calculation,* then Eq. (2.1) reduces to:

$$\left[-\sum_i^n \frac{h^2}{8\pi^2 m_0}\nabla_i^2 + V(Q_a', q_i)\right]\Psi(Q', q_i) = E(Q')\,\psi(Q', q_i)\,, \tag{2.5}$$

where $\Psi(Q', q_i)$ is written for $\Psi(Q_a', q_i)$. It is important to remember that $E(Q')$ *contains contributions from the kinetic and potential energies of the electrons as well as a (constant) additive term due to the internuclear potentials in the chosen* Q′.

Each choice of fixed internuclear geometry Q′ yields a particular wave function $\Psi_t(Q', q_i)$ and its associated energy $E_t(Q')$ for the electronic ground state of the molecule (or any other *particular* electronic excited state). It is thus possible to construct an *energy hypersurface* for the electronic ground state by plotting $E_t(Q')$ *versus* Q_{ax}, Q_{ay}, Q_{az} for a number of internuclear configurations Q′, where the co-ordinates are now referred to *molecule-fixed axes x, y* and *z* and care is taken to include only such Q′ which does not imply a translation or a rotation of the molecule (i. e. the usual process for splitting off the rotational and translational Schrödinger equations is implied, as usually discussed, using the Eulerian system of axes (14—18)). This means that $3\nu - 6$ (or $3\nu - 5$, in the case of linear molecules) co-ordinates are needed to specify the energy *hypersurface.* Such a variation of the energy $E_t(Q')$ with Q′ is exemplified by the variation of E(R) with internuclear distance $R = (Q_{1x} - Q_{2x})$ obtained from the solution of the Schrödinger equation of the H_2^+-system, as shown by *Burrau* (*19*) and extended to other diatomic molecules (see below); in this case only one internal co-ordinate is needed to specify the energy surface, i. e. it becomes a simple graph. The energy $E(Q')$ is a minimum at the equilibrium internuclear configuration Q_0, i. e.

$$\left(\frac{dE(R)}{dR}\right)_{Q_0} = 0\,; \tag{2.6}$$

the extension to the polyatomic case is obvious.

The classical significance of the E(R) *vs.* R curves for diatomic molecules (and the $E_t(Q')$ *vs.* Q′ hypersurfaces for polyatomic molecules) can be qualitatively seen as follows, if it is assumed that any infinitesimally small motion of the nuclei is instantaneously followed by the electrons (adiabatic principle): If R is changed from R_0, the equilibrium position, to R_0+dR, it is seen that the energy of the system increases to $E(R_0)+dE$. From this it follows that an *external force* F must be present to change the nuclear positions dR away from their positions of equilibrium. The work is equal to F. dR according to classical mechanics, where

$$F=\left(\frac{dE(R)}{dR}\right)_0 \tag{2.7}$$

which follows from

$$dE=\left(\frac{dE(R)}{dR}\right)_0 dR\,. \tag{2.8}$$

If the external constaint F is removed, the nuclei will move back to R_0, the equilibrium position. This means that a nett attractive force must be present (due to the displacement of the bonding electrons towards the bonding region between the nuclei) which overcomes the internuclear repulsion at distances R_0+dR [2]). This internal attractive force which will hold the system in equilibrium at R_0+dR against the external force F is equal to −F and will set the nuclei in motion if the external force is removed, and will return them to their equilibrium positions. The internal attractive force F_t which acts upon the nuclei (due to the bonding) is thus

$$F_{int}=-\left(\frac{dE(R)}{dR}\right)_0 \tag{2.9}$$

This shows that E(R) is very similar to an internuclear potential energy field, e. g. of the harmonic oscillator type (Hooke's law forces) with a potential energy V:

$$V=\tfrac{1}{2}k(R-R_0)^2\,, \tag{2.10}$$

where k is the *force constant*

$$k=\left(\frac{d^2V(R)}{dR^2}\right)_0, \tag{2.11}$$

[2]) This is not true at distances R_0—R where the electronic energy is not enough to offset the internuclear repulsion, but this does not affect the argument below. At $R=R_0$ the coulombic repulsion of the nuclei are exactly balanced by the binding effects.

and the *force* F is given by

$$F = -k(R - R_0) = -\left(\frac{dV(R)}{dR}\right)_0 . \tag{2.12}$$

By analogy, the variation of $E_t(R)$ *vs.* R can then be taken to represent an *internuclear potential energy curve* of the diatomic molecule in a specific electronic state *even though $E_t(R)$ contains kinetic contributions of the electrons*. If the various points on the E(R) *vs.* R curve are connected, the analytical form of the curve E(R) can be obtained and can then be used as a potential-energy function in a vibrating type Schrödinger equation:

$$\left[-\frac{h^2}{8\pi^2\mu} \nabla_R^2 + E_t(R)\right] \phi_E(R) = E' \phi_E(R), \tag{2.13}$$

where μ is the reduced mass of the diatomic molecule, *viz.* $\mu = M_1M_2/(M_1 + M_2)$. This argument can be easily extended to the polyatomic molecule, where

$$dE = -\sum_a F_a \, dQ_a' = -\sum_a \left(\frac{\partial E(Q_a')}{\partial Q_a'}\right) dQ_a' , \tag{2.14}$$

and where the summation goes over the discrete set of fixed-geometry position α chosen. The analytical form of the hypersurface is E(Q′), and the relevant vibrating Schrödinger equation is:

$$\left[-\frac{h^2}{8\pi^2} \sum_a^{\nu} \frac{1}{M_a} \nabla_a^2 + E(Q')\right] \phi_{vt'}(Q) = E' \phi_{vt'}(Q) , \tag{2.15}$$

where t' refers to the specific electronic state and v to the vibrational state. As this paper does not concern itself with vibration-electronic transitions, but only with vibration-transitions, the t's and t''s will be dropped, implying that only the electronic ground state is considered.

The assumption that E(Q′) can be used as an effective potential which regulates the vibratory internuclear motion in a molecule is known as the *Born-Oppenheimer assumption*. According to this assumption E′ of Eq. (2.15) is thus an eigenvalue of the total Schrödinger equation after the translation has been split off. This means that the total energy of a molecule is given by

$$E = E(Q') + E_v' , \tag{2.16}$$

where Q' is, naturally, the equilibrium value. Further, that the total molecular wave function Ψ (after translation has been split off) can be written as the product wave function of the type

$$\Psi = \psi(Q_a, q_i) \cdot \phi(Q_a)\,, \tag{2.17}$$

where, as stated above, only the electronic ground state is considered. This Born-Oppenheimer approximation is considered to be adequate if certain small terms can be neglected, as shown below for a diatomic molecule.

The Schrödinger equation which describes the internal motion of a diatomic molecule consisting of n electrons of masses m_0 and two nuclei with masses M_a and M_b, respectively, after the translational motion has been split off (*20*) is as follows (referred to centre of mass co-ordinates):

$$\left[-\frac{h^2}{8\pi^2\mu}\nabla_R^2 - \frac{h^2}{8\pi^2 M}\sum_i^n\sum_j^n \nabla_i\cdot\nabla_j - \frac{h^2}{8\pi^2 m_0}\sum_i^n \nabla_i^2 + V\right]\Psi = E\,\Psi \tag{2.18}$$

where $M = M_a + M_b$ and $\mu = M_a M_b/(M_a + M_b)$; V is a function of the relative co-ordinates $(r_j - r_i)$, r_i being the position vector of the ith particle, and R is the vector $A.B$. In the fixed-nucleus approximation the first term falls away and the second term is neglected because $M >>> m_0$, so that Eq. (2.18) reduces to Eq. (2.5) yielding eigenvalues $E(R')$ and eigenfunctions $\Psi(R', r_i)$ as discussed above.

The Born-Oppenheimer approximation can be tested by substituting

$$\Psi(R, r_i) = \psi(R, r_i) \cdot \phi(R) \tag{2.19}$$

into Eq. (2.18), multiplying by $\psi^*(R, r_i)$ from the left and integrating over all electron co-ordinates, keeping in mind that

$$\nabla_R^2\Psi = \phi\,\nabla_R^2\psi + 2\,\nabla\psi\,\nabla\phi + \psi\,\nabla_R^2\phi$$

$$\nabla_i\nabla_j\Psi = \phi\,\nabla_i\nabla_j\psi$$

$$\nabla_i^2\Psi = \phi\,\nabla_i^2\psi$$

and that

$$\int \psi^*\,\nabla_R\,\psi\,dr = 0$$

(because ψ^* and ψ belong to the same electronic state).

The result is, after rearrangement:

$$\left[-\frac{h^2}{8\pi^2\mu}\nabla_R^2 + \{E(R') + C(R)\}\right]\phi(R) = E_v'\,\phi(R)\,, \qquad (2.20)$$

where the constant C is given by

$$C(R) = -\frac{h^2}{8\pi^2\mu}\left[\int \psi^* \nabla_R^2 \psi\, dr\right] - \frac{h^2}{8\pi^2 M}\sum_i \sum_j \left[\int \psi^* \nabla_i \nabla_j \psi\, dr\right]. \qquad (2.21)$$

The nuclei thus move in an effective field

$$[E(R') + C(R)]\,. \qquad (2.22)$$

The Born-Oppenheimer approximation is thus valid provided the term $C(R)$ can be neglected. This term is indeed small, and is usually neglected in calculations of this type; one of the reasons is that these integrals depend upon $1/M$ or $1/\mu$ which is much smaller than $1/m_0$, e. g. one of the terms in the double summation of (2.22) is

$$\frac{m_0}{M}\left(\frac{h^2}{8\pi^2 m_0}\sum_i \int \psi^* \nabla_i^2 \psi\, dv\right) <<< E(R')\,,$$

viz. the electronic kinetic energy in the fixed-nucleus approximation which is found to be equal to $\frac{m_0}{M}E(R')$ using the Hellmann-Feynman theorem (see below) at the position of equilibrium where $\partial E(R')/\partial R$ is zero (see also Ref. (*14*), p. 925f.).

The Born-Oppenheimer approximation thus allows the separation of the electronic and nuclear co-ordinates. This in turn allows the construction of a potential-energy hypersurface for the internuclear vibrations of the molecule, using the fixed nuclear-geometry approach in which $E(Q')$ is calculated for various sets of internuclear configurations Q'. Once the $E(Q')$ variation is known (by curve-fitting, etc.), it is substituted in the vibrational Schrödinger equation which can then be solved for the eigenvalues so that the transitional frequencies can be calculated directly. The relevant force constants can then be calculated from the calculated frequencies or directly from the $E(Q')$ as discussed below. It is also possible to express the equation for the nuclear motion in terms of the 3 N–6 (or 3 N–5) internal co-ordinates by splitting off the rotational motion (*21*, *22*).

3. The Hellmann-Feynman Theorem

In the preceding section two types of potential energy functions were discussed, *viz.* (i) Eq. (2.3) which is the purely *classical electrostatic potential* V resulting from Coulombic forces between the charged particles, and (ii) $E(Q')$, the potential energy for a given nuclear configuration Q', as shown in Eq. (2.5); this equation can be written

$$H_e \psi = E(Q') \psi , \tag{3.1}$$

where H_e stands for the electronic Hamiltonian

$$H_e = \left[- \sum_i^n \frac{h^2}{8\pi^2 m_0} \nabla_i^2 + V \right] . \tag{3.2}$$

$E(Q')$ can be calculated from Eq. (3.1) by left-multiplication with ψ^*, using the normality of ψ:

$$E(Q') = \int \psi^* H \psi \, dQ . \tag{3.3}$$

The Hellmann-Feynman theorem states that the forces on the nuclei in a molecule can be calculated according to the classical electrostatic interaction formula

$$F = \sum_i \sum_j F_{ij} = \sum_i \sum_j \frac{e_i e_j}{r_{ij}^3} \mathbf{r}_{ij} , \tag{3.4}$$

once the distribution of the "electron cloud" has been established by the solution of the Schrödinger equation (3.1) to give the wave function ψ from which the potential energy hypersurface $E(Q')$ can be derived according to Eq. (3.3) (*23, 24, 25*).

According to the argument presented in the previous section, the force component acting on the α th nucleus in the direction $Q_a = x_a$ (see Eq. (2.2a)) is given by

$$F_{ax} = - \frac{\partial E(Q')}{\partial x_a} = - \frac{\partial}{\partial x_a} \int \psi^* H \psi \, dv \tag{3.5}$$

where $dv = \Pi_i \, dq_i$. This becomes

$$- \frac{\partial E(Q')}{\partial x_a} = - \int \psi^* \frac{\partial H}{\partial x_a} \psi \, dv - \int \frac{\partial \psi^*}{\partial x_a} H \psi \, dv - \int \psi^* H \frac{\partial \psi}{\partial x_a} dv . \tag{3.6}$$

The last integral in Eq. (3.6) cancels the second integral, as shown below. H is a real, Hermitean operator, so that

$$\int \psi^* H \frac{\partial \psi}{\partial x_a} dv = \int \frac{\partial \psi}{\partial x_a} H^* \psi^* dv , \tag{3.7}$$

as defined by *Margenau* and *Murphy* (*26*). This result, combined with $H \psi = \mathrm{E} \psi$ and with

$$H^* \psi^* = \mathrm{E}^* \psi^* = \mathrm{E} \psi \tag{3.8}$$

(i. e. E is real!) shows that the sum of the last two integrals in (3.6) is zero:

$$\int \frac{\partial \psi}{\partial x_a} H \psi dv + \int \psi^* H \frac{\partial \psi}{\partial x_a} dv = \mathrm{E} \int \left(\frac{\partial \psi^*}{\partial x_a} \psi + \frac{\partial \psi}{\partial x_a} \psi^* \right) dv$$

$$= \mathrm{E} \int \frac{\partial}{\partial x_a} (\psi^* \psi) dv = \mathrm{E} \frac{\partial}{\partial x_a} \int \psi^* \psi^* dv = \mathrm{E} \frac{\partial}{\partial x} (1) = 0 .$$

This means that

$$- \frac{\partial \mathrm{E}(\mathrm{Q}')}{\partial x_a} = - \int \psi^* \frac{\partial H}{\partial x_a} \psi dv = - \int \psi^* \frac{\partial \mathrm{V}}{\partial x_a} \psi dv , \tag{3.9}$$

because $H = T + V$ (Eq. 1.2), i. e. $\frac{\partial \mathrm{H}}{\partial x} = \frac{\partial \mathrm{V}}{\partial x}$ because T does not depend upon x; V is given by Eq. (2.3). From Eq. (3.9) it follows that

$$\mathrm{F}_{ax} = - \int \psi^* \left(\frac{\partial \mathrm{V}_{nn}}{\partial x_a} + \frac{\partial \mathrm{V}_{ne}}{\partial x_a} \right) \psi dv ,$$

because the interelectronic potential V_{ee} does not depend upon Q, the nuclear co-ordinates. This equation reduces to

$$\mathrm{F}_{ax} = - \left(\frac{\partial \mathrm{V}}{\partial x_a} \right) - \int \psi^* \left(\frac{\partial \mathrm{V}_{ne}}{\partial x_a} \right) \psi dv , \tag{3.10}$$

because $(\partial \mathrm{V} / \partial x_a)$ does not involve electronic co-ordinates over which the integral is executed, and $\int \psi^* \psi dv = 1$.

Eqs. (3.9) and (3.10) are, then, the mathematical statements of the Hellmann-Feynman theorem, i. e. once the wave functions are known, the forces on the nuclei can be calculated according to the principles of electrostatics. This Hellmann-Feynman theorem is very important for the calculation of force constants of polyatomic molecules, and can also be used to develop an electrostatic model of chemical bonds (*27—32*).

It should be noted that the Hellmann-Feynman theorem is only satisfied by exact wave functions or by certain types of approximate wave functions. This will be discussed below.

4. The Quantum-Mechanical Virial Theorem

The *virial theorem* of Clausius establishes the ratio of the average kinetic energy $\bar{K}$ to the average potential energy V of a mechanical system consisting of i particles (*28, 33—36*). The *virial of the force F* is equal to the total average kinetic energy $\bar{K}$:

$$-\tfrac{1}{2}\sum_i \overline{(r_i \cdot F_i)} = \sum_i \bar{K}_i = \bar{K}\,, \qquad (4.1)$$

where F_i is the force acting on the i th particle and r_i its position vector.

In *conservative systems* the virial theorem can be written in terms of the potential V, where V is a homogeneous function of degree t in the co-ordinates, using $F_i = -(\partial/\partial r_i)\,V$:

$$K = -\tfrac{1}{2}\sum_i \overline{(r_i \cdot F_i)} = \tfrac{1}{2}\sum_i \overline{\left[r_i \cdot \frac{\partial}{\partial r_i} V(r)\right]} = \tfrac{1}{2}\, t\, \bar{V}\,, \qquad (4.2)$$

using the theorem of *Euler* which states that if $V(x_1, x_2, ---, x_n)$ is homogeneous of degree t and has continuous first order partial derivatives, then:

$$\sum_i x_i \left(\frac{\partial V}{\partial x_i}\right) = t\,V\,. \qquad (4.3)$$

This means that the virial theorem states that the ratio of the average kinetic energy $\bar{K}$ and the average potential energy for a mechanical (conservative) system is just half the degree t of the total potential energy function

$$\frac{\bar{K}}{\bar{V}} = \tfrac{1}{2}\, t\,. \qquad (4.4)$$

For the harmonic oscillator

$$V = \frac{1}{2} kx^2 \tag{4.5}$$

with $t = 2$, i. e.

$$\bar{K} = \bar{V} \tag{4.6}$$

and for Coulombic interactions $V = e^2/r$, $t = -1$ and

$$\bar{V} = -2\,\bar{K} \tag{4.7}$$

Slater has shown that the virial theorem is also applicable in the case of quantum-mechanical systems if spatial averages replace the time averages of the classical system (*37–42*). The proof is as follows: The Schrödinger equation for a system of N particles is

$$\sum_i^{3N} \left(-\frac{h^2}{8\pi^2 m_i}\right) \frac{\partial^2 \psi}{\partial x_i^2} + (V - E)\,\psi = 0\,, \tag{4.8}$$

and for the complex conjugate of ψ

$$\sum_i^{3N} \left(-\frac{h^2}{8\pi^2 m_i}\right) \frac{\partial^2 \psi^*}{\partial x_i^2} + (V - E)\,\psi^* = 0\,. \tag{4.9}$$

If (4.8) is first differentiated with respect of x_j (i. e. operated upon by $\partial/\partial x_j$) and then multiplied by $x_j\,\psi^*$ from the left, the following equation is obtained:

$$\sum_i^{3N} \left(-\frac{h^2}{8\pi^2 m_i}\right) x_j\,\psi^* \frac{\partial^3 \psi}{\partial x_j\,\partial x_i^2} + \left(x_j \frac{\partial V}{\partial x_j}\right) \psi^*\,\psi + x_j (V - E)\,\psi^* \frac{\partial \psi}{\partial x_j} = 0\,. \tag{4.10}$$

From Eq. (4.9) it follows that

$$(V - E)\,\psi^* = -\sum_i^{3N} \left(-\frac{h^2}{8\pi^2 m_i}\right) \frac{\partial^2 \psi^*}{\partial x_i^2}\,,$$

i. e., when substituted in (4.10) and summed over j:

$$\sum_i \left(-\frac{h^2}{8\pi^2 m_i}\right) \sum_j x_j \left(\psi^* \frac{\partial^3 \psi}{\partial x_j\,\partial x_i^2} - \frac{\partial^2 \psi^*}{\partial x_i^2} \cdot \frac{\partial \psi}{\partial x_j}\right) + \psi^* \left(\sum_j x_j \frac{\partial V}{\partial x_j}\right) \psi = 0\,. \tag{4.11}$$

This equation can now be integrated over co-ordinate space to prove the required virial result, using the identities proved below. It follows that the following identity holds (which can be proved by executing the differentiations)

$$\psi^{*2}\frac{\partial}{\partial x_i}\left(\frac{1}{\psi^*}\cdot\sum_j x_j\frac{\partial\psi}{\partial x_j}\right)$$

$$=\sum_j\left(\psi^*\frac{\partial x_i}{\partial x_j}\cdot\frac{\partial\psi}{\partial x_j}+\psi^* x_j\frac{\partial^2\psi}{\partial x_i\,\partial x_j}-x_j\frac{\partial\psi}{\partial x_j}\cdot\frac{\partial\psi^*}{\partial x_i}\right) \quad (4.12)$$

$$=\psi^*\frac{\partial\psi}{\partial x_i}+\sum_j\left(\psi^* x_j\frac{\partial^2\psi}{\partial x_i\,\partial x_j}-x_j\frac{\partial\psi}{\partial x_j}\cdot\frac{\partial\psi^*}{\partial x_i}\right).$$

From this it follows, by differentiating and collecting terms, that

$$\frac{\partial}{\partial x_i}\left[\psi^{*2}\frac{\partial}{\partial x_i}\left(\frac{1}{\psi^*}\cdot\sum_j x_j\frac{\partial\psi}{\partial x_j}\right)\right]$$

$$=2\,\psi^*\frac{\partial^2\psi}{\partial x_i^2}+\sum_j x_j\left(\psi^*\frac{\partial^3\psi}{\partial x_i^2\,\partial x_j}-\frac{\partial\psi}{\partial x_j}\cdot\frac{\partial^2\psi^*}{\partial x_i^2}\right),$$

i. e. that

$$\sum_j x_j\left(\psi^*\frac{\partial^3\psi}{\partial x_i^2\,\partial x_j}-\frac{\partial\psi}{\partial x_j}\cdot\frac{\partial^2\psi}{\partial x_i^2}\right)$$

$$=-2\,\psi^*\frac{\partial^2\psi}{\partial x_i^2}-\frac{\partial}{\partial x_i}\left[\psi^{*2}\frac{\partial}{\partial x_i}\left(\frac{1}{\psi^*}\cdot\sum_j x_j\frac{\partial\psi^*}{\partial x_j}\right)\right], \quad (4.13)$$

and:

$$\int\sum_j x_j\left(\psi^*\frac{\partial^3\psi}{\partial x_i^2\,\partial x_j}-\frac{\partial\psi}{\partial x_j}\cdot\frac{\partial^2\psi}{\partial x_i^2}\right)$$

$$=-2\int_{-\infty}^{+\infty}\psi^*\frac{\partial^2\psi}{\partial x_i^2}\cdot dx_i+\left[\psi^{*2}\frac{\partial}{\partial x_i}\left(\frac{1}{\psi^*}\cdot\sum_j x_i\frac{\partial\psi^*}{\partial x_j}\right)\right]_{-\infty}^{+\infty}.$$

The term in brackets vanishes at the two limits because $\psi^{*2}=0$ at $x_i=\pm\infty$. This means that the integration of Eq. (4.11) yields

$$\sum_i \left(-\frac{h^2}{8\pi^2 m_i}\right) \int \psi^* \frac{\partial^2}{\partial x_i^2}\, \psi\, d x_i = -\tfrac{1}{2} \int \sum_j x_j \left(-\frac{\partial V}{\partial x_j}\right) \psi^* \psi\, d x_j . \quad (4.14)$$

It is clear that $(\partial^2/\partial x_i^2)(-h^2/8\pi^2 m_i)$ is just the kinetic energy operator ∇_i^2 for the i th particle, so that the average value of the kinetic energy is (left-hand side of Eq. (4.14):

$$\bar{K}_i = \int \psi \nabla_i^2 \psi\, d x \quad (4.15)$$

and

$$\bar{K} = \sum_i \bar{K}_i ,$$

the average (expectation) value of the kinetic energy, using normalized functions. If it is remembered that $F_{xi} = -(\partial V/\partial x_i)$, then it is easily seen that the right-hand expression of Eq. (4.14) is just the *expectation value* of the virial:

$$\tfrac{1}{2} \int \psi^* (\sum_j x_j F_{x_j})\, \psi\, d x .$$

This is the quantum-mechanical analog of the virial, i. e. it is the *average (expectation) value* of $\sum x_j F_{xj}$ because $\int \psi^* \psi\, d\varkappa = 1$. This means that

$$\bar{K} = -\tfrac{1}{2} \overline{\sum x_j F_{x_j}} . \quad (4.16)$$

This result is the same as that of the classical case, *viz.*

$$\bar{K} = -\tfrac{1}{2} \bar{V} \quad (4.17)$$

for an electrostatic potential V, i.e. by using

$$\bar{K} = -\tfrac{1}{2} t\bar{V} , \quad (4.18)$$

which is obtained by applying Euler's theorem (see above) to $\Sigma_j x_j (\partial V/\partial x_j)$ in order to obtain tV where t is the order of V.

It is important to obtain expressions for $\bar{K}_e$, the average kinetic energies of the electrons, and $\bar{V}_e$ the average potential energy in terms of V_n, the internuclear potential energy. This can be done through the virial theorem and the Born-Oppenheimer approximation which allows the separation of electronic and nuclear co-ordinates — even in non-equilibrium positions.

If the ν nuclei α and the n electrons i are specified separately, the virial becomes

$$\tfrac{1}{2}\,\overline{V}_e=\tfrac{1}{2}\sum_{i}^{n}(\overline{r_i\cdot F_i})+\tfrac{1}{2}\sum_{\alpha}^{\nu}(\overline{r_\alpha\cdot F_\alpha})=\overline{K}_e+\tfrac{1}{2}\sum_{\alpha}^{\nu}(\overline{r_\alpha\cdot F_\alpha})\,, \qquad (4.19)$$

because V (Eq. 2.3) is a homogeneous function of the co-ordinates of degree -1, where $\overline{K}_e$ is just the virial of the *electrons*. If the nuclei are considered to be *stationary* (Born-Oppenheimer approximation), then the average value $\overline{r_\alpha\cdot F_\alpha}$ becomes just $r_\alpha\cdot F_\alpha$ for a particular choice of nuclear co-ordinates, i.e.

$$\tfrac{1}{2}\,\overline{V}_e=\overline{K}_e+\tfrac{1}{2}\sum_{\alpha}^{\nu}(r_\alpha\cdot F_\alpha)\,. \qquad (4.20)$$

The force on the αth nucleus in this fixed-nucleus approximation is given by

$$F_\alpha=-\frac{d\mathrm{V}_n}{dr_\alpha}\,, \qquad (4.21)$$

where V_n is the *internuclear potential energy* which was previously defined as $\mathrm{E}(Q')$ in Eq. (2.5) and (2.15); this means that

$$\overline{V}_e+\overline{K}_e=\mathrm{V}_n\,, \qquad (4.22)$$

because in the fixed-nucleus approximation of Eq. (2.5) $\mathrm{V}_n(\equiv\mathrm{E}(Q'))$ plays the rôle of the total electronic energy. From (4.21) and (4.20) it follows that

$$\overline{K}_e=-\tfrac{1}{2}\,\overline{V}_e-\tfrac{1}{2}\sum_{\alpha}^{\nu}\left(r_\alpha\cdot\frac{\partial\mathrm{V}_n}{\partial r_\alpha}\right). \qquad (4.23)$$

In many cases the energy $\mathrm{V}_n(\equiv\mathrm{E}(Q'))$ ist just a function of the distances $r_{\alpha\beta}$ between the nuclei, so that

$$\sum_{\alpha}^{\nu}\left(r_\alpha\cdot\frac{\partial\mathrm{V}_n}{\partial r_\alpha}\right)=\sum_{\alpha}^{\nu}\sum_{>\beta}^{\nu}r_{\alpha\beta}\left(\frac{\partial\mathrm{V}_n}{\partial r_{\alpha\beta}}\right). \qquad (4.24)$$

This result, when combined with Eq. (4.22) gives the required values of $\bar{K}_e$ and $\bar{V}_e$:

$$\bar{K}_e = -V_n - \sum_{a>\beta}^{\nu} \sum^{\nu} r_{a\beta} \left(\frac{\partial V_n}{\partial r_{a\beta}} \right) \tag{4.25}$$

$$\bar{V}_e = +2\, V_n + \sum_{a>\beta}^{\nu} \sum^{\nu} r_{a\beta} \left(\frac{\partial V_n}{\partial r_{a\beta}} \right), \tag{4.26}$$

i.e. the average electronic kinetic energy and the average potential energy. These expressions become particularly simple for the case of the diatomic molecule, *viz.*

$$\bar{K}_e = -E(R) - R \left(\frac{dE(R)}{dR} \right) \tag{4.27}$$

$$\bar{V}_e = +2\, E(R) + R \left(\frac{dE(R)}{dR} \right) \tag{4.28}$$

At the point of equilibrium $(\partial V_n / \partial r_{ab})$ is zero, so that $\bar{K}_e = -V_n$ and $\bar{V}_e = 2\, V_n$, as well as $\bar{K}_e = -\frac{1}{2} \bar{V}_e$.

It is very instructive to plot graphs of the variation of $\bar{K}_e$, $\bar{V}_e$ and V_n for the diatomic molecule, using the Morse potential-energy function

$$E(R) = D\{1 - e^{-a(R-R_e)^2}\} = D\{-2\, e^{-a(R-R_e)} + e^{-2a(R-R_e)}\}, \tag{4.29}$$

to approximate the actual E(R′) which can be computed from the fixed-nucleus approximation, where D is the dissociation energy and $E(R) \to 0$ as $R \to +\infty$. If, e.g. D = 100 k cal/mole and $a = R_e = 1$ are used it is seen that $\bar{K}_e$ decreases gradually going from $r = \infty$ to $\sim R = 2.5$ and then increases sharply at shorter distances. $\bar{V}_e$ first increases (i.e. it becomes >0) gradually at smaller distances till it reaches a maximum at ~ 3.2 R and then decreases sharply till it reaches a minimum at ~ 0.7 R. The actual potential-energy curve E(R) at all distances R is simply the arithmetical sum of the $\bar{K}_e$ and $\bar{V}_e$-curves, as predicted by Eq. (4.22). At the point of equilibrium $\bar{V}_e = -2\, \bar{K}_e$, so that the increase in kinetic energy as the point of equilibrium is approached is more than compensated by the decrease in potential energy, resulting in a stable system. At distances smaller than R_e the ratio $\bar{V}_e / \bar{K}_e < 2$, so that the molecule becomes unstable with respect to the equilibrium position. (Note that it is usual in constructing curves such as this to add energy constants so that each curve goes to zero at $R = +\infty$ — See Ref. (*12*), p. 35ff.)

5. The Calculation of the Molecular Frequencies: Diatomic Molecules

In the theory of the vibrations of the anharmonic diatomic molecule the potential $V(R)[\equiv E(R')]$ of Eq. 2.5 is expanded as a Taylor series around the equilibrium position (*43*) to give

$$V(R) = \text{constant} + (R - R_e)\left(\frac{dV}{dR}\right)_{R_e} + \frac{(R-R_e)^2}{2!}\left(\frac{d^2V}{dR^2}\right)_{R_e} + \frac{(R-R_e)^3}{3!}\left(\frac{d^3V}{dR^3}\right)_{R_e} + \cdots. \tag{5.1}$$

The constant is zero when V is measured relative to the minimum, and at $R = R_e$ the tangent is zero, so that the first two terms of Eq. (5.1) drop out. The third term gives a parabolic (quadratic) potential which causes a harmonic oscillation of the molecule, and where the force constant is

$$k = \left(\frac{d^2V}{dR^2}\right)_{R_e} = 4\,\pi^2 c^2 \mu \omega_e^2 = 4\,\pi^2 \mu \nu_e^2\,, \tag{5.2}$$

Where ω_e is the corresponding frequency of vibration (in cm^{-1}). The remaining terms give the cubic, quartic, etc. contributions of the potential, with the cubic, quartic, etc. contributions of the potential, with the cubic, quartic, etc., force constants, respectively. The energy levels of an anharmonic oscillator follow from the term values

$$G_n = \omega_e(n+\tfrac{1}{2}) - \omega_e x_e (n+\tfrac{1}{2})^2 + \omega_e y_e (n+\tfrac{1}{2})^3 - \cdots, \tag{5.3}$$

where n is a quantum number and $\omega_e >> \omega_e x_e >> \omega_e y_e$ and

$$D_e = \frac{\omega_e}{4(\omega_e x_e)}\,, \tag{5.4}$$

where $D_e = (D_0 + \text{zero-point energy})$, or alternatively,

$$\omega_e x_e = \frac{h}{8\,\pi^2 \mu c}\cdot \left\{\frac{1}{24}\left\langle 5\left[\left(\frac{d^3V}{dR^3}\right)_{R_e} \Big/ \left(\frac{d^2V}{dR^2}\right)_{R_e}\right]^2 - 3\left[\left(\frac{d^4V}{dR^4}\right)_{R_e} \Big/ \left(\frac{d^2V}{dR^2}\right)_{R_e}\right]\right\rangle\right\}. \tag{5.4}$$

Once the potential energy curve E(R′) vs R of a diatomic molecule has been determined from the Schrödinger equation of the electronic problem in the fixed-nucleus approximation (Born-Oppenheimer), there are various methods to determine the force constants and vibrational frequencies non-empirically. These methods will now be described below.

A. The Parabolic Approximation

The effort to compute the values of E(R′) for various R′ is great, especially if multi-electron molecules are considered. Computer time is, therefore, saved by establishing the position $R=R_e$ and then calculating the values of E(R′) for a few values of R around R_e. The value of the potential is obtained by fitting a quadratic expression to the points, mostly using a least-squares procedure. The basic uncertainty in this type of calculation is caused by the truncation of Eq. (5.1) after the third (harmonic) term: very few vibrations, if any, are harmonic, even in the "vibrationless" state with $n=0$.

Menna, *Moccia* and *Randaccio* (*44*), using the molecular-orbital approach in the one-centre, self-consistent-field approach (OCE—SCF—MO) developed by *Moccia* (*45—47*), reports the calculated vibrational frequencies of HF and HCl (using the *FG*-method (*7*)):

	HF	HCl
r_e (exp.) Å	1.733	2.409
r_e (calc.) Å	1.730	2.406
ω (exp.) cm^{-1}	4137	2990
ω (calc.) cm^{-1}	4518	3446

Bishop (*48*) fitted the values of E(R′) vs R to a polynomial of the type

$$a+bR+CR^2+dR^3+eR^4$$

for the hydrogen molecule, differentiated it twice to R to give the force constant $k=6.33$ md/Å (exp.: 5.75 md/Å) ($R_e=1.38$ a.u. calculated; 1.40 a.u. exp.); the one-centre approach with non-integral principal quantum numbers in Slater atomic orbitals. This calculation does not compare favourable with the extended *McLean*, *Weiss* and *Yoshimine* calculation (*49*) who obtained a five-term MO wave function and fitted the E(R′) points (9 points in all) to an eight degree polynomial; the calculated value is $k = 5.707$ md/Å, which nearly coincides with k exp. (ω_e calc. $= 4385$ cm^{-1}/exp. 4400 cm^{-1}).

It is thus seen that calculations of this type are prone to great uncertainties, especially if the potential-energy curves are to be fitted to very few points around the potential-energy minimum (see e.g. the calculations of *Bishop* and *Hoyland* (*50*) on LiH, BeH, BH, OH, etc.).

This type of calculation can also be applied to the 'breathing' force constant of symmetrical molecules where all the atoms move along *main symmetry axes*[3]), e.g. for CH_4 (*44, 48, 52*), SiH_4 (*44, 53*), NH_4^+ (*44*), PH_4^- (*44*) and AlH_4^- (*54*); it is relatively easy to obtain E(R') around the minimum in this type of molecules by keeping all protons in the tetrahedral arrangement and altering the X—H-distance by (fixed) amounts. The vibrational frequencies calculated by these methods are of the correct *order*, as exemplified by the result of *Menna, Moccia* and *Randaccio* (*44*)

	CH_4	SiH_4	NH_4^+
ω_1	3709 (3137)	2542 (2237)	3634 (3252)

where the experimental values are given in brackets. These values compare favourably for CH_4 to the values obtained by *Allavena* (*52*) viz. 3350 cm^{-1} (one-centre configuration-interaction calculation; FG matrix V taken to quadratic terms in the symmetry co-ordinates) and *Hartman* and *Glieman* (*55*) who obtained 3009 cm^{-1} theoretically[4]). These calculations can be compared with the empirical force constants (general and Urey-Bradley) of these tetrahedral molecules as calculated by *Shimanouchi et al.* (*56*).

McLean and *Yoshimine* (*58, 59*) published an extensive list of "Tables of Linear Molecule Wave Functions" which include the diatomic molecules e.g. the hydrogen halides, the oxides (e.g. BeO, MgO), the alkali halides, PN, as well as HCN, HF_2^-, CO_2, N_2O, SCN, C_2H_2, each for a number of internuclear configurations. These Tables represent the "best"

[3]) It can thus not be applied to the "breathing" force constant of water ν_1 where only the O-Atom moves along the main C_2-axis of the molecule (see *Herzberg* (*51*), p. 171) unless the R' are taken such that the atoms are placed along the lines in space along which the normal mode takes place (see (*57*)).

[4]) In Refs. (*44, 52, 55*) the same potential function used to calculate ω_1 is also used to calculate the other tetrahedral frequencies ω_2, ω_3 and ω_4. It is not clear whether V was redetermined for each type of normal mode: *V(Q') should be calculated in the fixed nucleus approximation, shifting the atoms along their actual normal-mode trajectories for the various normal modes;* then a quadratic potential curve should be fitted. This could be explanation for the bad correlation between ω calc and ω exp which involve motion of the central x-atom and/or bending modes of the H-atoms (see later).

calculations, and these E(R′) points can be used to fit a V-polynomial and thus to yield $k = (\partial^2 V/\partial R^2)$ and the harmonic frequencies.

B. The Complete E(R′) Curve: Harmonic Approximation

It is possible to obtain an analytical expression for the variation of V(R) = (E(R′)) with R in the case of simple diatomic molecules. *Sugiura* (*60*) obtained an analytical expression for the variation of the internuclear potential V(R) in H_2 for both the bonding and antibonding states of the electronic ground state, in terms of the *Heitler* and *London* model (*61*), using the Born-Oppenheimer approximation (see Section 2 above). The result is an involved expression in ϱ, where $\varrho = R/a_0$:

$$V_s = \frac{e^2}{a_0(1+S)}\left[\frac{e^{-2\varrho}+S}{\varrho} - e^{-2\varrho}\left(\frac{3}{4}+\frac{59}{10}\varrho+\frac{103}{30}\varrho^2+\frac{11}{15}\varrho^3\right)\right.$$
$$\left.+\frac{6}{5\varrho}\{S(C+\log\varrho)+S'\,\mathrm{Ei}(-4\varrho)-2\sqrt{SS'}\,\mathrm{Ei}(-2\varrho)\}\right]$$

where

$$S = (1+\varrho+\tfrac{1}{3}\varrho^2)^2 e^{-2\varrho}\,; \qquad S' = (1-\varrho+\tfrac{1}{3}\varrho^2)^2 e^{-2\varrho}$$

$$C = 0.57722 \qquad \text{(Euler constant)}$$

$$\mathrm{Ei}(\varkappa) = \frac{e^{-\varkappa}}{\varkappa}\left(1-\frac{1!}{\varkappa}+\frac{2!}{\varkappa^2}-\frac{3!}{\varkappa^3}+\cdots\right) \qquad \text{(integral logarithm).}$$

The curve V_s(= E(R′)) has a minimum around $\varrho = 1.5$, as can be seen from (dV_s/dR), with calculated dissociation energy $D = -(V_s)_{R=R_0}$ (which is about 25% too low in this approximation!). The vibrational frequency is obtained from

$$\nu_0 = \frac{1}{2\pi}\sqrt{\left(\frac{1}{m_H}+\frac{1}{m_H}\right)\left(\frac{d^2V_s}{dR^2}\right)_{R=R_0}} = \frac{1}{2\pi}\sqrt{\frac{2k_e}{m_H}},$$

where $m_H = 1.65 \times 10^{-24}$ g. This gives $\omega = \nu/c \approx 4800$ cm^{-1}, compared with the experimental value of ~4400 cm^{-1}. (The more recent evaluations of the energy of the lowest state are discussed by *Parr* (*62*), and should be referred to.)

C. The Morse Approximation

The Morse potential (Eq. 4.29) with its two disposable constants a and D gives very good approximation to the potential of actual molecules, as, e.g. an examination of the actual and Morse values of H_2^+ shows (*63*), that the correlation is good, especially near the point of equilibrium $r_e = 2.00$ au, $a = 0.6678$ au^{-1} and D $= 0.2053$ rydbergs.

The *vibrational terms* of an anharmonic osciallator which moves under a Morse potential are given by (*63, 64*)

$$G_n(\text{cm}^{-1}) = -D_e + \frac{a}{\pi c}\sqrt{\frac{D_e}{2\mu}}\,(n+\tfrac{1}{2}) - \frac{ha^2}{8\pi^2 c\mu}\,(n+\tfrac{1}{2})^2, \qquad (5.5)$$

where

$$\omega_e = \frac{1}{2\pi c}\sqrt{\frac{k_e}{\mu}} = a\sqrt{\frac{hD_e}{2\pi^2 c\mu}} \quad \text{and} \quad \omega_e x_e = \frac{ha^2}{8\pi^2 c\mu}. \qquad (5.6)$$

Both D_e and R_e are known from any advanced calculation of the structure of a diatomic molecule. If the value of the disposable Morse constant a were known, Eq. (5.5) can be used to calculate the vibrational levels, and thus the value of ω_e, where $k_e = (d^2V/dR^2)_{R_e} = 2\,a^2 D_e$; the value of a at $R = R_e$ gives a compromise between the repulsive term *exp*. $(-2\,a(R-R_e))$ and the attractive term 2 *exp*. $(-a(R-R_e))$ which operate at small and large distances R, respectively. It is easy to obtain an optimum value of a by fitting Eq. (4.29) to the actual variation of $E(R')$ vs R, using the theoretically determined values of R_e and D_e.

This method can be tested using the hydrogen molecule-ion H_2^+ for which D $= 0.2053$ Rydberg ($\sim 22{,}000$ cm^{-1}), R $= 2.00$ atomic units (≈ 1.06 Å) and $a = 0.6678$ atomic unit^{-1} (1 au$^{-1} \approx 1.885 \times 10^8$ cm^{-1}) cited above. If μ is written in terms of atomic weights (referred to carbon 12.000) Eq. (5.6) can be written

$$\omega_e = \frac{a}{1\cdot 2177 \times 10^7 \sqrt{2/D_e}}, \qquad (5.7)$$

where a is in cm^{-1}, D_e in cm^{-1} and $\mu_H = 2/m_H$ (*65*). Substitution of the above values yields $\omega_e = 2170$ cm^{-1}, whereas the value obtained from the *exact solution* of the H_2^+ vibrational-rotational problem by *Wind* (*66*) yields $(E_{1.0} - E_{0.0}) \approx 0.02$ Rydberg $= 2194$ cm^{-1} for the transition with n, J values [0.0 → 1.0]. If the $\omega_e x_e$-correction (see Eq. 5.6(b)) is taken into account, $\omega_e x_e \approx 10$ cm^{-1} (using Eq. (5.7)), the approximate Morse potential is seen to yield *adequate values* for the vibrational transitions. (To avoid confusion, attention must be drawn to the fact that the atomic units used by *Wind* (*66*) are just twice as big as those used by *Slater* (*63, 67*); the actual energy is ≈ -1.2 Rydbergs at 2.00 atomic unit, i.e, -0.5971 au in terms of Wind's units. In computing D above, one Rydberg was added to V so that the curve goes to zero at infinite separation, as required by the Morse potential (see Ref. (*12*), p. 36).)

This method is, naturally, very sensitive to the value of the computed binding energy D_e; the more accurate D_e and $E(R')$ around the equilibrium position, the better would the calculated ω_e be. The results of *Bates*, *Ledsham* and *Stewart* (*68*) show how far off the simple L.C.A.O. computed values of $E(R')$ could be (see also (*69, 70, 18c*)) for the case of H_2^+; in these results V_{nn} was not absorbed into $E(R')$ as implied by Eq. (2.5).

This method should lead to results which are just as accurate as the results of the methods described in the previous sections, and can be used as a check on the computed potential-energy minimum $E(R')$ at $R = R_e$: if a is determined from curve-fitting of the Morse potential with the computed R_e and D_e and this leads to a wrong ω_e and/or $\omega_e x_e$, then it can be assumed that D_e and/or R_e are/is wrong. It is to be emphasized (*12*) that the Morse curve can mostly not be used with essentially ionic compounds like NaF because the attraction given by the Coulomb term extends out in space to greater distances than the Morse exponential part; for these compounds many other types of potential have been postulated (e.g. the Hellmann-potential or the Born-Landé potential (*71*)). The reader can try to calculate ω_e, etc. of NaF from the SCF—LCAO—MO calculation of *Matcha* (*72*) in the Roothaan-Hartree-Fock approximation, using the Morse curve ($E_e = -261.38$ au, $R_e = 3.628$ au; experimental values $R_e = 3.639$ au, $\omega_e = 536\ cm^{-1}$, $\omega_e x_e = 3.83\ cm^{-1}$).

Morse curves, however, seem to fit both the calculated and experimental potential energy curve of LiH as exemplified by the *Karo* and *Olson* calculations (*73*) which use the determinantal (case 1) and SCF—LCAO—MO (case 2) approximations to study the effect of configuration-interaction. Their results are given by:

case 1:

$$U(R) = 1.324\{[e^{-0.682(R-3.238)} - 1]^2 - 1\}$$
$$\omega_e = 1.165 \times 10^3\ cm^{-1}$$
$$D_e = 1.324\ eV$$
$$r_e = 3.238\ a.u.$$

case 2:

$$U(R) = 1.669\{[e^{-0.632(R-3.245)} - 1]^2 - 1\}$$
$$\omega_e = 1.212 \times 10^3$$
$$D_e = 1.669\ eV \approx 0.1227\ Ry$$
$$r_e = 3.245\ a.u.$$

exp.:

$$\omega_e = 1406\ cm^{-1}\ (65, 71a)$$
$$D_e = 0.18493\ Ry.$$

The Karo-Olson calculation does not give the correct D_e, so that the calculated values of ω_e are far from the experimental value.

This approach was used for the first time by Hylleraas in his extension of the *Burrau* calculations (*19*) on H_2^+, to give $\omega_0 = 0.0206$ Ry and $\omega_e x_e = 0.00051$ Ry, with $E_{min} = -1.2053$ Ry at $R = 2.000$ (*74*). The Hylleraas levels for H_2^+ are thus given by:

$$E^n_{H_2^+} = -1.2053 + 0.0206\,(n+\tfrac{1}{2}) - 0.00051\,(n+\tfrac{1}{2})^2.$$

The value of ω_e is nearly the same (2260.8 cm^{-1}) as that determined above, and if $\omega_e x_e$ is subtracted (for the transition ($n=0 \rightarrow n=1$)) it gives the value of ≈ 2204 cm^{-1} for the actual transition. This compares favourably with the value of the predicted transition from the *Wind* rotationless energy levels (*66*), viz. 2194 cm^{-1}; *Herzberg* (Ref. (*65*), p. 534) quotes the experimental values of $\omega_e = 2297$ cm^{-1} and $\omega_e x_e = 62$ cm^{-1}, which were "indirectly observed" (see also (*75*)). (In the same paper Hylleraas gave a variational calculation of H_2 and calculated ω_e and $\omega_e x_e$ for various electronic states; the values for the ground state agreed satisfactorily with the experimental values.)

D. The Exact Calculation

The simplest molecular problem, viz. H_2^+, can be attacked by numerical methods to yield accurate vibrational and rotational energy levels in the electronic ground state. This calculation was done by *Wind* (*66a*) using the electronic energy $E_e(R)$ determined by him (*66b*) to a very high degree of accuracy and the rotation-vibration wave equation derived by *Cohen*, *Hiskes* and *Riddell* (*76*)

$$\nabla^2\phi(R) = \left\{\frac{M(1+2/2m)}{m}\,[E - E_e(R) - R^{-1}] - \right.$$
$$\left. - \frac{J(J+1)}{R^2} - g_{00}(R)\right\}\phi(R) = 0\,. \qquad (5.8)$$

R^{-1} is just the coulomb internuclear repulsion (in atomic units), so that $(E_e + R^{-1})$ is $E_t(R)$ of Eq. (2.13); g_{00} is a first-order correction to the Born-Oppenheimer approximation, and J is the rotational quantum number. Wind determined the vibratory levels up to the *dissociation limit* (from $n=0$ to $n=19$) and with J values 0, 1, -----, 8.

From these calculations on H_2^+ it is clear that the Morse approximation, although it is *adequate* in the lowest vibrational energy levels, is far from satisfactory because the graph ΔG (the spacing between the successive vibrational term levels) is definitely not linear, as the Morse approximation assumes (Eq. 5.5):

n	E_n	ΔG	$\Delta^2 G$
0	−0.59713932		
		.00998449	
			.00057971
1	−0.58715483		
		.00940478	
			.00056046
2	−0.57775005		
		.00884432	
			.00054424
3	−0.56890573		
		.00830008	
			.00053088
4	−0.56060565		
		.00776920	
5	−0.55283645		

The second differences $\Delta^2 G$, which should be equal to $-2\,\omega_e x_e$ (*77*) are not constant (compare with the 0.00051 Ry calculated by Hylleraas with the Morse Curve, as discussed above, if it is remembered that an energy of *a* on the Wind scale is just half that on the Hylleraas scale). The linear Birge-Sponer extrapolation (77) to determine D can thus not be used with molecules having E(R′) vs. R curves which deviate from the Morse curves, because the energy terms are given by

$$G_n = \omega_e(n+\tfrac{1}{2}) - \omega_e x_e(n+\tfrac{1}{2})^2 + \omega_e y_e(n+\tfrac{1}{2})^3 - \omega_e z_e(n+\tfrac{1}{2})^4 + \cdots \quad (5.9)$$

where $\omega_e >> \omega_e x_e >> \omega_e y_e >> \omega_e z_e$ etc. (this equation can also be derived from that of the harmonic oscillator using first-order perturbation theory where the perturbing hamiltonian is taken to be due to the addition of successive terms to the potential energy of the harmonic oscillator: introducing a cubic term x^3 does not influence the energy, while the quartic term x^4 gives an approximate value of $\omega_e x_e$, etc. (*78*)).

E. The United-Atom Approximation

Bingel (*79*) has given a rigorous derivation of the potential energy variation at *small internuclear separations*, using the united-atom method.

In principle (d^2V/dR^2) can be evaluated, e.g. from the curves given by *Preuss* for H_2^+ and H_2 (*80*).

F. From the Virial and Hellmann-Feynman Theorems

Eq. (4.27) gives the variation of the virial of the electrons $\mathfrak{K}_e$ with R for diatomic molecules; at $R = R_e$ it is found that $\mathfrak{K}_e = -V_n$, where is the internuclear potential E(R'), so that $\mathfrak{K}_e/T = -1$. By rearranging Eq. (4.27), it follows that

$$\frac{\mathfrak{K}_e}{E(R)} = -1 - \frac{R}{E(R)} \left(\frac{dE(R)}{dR}\right),$$

i.e. that

$$\left(\frac{d\left(\frac{\mathfrak{K}_e}{E}\right)}{dR}\right)_{R_0} = -\frac{R_0}{E_0} \left(\frac{d^2E}{dR^2}\right)_{R_0},$$

and hence that

$$k = -\frac{E_0}{R_0} \left(\frac{d\left(\frac{\mathfrak{K}_e}{E}\right)}{dR}\right)_{R_0}. \tag{5.9}$$

The variation of $\mathfrak{K}_e/E$ with R can be determined by plotting $\mathfrak{K}_e/E$ vs. R and then fitting a polynomial

$$\frac{\mathfrak{K}_e}{T} = -1 + aR + bR^2 + cR^3 + \cdots. \tag{5.10}$$

R_0 is known from the variation of E with R, as calculated. This means that

$$k = -\frac{E_0}{R_0} \left(a + 2bR_0 + 3cR_0^2 + \cdots\right), \tag{5.11}$$

The accuracy of this force-constant calculation depends primarily upon the accuracy of the computed potential-energy curve E(R') in the Born-Oppenheimer approximation, and thus upon the accuracy of the computed wave function ψ. This method was, e. g. used by *Bishop* (*48*) to determine the force constant of methane CH_4 using a one-centre of calculation, to give $k = 22.4$ md/Å, compared with the experimental value of 23.47 md/Å.

The calculated potential-energy curve E(R′) for diatomic molecules in the Born-Oppenheimer approximation does not, in general satisfy the virial theorem, if approximate wave functions are used. There are two methods which can be used to compute force constants and frequencies, *viz.* the methods of *McLean* (*81*) and *Phillipson* (*82*).

McLean (*81*) has demonstrated that scale factors (*83*) can be incorporated in the electronic wave functions $\psi(Q', q_i)$ of Eq. (2.5) in such a way that the potential-energy surface of the inter-nuclear motion is lowered (optimized with respect to the scale factor through the variation principle). This optimization, in the case of diatomic molecules, yields electronic wave functions from which $\bar{K}_e$ and V_n *can be computed which satisfy the virial theorem.* The scaled wave functions are of the type

$$\phi_{a,\lambda}(X) = \lambda^{3n/2}\, \phi_a(\lambda x)\,, \tag{5.12}$$

where x is a point in the 3n electronic space. The condition that $E_{a,\lambda}$ be minimized with respect to λ is

$$\lambda_a \min = -\frac{\bar{V}_e}{2\,K_e}\,. \tag{5.13}$$

Each unscaled ϕ at a specific R, in the case of diatomic molecules, gives rise to a scaled energy curve; the *envelope* of a family of such curves can be taken to be the optimal internuclear potential energy. Exact wave functions, minimal STO-SCF wave functions optmized with respect to coefficients and orbital exponents, and Hartree-Fock wave functions yield the optimal internuclear potential-energy curves which satisfy the virial. At the minimum in the optimal PE-curve $(-\bar{V}_e/2\,\bar{K}_e)$ will be unity. *McLean* uses the *Dunham* analysis (*3b*) to determine the spectroscopic constants through the potential curve

$$(E - E_{\min}) = a_0\, \xi^2 (1 + a_1\, \xi + a_2\, \xi^2 + \cdots)\,, \tag{5.14}$$

where $\xi = (R - R_0)/R_0$. ξ is a dimensionless variable, as well as the constants a_i; the dimensions of a_0 is cm^{-1} (energy). This is differentiated twice to give the force constants, or calculated in terms of Y_{ij} of *Dunham* to give ω_e, etc. directly. For details the reader is referred to the original papers (*36, 81*). The computed values for the spectroscopic constants are fairly good for H_2 (using the *Kolos* and *Roothaan* computation (*84*)), while the $Li^{19}F$ results of *McLean* (*85*) diverge from the experimental values: $\omega_e = 1134$ cm^{-1} (exp.: 964 cm^{-1}), $\omega_e x_e = 9.4$ cm^{-1} (exp.: 8.9 cm^{-1}). The differences between the scaled and unscaled curves are small.

Phillipson (*82*) showed that it is important which co-ordinate representation is used to describe the electronic wave function. He investigated the force constant calculation of *diatomic molecules*, using confocal elliptic co-ordinates (*86, 87*) (ξ_i, η_i, ϕ_i) where ϕ_i is the azimuthal angle of the i th electron around the internuclear axis, R and

$$
\begin{aligned}
r_{ai} &= \tfrac{1}{2}(\xi_i + \eta_i)\,\mathrm{R} \\
r_{bi} &= \tfrac{1}{2}(\xi_i - \eta_i)\,\mathrm{R}
\end{aligned}
\tag{5.14}
$$

where r_{ai} and r_{bi} are the distances of the nuclei, a, b to electron i. The N-electron wave function is $\Phi = \Phi(\xi, \eta, \phi; \alpha)$ where ξ stands collectively for the set, $\xi_1, \xi_2, \xi_3, \cdots$, etc., and α is a set of variational parameters, $\alpha, \alpha_2, \alpha_3, \cdots$ where $\alpha_i = f_i(\mathrm{R})$. The normalized wave function is

$$
\psi = \Phi(\xi, \eta, \phi; \alpha)/\sqrt{\mathrm{S}}; \quad \mathrm{S} = \int \Phi^* \Phi \, d\tau \,, \tag{5.15}
$$

and ψ is a normalized eigenfunction of the Hamiltonian operator $H(\xi, \eta, \phi; \alpha) = \bar{\mathrm{K}}_e + \bar{\mathrm{V}}_e$. If E is the total *electronic energy* of the molecule, then

$$
\mathrm{E}(\alpha, (\mathrm{R})) = \int \psi^*(\alpha)\, H(\mathrm{R})\, \psi(\alpha)\, d\tau \,. \tag{5.16}
$$

Differentiation of Eq. (5.16) on the left and right to R yields the equivalent of Eq. (3.6) with, in the case of exact wave functions, the *Hall* condition of stability (*88*) of Eq. (3.8a). The general expression for $d\mathrm{E}/d\mathrm{R}$ is

$$
\begin{aligned}
\frac{d\mathrm{E}}{d\mathrm{R}} = \frac{\partial \mathrm{E}}{\partial \mathrm{R}} &= -\frac{1}{\mathrm{R}} \int \psi^* \left[2\,\mathrm{K}_e + \mathrm{V}_e + \mathrm{R}\,\frac{d\mathrm{V}}{d\mathrm{R}}\right] \psi \, d\tau \\
&+ \int \psi^* \frac{d\mathrm{V}_e}{d\mathrm{R}}\, \psi \, d\tau \,.
\end{aligned}
\tag{5.17}
$$

In the case of exact wave functions the first integral vanishes, as proved by *Phillipson*, so that

$$
\frac{d\mathrm{E}}{d\mathrm{R}} = \frac{\partial \mathrm{E}}{\partial \mathrm{R}} = \int \psi^* \frac{d\mathrm{V}_e}{d\mathrm{R}}\, \psi \, d\tau = -\bar{\mathrm{F}}, \tag{5.18}
$$

i. e. just the Hellmann-Feynman result (Eq. 3.9).

If, however, the wave functions are *approximate*, then the virial theorem does not hold because the first integral in Eq. (5.17) does not vanish.

The last integral of Eq. (5.17) then cancels with the dV/dR integral in the first term, so that

$$\frac{dE}{dR} = \frac{\partial E}{\partial R} = -\frac{1}{R} \int \psi^* [2\,K_e + V_e]\, \psi \, d\tau \,. \tag{5.19}$$

This equation holds also for the case of exact wave functions because it is just a rearrangement of Eq. (5.17), but Eq. (5.18) holds only for the exact wave functions; this means that

$$\left[\int \psi^* \frac{dV_e}{dR} \psi \, d\tau = -\bar{F}\right] = \left[-\frac{1}{R} \int \psi^* [2\,K_e + V_e]\, \psi \, d\tau = \frac{\partial E}{\partial R}\right], \tag{5.20}$$

where both brackets hold for exact wave functions and only the right-hand bracket for approximate wave functions. It is to be noted that dE/dR is not equal to $-\bar{F}$ for approximate wave functions, so that the Hellmann-Feynman force method is then inapplicable, as stated in Eq. (5.18)—Eq. (5.20) must be used.

For *exact wave functions,* differentiation of the Hellmann-Feynman force (5.18) yields

$$\begin{aligned} \frac{d^2E}{dR^2} &= \int \frac{d\psi^*}{dR} \cdot \frac{dV_e}{dR} \psi \, d\tau + \int \psi^* \frac{dV}{dR} \cdot \frac{d\psi}{dR} d\tau \\ &+ \int \psi^* \frac{d^2V}{dR^2} \psi \, d\tau = \frac{a}{R} \bar{F} - \frac{\partial F}{\partial R} \,. \end{aligned} \tag{5.21}$$

If (5.21) is evaluated at R_e as required, $\bar{F} = 0$ because the system is in equilibrium, then

$$k = \left(\frac{d^2E}{dR^2}\right)_{R_e} = -\left(\frac{\partial \bar{F}}{\partial R}\right)_{R_e} = -\sum_i \left(\frac{\partial \bar{F}}{\partial \alpha_i} \cdot \frac{\partial \alpha_i}{\partial R}\right)_{R_e}, \tag{5.22}$$

where $\bar{F}$ is given by Eq. (5.18). *The harmonic force constant k can thus be evaluated from the gradient of the force vs. R curve at $R = R_e$, provided the derivative acts only upon the electronic charge distribution via the parameters* α_i (valid only in confocal elliptic co-ordinates; for expressions for the k of one-centre calculations, see *Phillipson* (*82*)).

Goodisman (*89*) used the Hartree-Fock method to calculate the variation of the force on the nuclei in H_2 with R, where the variation wave function is of the James-Coolidge type in confocal elliptic co-ordinates. The normal Hartree-Fock method (*without* configuration interaction) satisfies the Hellmann-Feynman theorem (*90*), and the true H—F method should satisfy the virial theorem, so that Eq. (5.22) can be used. The variation of E with R near R_e was fitted to a parabola through 3 points which yielded $(d^2E/dR^2)_{R_0} = 0.3795$ au$/a_0^2$ (i. e. $\omega_e = 4458$ cm^{-1}; a fit through 5 points yielded a worse result!). The gradient of the $\bar{F}$ vs. R curve at R_e yielded $k_e = 0.3727$ au$/a_0^2$, i. e. $\omega_e = 4417$ cm^{-1}, which is near the experimental value of 4400 cm^{-1}. According to Eq. (5.22) both k_e's should be equal, indicating, perhaps that the variation principle or the virial theorem was not satisfied, so that the wave functions can not be used in a Hellmann-Feynman computation. This aspect is further emphasized by the calculations of *Ross* and *Phillipson* (*91*) who used a series of wave functions for H_2^+ and H_2 to calculate k with Eq. (5.22). For most wave functions $\bar{F} = 0$ at $R = R_e$, so that k_e gives an ω_e which is far too big; the more $\bar{F}$ deviates from 0, the greater is ω_e (see Table 1 below).

Table 1

Molecule	ω_e cm^{-1} (exp.)	ω_e approximate cm^{-1}	ω_e exact cm^{-1}	$\bar{F}$ au	D ev	Wave function
H_2^+	2297	2856	3213	−0.044	2.214	$\psi = \exp(-\delta\xi)$
H_2^+	2297	2326	2407	−0.007	2.786	$\psi = \exp(-\delta\xi)(1 + c\eta^2)$
H_2^+	2297	2221	1474	0.087	2.354	$\psi = 1s_a + 1s_b$ (*Finkelstein* and *Horowitz*)
H_2	4400	4240	6723	1.450	3.784	$\psi = 1s_a \cdot 1s_b + 1s_b \cdot 1s_a$ (*Wang*)
H_2	4400	5365	5390	1.842	3.488	$\psi = (1s_a \cdot 1s_b + 1s_b \cdot 1s_a) + (1s_a \cdot 1s_a + 1s_b \cdot 1s_b)$ (*Coulson*)
H_2	4400	4772	6012	1.602	4.024	$\psi = (1s_a \cdot 1s_b + 1s_b \cdot 1s_a) + (1s_a \cdot 1s_a + 1s_b \cdot 1s_b)$ (*Weinbaum*)
H_2	4400	–	4417 (from $\bar{F}$)	0.0025		
			4458 (from E)			$\psi =$ *James* and *Coolidge*, by *Goodisman* (*89*)
					$D_{H_2\,\text{exact}} = 4.75$ eV	

There is thus a need to obtain better wave functions which satisfy Hellmann-Feynman theorem, or another force-constant expression must be obtained which is valid for non-exact wave functions. The second procedure was adopted by *Phillipson* (*82*) who differentiated Eq. (5.17) to R in the form

$$\frac{dE}{dR} = -\frac{2E}{R} + \frac{1}{R} \int \psi^* \, V \, \psi \, d\tau \tag{5.23}$$

which follows from $H = \bar{K}_e + \bar{V}_e$ and Eq. (5.16), to obtain

$$\frac{d^2E}{dR^2} = -\frac{3}{R}\frac{dE}{dR} + \frac{1}{R}\frac{d}{dR} \int \psi^* \, V_e \, \psi \, d\tau \, , \tag{5.24}$$

where (5.23) is kept in mind. The above integral reduces to

$$\frac{\partial}{\partial R'} \int \psi^* \, V_e \, \psi \, d\tau - \frac{1}{R} \int \psi^* \, V_e \, \psi \, d\tau \, ,$$

where the $\partial/\partial R'$ shows that it applies only to the α-parameter, i. e.

$$k_e = \frac{d^2E}{dR^2} = \frac{1}{R}\frac{\partial V_e}{\partial R'} - \frac{V_e}{R^2} = \frac{1}{R}\frac{\partial \alpha_i}{\partial R'} \cdot \frac{\partial \bar{V}}{\partial \alpha_i} - \frac{\bar{V}}{R^2} \, , \tag{5.25}$$

where

$$\bar{V} = \int \psi^*(\alpha) \, V \, \psi(\alpha) \, d\tau \, . \tag{5.26}$$

Eq. (5.25) was used by *Ross* and *Phillipson* (*91*) to compute k_e, and hence ω_e, for the H_2^+ and H_2 molecules mentioned above, using various wave functions; the results depend upon the wave function, as shown below in Table 1.

These studies by *Phillipson* and *Goodisman* show that the computation of vibrational frequencies using the Hellmann-Feynman theorem and the virial theorem is very much dependent upon the wave functions used. It also show that many cases are known where a specific type of wave function yields a poor binding energy but an accurate ω_e: this occurs because the computed curve is "parallel" to the actual curve, especially near the minimum, so that the curvatures (d^2E/dR) at R_e for these curves are identical, or nearly identical, e. g. the wave function of H_2 which yields a binding energy which is $\sim$1 eV too low ($\sim$20%!), but yields an acceptable value of ω_e (Table), or the *McLean* calculations of the PE-curves of LiF (*85*). It is further very important that at equilibrium $\bar{F}=0$, and also that the Hellmann-Feynman theorem must be satisfied if reasonably accurate values of ω_e are to be obtained from Eq. (5.25). In addition attention must be paid to the co-ordinate representation of the problem, because the form of the force-constants depends upon it. The method suffers from the absence of higher derivatives of E to account for $\omega_e x_e$, etc.

It is perhaps worthwhile to return again to the poor results which *Ross* and *Phillipson* (*91*) obtained when certain "popular" wave functions which appear in the literature are used to calculate the force constants through the virial theorem Eq. (5.18). In these cases a "fixed" function of a certain form is used instead of the correct wave function to determine the energy, e. g. the familiar *Heitler-London function* (*92*) is used:

$$\psi=\{2\pi(1+S)\}^{-\frac{1}{2}}\left[\exp(-r_{a1}-r_{b2})+\exp(-r_{b1}-r_{a2})\right]. \tag{5.27}$$

This function is not "fixed" in the sense that it not only depends upon r_{a1}, r_{a2}, r_{b1} and r_{b2}, but also explicitly on the internuclear separation R. This can be seen by rewriting Eq. (5.27), following *Hurley* (*93*), using a co-ordinate representation situated at the centre of symmetry and the z-axis along the molecular axis:

$$\begin{aligned}\psi=&[2\pi\{1+S^2(R)\}]^{-\frac{1}{2}}\left[\exp\{-(x_1^2+y_1^2+\langle z_1-\tfrac{1}{2}R\rangle^2)^{\frac{1}{2}}\right.\\ &-(x_2^2+y_2^2+\langle z^2+\tfrac{1}{2}R\rangle^2)^{\frac{1}{2}}\}+\exp\{-(x_1^2+y_1^2+\langle z_1+\tfrac{1}{2}R\rangle^2)^{\frac{1}{2}}\\ &\left.-(x_2^2+y_2^2+\langle z_2-\tfrac{1}{2}R\rangle^2)^{\frac{1}{2}}\}\right],\end{aligned} \tag{5.28}$$

where $(x_1y_1z_1)$ and $(x_2y_2z_2)$ are the co-ordinates of electrons 1 and 2, respectively, and $R=(Z_1-Z_2)$. This means that

$$\psi=\psi(x_a;R), \tag{5.29}$$

where the $x_a\,(a=1, 2, \text{-----}\ 4n)$ are the zn space co-ordinates of the two electrons. Then

$$\begin{aligned}E(X_a, Y_a, Z_a)&=\varepsilon(X_a, Y_a, Z_a; R)\\ &=\int\psi^*(R;x_a)\,H(X_a, Y_a, Z_a)\,\psi(R;x_a)\,d\tau,\end{aligned} \tag{5.30}$$

where X_a, etc. refer to the nuclear co-ordinates of the α th atom, and H is total electronic Hamiltonian ($K_e + V_{nn}$). The force on the 1st nucleus in the Z-direction is $F_{az} = -(\partial E/\partial Z_a)$, where

$$\frac{\partial E}{\partial Z_1} = \frac{\partial \varepsilon}{\partial Z_1} + \frac{\partial \varepsilon}{\partial R} \cdot \frac{\partial R}{\partial Z_1}. \tag{5.31}$$

Eq. (5.31) can be used to calculate F_{az} using the Hellmann-Feynman theorem if the last term is equal to zero. It is clear that $\partial R/\partial Z_1$ is not equal to zero, while $\partial E/\partial R$ is also not equal to zero because the values of R were not determined by the variational method (only the c_i's are variationally determined in the linear,combination $\psi = c_1 \phi_1 \phi_2 + c_2 \phi_2 \phi_1$ in the Heitler-London method; the values with respect to c must be included in the last term of Eq. 5.31, but as $(\partial E/\partial c_i) = 0$ both in cases they were dropped). This means that R in (5.28) must be replaced by a function $\mu(R)$, the value of which, for each value of R is determined by the variation method, so that

$$\frac{\partial \varepsilon}{\partial \mu} = 0 \tag{5.32}$$

at each R, and then

$$F_{az} = -\frac{\partial \varepsilon}{\partial Z_a} = -\int \psi^* \frac{\partial V}{\partial Z_a} \psi \, d\tau . \tag{5.33}$$

The reason why the Heitler-London method gives such a bad Hellmann-Feynman force-constant is thus that R is "attached" to the nuclei so that $\partial \varepsilon / d R$ is not zero; the same applies to the *Weinbaum* function (*94*), to the *Wang* function (*95*), and to the *Coulson* function (*96*) for H_2. To yield better force results, the variable parameters[5]) must be "detached" from the nuclei and their values determined at each internuclear configuration. A wave function in which *all* the parameters are determined by the variational procedure is called a *floating function* by *Hurley* (*93, 97, 30, 31, 32*) (this is equivalent to Hall's *stable wave functions* (*88*)). This procedure can be extended to the scale factors, as discussed by *McLean* (*81*) and *Löwdin* (*83*). The vibrational frequency of H_2 determined by *Ross* and *Phillipson* using the differentiation of the virial theorem (which assumed that all the variable parameters are variationally

[5]) *Coulson* and *Hurley* (*31*) suggest that a *variable parameter* is one whose value changes with that of the parameter being varied in the Hamiltonian.

determined for each R) with the three Heitler-London type wave functions discussed, is also considerably in error because the μ(R)-dependence does not satisfy the variational principle. For floating functions which contain variable scale parameters the forces determined from $\partial E/\partial Z$, from the Hellmann-Feynman theorem and from the virial theorem must agree (*93*).

G. Platt's Approximation

The properties of the diatomic hydrides MH can be derived from the density distribution of the united-atom approximation (see e. g. (*79, 80*) for the potential-energy-curves) of the neutral united atom (M+1). *Platt* (*28, 98*) used this approximation to calculate the frequencies via the force constants of these diatomic hydrides through purely electrostatic arguments. To make the hydride from the united atom one *proton* is removed from the nucleus without distortion of the charge cloud of the united atom (i. e. the removal of the proton is considered to be a small perturbation). If r_0 is the radius at which the effective nuclear charge is *unity*, the proton should be moved to r_0, i. e. to the radius beyond which lie a total number of 1.00 electrons. If the proton is replaced outside r_0, the excess electrons will attract it, and it will be repelled if it is situated inside the nucleus. This is just the situation for the vibrating motion of a proton attached to another atom through a stable chemical bond: a small displacement $(+dr)$ from r_0 will subject the proton to an added force

$$dF = -4\,\pi^2\, r^2\, \varrho_{1.0}\, e^2 \frac{dx}{r^2} \tag{5.34}$$

giving the force constant

$$k = \frac{dF}{dx} = 4\,\pi\, e^2\, \varrho_{1.0}\,, \tag{5.35}$$

where $\varrho_{1.0}$ is the density of electrons in the undisturbed united-atom distribution at radius r_0. Eq. (5.35) can also be written as

$$k = \frac{e^2}{a_0^3} \cdot \frac{q_{1.0}}{z_{1.0}^2} = 15.62 \times 10^5\, \frac{q_{1.0}}{z_{1.0}^2}\ \text{dynes/cm}\,, \tag{5.36}$$

where q is the radial electron density $q = 4\,\pi^2\, r_{1.0}^2\, \varrho_{1.0}\, a_0$ and $z_{1.0} = r_{1.0}/a_0$. Mostly $r_{1.0}$ is not equal to r_e, so that two force constants can be calculated, *viz.* k_0 at $r_{1.0}$ and k_e at r_e. This calculation contains no adjustable semi-empirical parameters depending upon experimental frequencies, and give very good results, e. g. an average error of 9% in predicting r_0 and 7% in predicting k_0 (compared with observed values at r_e) when Hartree united-atom functions are used.

Clinton (*99*) showed that *Platt's* approximation is valid if the last term in a derived expression is neglected. *Bratoz et al.* (*100*) also derived *Platt's* formula using a perturbation procedure, and showed that the *Platt* model corresponds to the leading term of the perturbation analysis based upon the united atom method and is valid only for a spherical charge distribution; they also show how the electrons in the various orbitals effect the force constants.

H. The Empedocles Approximation

Empedocles (*101*) proposed a general semi-empirical approach using data from the united-atom and separated-atom models and calculate the quadratic (harmonic), cubic and quartic force constants k_2, k_3 and k_4 from the differentiation of the virial theorem electronic kinetic energy expression Eq. (4.27), where E is the total energy

$$\bar{K}_e = -E - R_0 \left(\frac{dE}{dR}\right)_{R_0} \tag{5.37}$$

evaluated at $R = R_0$, the equilibrium position. Then

$$k_2 = \left(\frac{d^2E}{dR^2}\right)_{R_0} = \frac{1}{R_e}\left(\frac{d\bar{K}_e}{dR}\right)_{R_0}; \tag{5.38}$$

$$k_3 = \left(\frac{d^3E}{dR^3}\right)_{R_0} = \frac{3}{R_0^2}\left(\frac{d\bar{K}_e}{dR}\right)_{R_0} - \frac{1}{R_0}\left(\frac{d^2\bar{K}_e}{dR^2}\right)_{R_0}; \tag{5.39}$$

$$k_4 = \left(\frac{d^4E}{dR^4}\right)_{R_0} = \frac{12}{R_0^3}\left(\frac{d\bar{K}_e}{dR}\right)_{R_0} + \frac{4}{R_0^2}\left(\frac{d^2\bar{K}_e}{dR^2}\right) - \frac{1}{R_0}\left(\frac{d^3\bar{K}_e}{dR^3}\right). \tag{5.40}$$

Empedocles argues that $(d\bar{K}_e/dR)$ is relative insensitive to "crudities" in the wave function. In addition, $\bar{K}_e$ is a one-electron quantity, a thing which simplifies multi-electron calculations considerably. Further, it must be pointed out that Eq. (5.37) (see Eq. 4.27) and Eq. (5.41) (see Eq. 4.28)

$$\bar{V}_e = 2E + R\left(\frac{dE}{dR}\right), \tag{5.41}$$

partition the kinetic energy and the potential energy correctly only when the exact wave functions are used, so that at equilibrium or at $\mathrm{R}=\infty$, where $d\mathrm{E}/d\mathrm{R}=0$, the relation

$$\bar{\mathrm{K}}_e = -\tfrac{1}{2}\,\bar{\mathrm{V}}_e \tag{5.42}$$

holds, as already discussed. *Properly-scaled wave functions* can be also used (see *Löwdin* (*83*)) at $\mathrm{R}=\mathrm{R}_0$. If a properly scaled wave function for a particular molecule is available at $\mathrm{R}=\mathrm{R}_0$, then the force constants (5.38 to 40) can be calculated using the derivatives $d\bar{\mathrm{K}}_e/d\mathrm{R}$, etc. The expectation value of the *kinetic-energy operator* K' for a properly scalled wave function $\psi_\eta(\eta r_1, \eta r_2, \cdots)$ is equal to

$$\mathrm{K}(\eta) = \int \overset{*}{\psi}_2(\eta r)\, K_e'\, \psi_\eta(\eta r)\, dr = \eta^2 \int \psi^*(r)\, K_e'\, \psi(r)\, dr = \eta^2\, \mathrm{K}(1)\,, \tag{5.43}$$

where ψ is the unscaled wave function $\psi(r)$, and $\mathrm{K}(1)$ the expectation value of the unscaled kinetic energy

$$\mathrm{K}(1) = \int \psi^*(r)\, K\, \psi(r)\, dr \tag{5.44}$$

since K is a homogeneous function of degree -2. The *Hoare* and *Linnett* (*102*) approach allows the separation of K into its three cartesian components, viz.

$$K = \frac{d^2}{dx^2} + \frac{d^2}{dy^2} + \frac{d^2}{dz^2}\,, \tag{5.45}$$

and each component is homogeneous and of degree -2, so that (5.43) is satisfied for each of the three components along the *1, 1, 1,* directions, respectively. This yields

$$\bar{\mathrm{K}}(1,1,1) = \bar{\mathrm{K}}_x(1) + \bar{\mathrm{K}}_y(1) + \bar{\mathrm{K}}_z(1) = \sum_{i=x,y,z} \mathrm{T}_i(1) \tag{5.46}$$

$$\bar{\mathrm{K}}_i(\eta_i) = \eta_i^2\, \bar{\mathrm{K}}_i(1) \tag{5.47}$$

$$\bar{\mathrm{K}}(\eta_x, \eta_y, \eta_z) = \sum_{i=x,y,z} \eta_i^2\, \bar{\mathrm{K}}_i(1)\,. \tag{5.48}$$

If Eq. (5.48) is differentiated the force constants are obtained, e. g.

$$\frac{d\mathrm{K}}{d\mathrm{R}} = \sum_{i=x,y,z} 2\,\eta_i\, \bar{\mathrm{K}}_i(1)\, \frac{d\eta_i}{d\mathrm{R}}\,, \tag{5.49}$$

etc., where the relations:

$$\frac{d\,\mathfrak{K}_i(\eta_i)}{d\eta_i} = 2\,\eta_i\,\mathfrak{K}_i(1)\;; \qquad \frac{d^2\,\mathfrak{K}_i(\eta_i)}{d\eta_i^2} = 2\,\mathfrak{K}_i(1)\;; \qquad \frac{d^3\,\mathfrak{K}_i}{d\eta_i^3} = 0 \tag{5.50}$$

are used. Empedocles uses the following approach to calculate the $\mathfrak{K}(1)$ and by choosing a distortion operator D he calculates η_i and $d\eta_i/d\mathrm{R}$ at $\mathrm{R} = \mathrm{R}_0$. Further, a scale is chosen which defines $\eta_x = \eta_y = \eta_z = 1$, leading to a simpler evaluation of $\mathfrak{K}(1)$. The distortion operator D is chosen in such a way that it expresses the wave function $\psi(\mathrm{R}_0 + \Delta\mathrm{R})$ for the molecule after an adiabatic increase in internuclear distance $\Delta\mathrm{R}$ away from the R_0-position, in terms of $\psi(\mathrm{R}_0)$, where $\Delta\mathrm{R} = \mathrm{R} - \mathrm{R}_0$, i. e.

$$D(\Delta\mathrm{R})\,\psi(\mathrm{R}_0) = \psi(\mathrm{R}_0 + \Delta\mathrm{R})\,. \tag{5.51}$$

The force constants, from the considerations above, are functions of $\psi(\mathrm{R}_0)$ and depend on the way in which $D(\Delta\mathrm{R}) \to 1$ as $\mathrm{R} \to 0$. The operator D is obtained by distorting the molecular orbitals of the molecule to the shape which closely resembles the united-atom orbitals from which they were derived — e. g. an $1s\sigma_g$ — MO becomes an $1s$-united atom orbital and an $1s\sigma_u$-MO becomes a $2p$-united atom orbital. The origin of the scale is then fixed so that for this wave function $\eta_x = \eta_y = \eta_z = 1$. The fundamental Empedocles-assumption is now that the components of the kinetic energy of the distorted MO are approximately equal to those of the actual atomic orbital, i. e.

$$\mathfrak{K}_x(1_{\mathrm{MO}}) \approx \mathrm{T}_x(\eta_u) \tag{5.52}$$

and identical equations for $\mathfrak{K}_y$ and $\mathfrak{K}_z$, where n_u is the Slater orbital exponent of the united atom. The only problem which is left, is the variation of the scale factors η with R; these are obtained from the scale factors of the united atoms η_u and the separated atoms η_5 by extrapolation using an empirical relationship. This empirical relationship destroys the *ab-initio*-calculation of the k's or ω's; it yields reasonably good results for all three force constants k_2, k_3 and k_4 for the homonuclear diatomic molecules of the 1st and second periods, but deviates rather badly from the experimental values for heteronuclear diatomics of the same periods. *Empedocles* concludes from his analysis of the contributions to k_2 from each occupied molecular orbital, that the electrons in the valence shells of the atoms which form the M.O.'s contribute practically all of the force constant.

I. From Perturbation Theory

The Born-Oppenheimer Hamiltonian operator $H=(T_e+V)$ for the fixed-nucleus approximation (Eq. 2.5) is expanded in powers of the displacement $(\mathrm{R}-\mathrm{R}_e)$ near R_e and second-order perturbation theory is used to calculate E correct to second-order. Then k is given by

$$k=\left(\frac{d^2\mathrm{E}}{d\mathrm{R}^2}\right)_{\mathrm{R}_e}, \tag{5.52}$$

as usual. Then near R_e:

$$\begin{aligned} H(\mathrm{R}, r_i) = H(\mathrm{R}_e, r_i) &+ \left(\frac{\partial H}{d\mathrm{R}}\right)_{\mathrm{R}_e}(\mathrm{R}-\mathrm{R}_e) \\ &+\tfrac{1}{2}\left(\frac{\partial^2 H}{\partial \mathrm{R}^2}\right)_{\mathrm{R}_e}(\mathrm{R}-\mathrm{R}_e)^2+\cdots, \end{aligned} \tag{5.53}$$

(the partial derivatives $\partial/\partial\mathrm{R}$ imply that the electronic co-ordinates and some point S on the molecular axes are held fixed). But

$$\frac{\partial H}{\partial \mathrm{R}}=\frac{\partial \mathrm{V}}{\partial \mathrm{R}}, \tag{5.54}$$

because T_e involves no internuclear co-ordinates, hence

$$\begin{aligned} &\mathrm{E}^0(\mathrm{R}) = \\ &\left\langle \psi^{*0}\left| H(\mathrm{R}_e)+\left(\frac{\partial \mathrm{V}}{\partial \mathrm{R}}\right)_{\mathrm{R}_e}(\mathrm{R}-\mathrm{R}_e)+\tfrac{1}{2}\left(\frac{\partial^2 \mathrm{V}}{\partial \mathrm{R}^2}\right)_{\mathrm{R}_e}(\mathrm{R}-\mathrm{R}_e)^2\right|\psi^0\right\rangle \\ &\quad -\sum_{n'}\left\langle \psi^{*0}\left|\left(\frac{\partial \mathrm{V}}{\partial \mathrm{R}}\right)_{\mathrm{R}_e}(\mathrm{R}-\mathrm{R}_e)_{0n}\right|\psi^0\right\rangle^2\Big/(\mathrm{E}_n-\mathrm{E}_0) \end{aligned} \tag{5.55}$$

where $<||>$ indicates the expectation value of ψ^0. It is clear that

$$\left\langle \psi^{*0}\left|\left(\frac{\partial^2 \mathrm{V}}{\partial \mathrm{R}^2}\right)_{\mathrm{R}_e}\right|\psi^0\right\rangle=\left\langle \psi^{*0}\left|\left(\frac{\partial^2 \mathrm{V}}{\partial \mathrm{R}^2}\right)\right|\psi^0\right\rangle_{\mathrm{R}_e} \tag{5.56}$$

because R_e is a parameter of integration over *electronic co-ordinates*, so that

$$k_0=\left\{\left\langle \psi^{*0}\left|\left(\frac{\partial^2 \mathrm{V}}{\partial \mathrm{R}^2}\right)\right|\psi^0\right\rangle-2\sum_{n'}\frac{\left\langle \psi^{*0}\left|\frac{\partial \mathrm{V}}{\partial \mathrm{R}}\right|\psi^0\right\rangle^2}{(\mathrm{E}_n-\mathrm{E}_0)}\right.. \tag{5.57}$$

This equation gives the force constant k_0 (*103—108*) but it is rather "intractable", to use *Salem's* description (*106*). *Bader* (*103*), however, used it to predict the sign of the interaction constants in polyatomic molecules.

Gerratt and *Mills* (*109a*) developed a perturbation-type Hartree-Fock calculation within the Born-Oppenheimer framework for force constants, and applied it to LiH, BH and HF using a limited basis-set of SCF-MO wave functions. They chose a co-ordinate system external to the molecule; the positions of the N nuclei relative to this co-ordinate system are denoted vectorially as $R_J\,(J=1, ---, N)$ with components $R_J^\alpha\,(\alpha=x,y,z)$ and the positions of the n electrons by $r_\mu\,(\mu=1, ---, n)$. The choice of the co-ordinate system then influences the force constants $f_{JK}^{\alpha\beta}$ which are defined by the harmonic potential expression.

$$V=\tfrac{1}{2}\sum_{J,K=1}\sum_{\alpha,\beta} f_{JK}^{\alpha\beta}\,\delta R_J^\alpha\,\delta R_K^\beta \tag{5.58}$$

i. e. by

$$f_{JK}^{\alpha\beta}=\left(\frac{\partial^2 E}{\partial R_J^\alpha\,\partial R_K^\beta}\right)_0 \tag{5.59}$$

Proceeding with an argument similar to that used by *Phillipson* (*82*, see also § 5F above), except that in this case the potential energy of internuclear interaction Ω is explicitly included, *Gerratt* and *Mills* derive the following equation for the force constants

$$\begin{aligned} f_{JK}^{\alpha\beta}=&\frac{\partial^2\Omega}{\partial R_J^\alpha\,\partial R_K^\beta}+\int\varrho(r,R)\,\frac{\partial^2 v}{\partial R_J^\alpha\,\partial R_K^\beta}\,dr\\ &+\int\frac{\partial\varrho}{\partial R_K^\beta}\cdot\frac{\partial v}{\partial R_J^\alpha}\,dr \end{aligned} \tag{5.60}$$

where $\varrho(r,R)$ is a one-electron density function

$$\varrho(r,R)=n\int\cdots\int\psi^*\,\psi\,dr_2\,dr_3\cdots\cdots dr_n\,dS_1\,dS_2\cdots dS_n \tag{5.61}$$

the space integration excludes electron 1 although its spin is included; v is the potential experienced by the electron in the field of all the N nuclei. The force constant is thus the sum of three distinct contributions, viz. the internuclear contribution, while the second term gives the contribution to f_{JK} if the nuclei are moved while $\varrho(r,R)$ remains unaltered (Hellmann-Feynman term) and the third term represents the energy which is "gained" when the one-electron density adjusts according to the internuclear motion ("backlash"-effect), and is called the relaxation term. This last term is always negative for f^{aa}_{KK}, and for $f^{a\beta}_{JK}$($J \neq K$) it represents the electronic contribution. The second and third terms in equation (5.60) are not individually invariant because they depend, just as in *Phillipson's* case, upon the co-ordinate system chosen; if the co-ordinate system is properly chosen, the Hellmann-Feynman term becomes zero (see for instance *Murrell* (*104*)). Using the standard Roothaan-approach (*110*), Eq. (5.60) is transformed into the Hartree-Fock form; a perturbation approach is used to evaluate *f*. The results with HF, LiH and BH are encouraging, and show that Hartree-Fock calculations of the force constants are feasible, and that it might be possible to localise the origins of the forces which give rize to the force constants (i. e. using Eq. 5.60).

Gerratt and *Mills* showed that the perturbation-approach circumvent the errors involved in the numerical differentiation of the potential curve (see also (*89*)). However, *Freed* (*111*) showed that the calculation of force constants by either the numerical differentiation of the Hartree-Fock potential energy curve or by the Hartree-Fock preturbation theory which evaluates the relaxation contribution at a fixed internuclear distance should formally yield the samc answcr. The method of *Gerratt* and *Mills* is essentially the perturbation analogue of the *Bishop* and *Randic* (*112*) method of direct differentiation which goes back to the work of *Bratoz* (*112*). *Bader* and *Bandrauk* (*114*) also discuss the nature of the force constant in terms of the relaxation of the molecular charge distribution during nuclear motion, using the Hartree-Fock wave functions of N_2, CO, BF, BeO and LiF; it is, for instance, found that the charge distribution of a covalently bonded molecule AB relaxes in such a manner that it "helps" the motion of the displaced nuclei, whilst the ionic molecules A^+B^- the cation-charge relaxation opposes the nuclear motion and the anionic charge relaxation helps the displacement.

J. The Parr "Bond-Charge" Model

A simple bond-charge model of both homonuclear (*114a*) and heteronuclear (*114b*) neutral diatomic molecules has been developed by *Parr*, using a Fues-type vibrational potential (*115*)

$$W = W_0 + (W_1/R) + (W_2/R^2)$$

The coefficients W_1 and W_2 are expressed in terms of the empirical parameters q (bond charge) and v (a one-dimensional box length). It is shown that q and v are related to R_e and K_e, using the virial theorem; thus no wave function enters in the discussion. This simple model adequately describe the vibrations of neutral diatomic molecules, but the relation between the operationally defined parameters q ($\approx$ bond orders) and ν ($\approx$ core radii for atoms in molecules) and actual charges in bonds and actual lengths of charge delocalisations in bonds, repectively, need further investigation.

K. The Ab-Initio Wave Functions

Although the actual calculation of the wave functions themselves, falls outside the scope of this paper, it is perhaps the best to give an indication where to find the information. The papers referred to give, of course, descriptions of the methods followed, but reference to the basic papers of *Allen* and *Karo* (*116*) and *Kotani et al.* (*117*) will place the methods in general perspective. The scope and summary of diatomic computations are given by *Ransil* (*118*), as well as by *McLean* and *Yoshimine* (*58, 59*). *Wahl et al.* (*119*) give a review of recent progress beyond the Hartree-Fock method for diatomic molecules as well as a review of the Hartree-Fock calculations during the past six years (e.g. they conclude that Hartree-Fock ω_e's are generally poor, probably because the accuracy is $\sim$10 %); the emphasis is on the method of optimized valence configurations as an alternative for Hartree-Fock calculations ((*119*) has a comprehensive list of references!). *Pople et al.* (*120*) review the results of approximate SCF theory (zero-differential overlap), and include references to the calculation of both ω_e and r_e for diatomics, and XY_3 and XY_2 molecules. The energy hypersurface of molecules is further discussed by *Preuss* (*121*) and applied to H_2 and (H_2He) to show how good E vs. R compares with *ab-initio* calculations. *Nesbet* (*122*) reviews approximate Hartree-Fock calculations on small molecules, and tabulates a whole series of ω_e, r_e and K_e calculations for these molecules, e.g. HF, HCl, N_2, CO, CO^+, BF, BF^+, LiF, BeO, F_2; the agreement between ω_e calc. and ω_e exp. is "fair", but fluctuates between wide margins. *Bishop* (*123*) gives an extensive review of single-centre molecular wave functions (see also *44—58, 50, 52, 53*), with special reference to polyatomic molecules, using a corrected form of the *Bratoz-Allavena* (*124*) evaluation of (d^2E/dR^2). *Bratoz* attacks the hydrogen bond via an electronic theory, and discusses the infra-red implications (*125*). *Joshi* (*126*) calculates the potential energy curve for the inversion mode of NH_3 (a one-dimensional mode!) using SCF—M.O.-single centre approximation (and gives a complete biography of the method, see (*123*); reasonable results are obtained.

H. Preuss has recently given a systematic review of ab-initio calculations of molecules and atomic systems, with emphasis on the construction of energy hypersurfaces (*127*). *Kouba* and *Öhrn* (*128*) give a natural orbital valence shell configuration interaction study of diatomic molecules and apply the theory to NH, obtaining reasonable results for the various electronic states, e.g. $X^3\Sigma^-$I, $a^1\Delta$ etc. *Ellison* developed a scaled atoms in molecules theory for predicting diatomic potential energy curves, and applied it to the $X^1\Sigma_g^+$ and $E^1\Sigma_g^+$ states of H_2 (*129*).

Kolos and *Wolniewicz* (*130*) published some very accurate absolute calculations on hydrogen. The SCF—M.O. method was used by *Sahni, Sahwney* and *Hanley* (*131*) to calculate ω_e, $\omega_e x_e$, B_e, α_e, R_e and D_e of LiH. The unrestricted Hartree-Fock method as a tool for the computation of potential hypersurfaces was investigated by *Salotto* and *Burnelle* (*132*).

Pack and *Hirschfelder* (*133*) discuss the energy corrections to the Born-Oppenheimer approximation, while *Bunker* (*134*) discusses the partial breakdown of the Born-Oppenheimer approximation. The Hellmann-Feynman theorem is also discussed by *Tuan* (*135*) with reference to multiconfiguration SCF-theory, while *Loeb* and *Rasiel* (*136*) discuss constraints upon the LiH molecule with reference to the Hellmann-Feynman theorem. *King* (*137*) developed a theory of effective Cartesian force constants which relate to the sum of the squares of the normal frequencies; these effective force constants are independent of the internuclear repulsion of the nuclei in a molecule, and thus removes one of the indeterminacy of the Hellmann-Feynman method described above.

The general approach to SCF—MO theory on an elementary level is given by *Simonetta* and *Gianinetti* (*138*), and should be consulted.

The eigenfunctions obtained from a Morse Potential were used by *Greenawalt* and *Dickenson* (*139*) in a variational calculation of bound states of diatomic molecules; a basis set of Morse functions was chosen for the *Wolniewicz* calculation of the $X^1\Sigma_g^+$-state of H_2 with $J=0$, and the calculated vibrational levels were compared with those obtained by the accurate integration procedure of *Le Roy* and *Bernstein* (*140*).

L. Conclusions

The various methods to calculate the vibrational frequencies and force constants from *ab-initio* data on diatomic molecules is represented in Sections 5 A to K. It is seen that the various approximations yield results which fluctuate from molecule to molecule, although the order of magnitude is mostly correct. It is clear, however, that it is not at the present moment possible to calculate ω_e and k_e of molecules to such a degree of accuracy that the factors which contribute to the internuclear forces in molecules can be pinpointed and compared. This is perhaps the reason why semi-empirical models continue to be exploited, e.g. the simple bond-charge model (electrostatic) model for P.E.-curves of homonuclear diatomic molecules of *Parr* and *Borckmann* (*114*) based upon the Fues potential from which the famous Birge-Mecke relation is derived:

$\omega_e R_e^2 = \text{constant}$. Another approach is that of *Cashion* (*141*) who gives simple formulae of the vibrational and rotational eigenvalues of the 6—12-Lennard Jones potential. But in spite of the difficulties of the *ab-initio* method, careful studies and exact computations appear, e.g. the *Kolos* and *Wolniewicz* work on H_2 (*130*).

6. The Calculation of the Molecular Frequencies: Polyatomic Molecules

The exact wave-mechanical calculation of the various ψ's und E's of polyatomic molecules still poses a formidable computational problem in *ab-initio* studies, although promising results have been obtained (see e.g. (*120—122*), *142, 143*)) in the various approximations; the calculations are too numerous to refer to specifically here. A few general observations can, however, be made, especially with reference to polyatomic molecules and ω_e-calculations.

It has already been observed in Section **5** that the frequency of the A_1-normal mode of *certain* polyatomic molecules in which the centre atom does not move and the peripheral atoms move along a set of the main symmetry axes, can be directly computed from the (effective one-dimensional) potential energy curve. This potential-energy curve is usually obtained (within the Born-Oppenheimer approximation) by keeping the centre atom rigid and bringing the peripheral atoms adiabatically in along the main symmetry axes. Differentiation of this curve at $R = R_e$ should yield k, and thus ω_e, as described.

If, however, the *ab-initio* calculation of a specific molecule yields the *complete* potential energy hypersurface (as discussed above) the *FG*-matrix method (*7*) can be used. This was done by, e.g. *Allavena* (*52*), *Hartmann* and *Gliemann* (*55*) and *Menna, Moccia* and *Randaccio* (*44*). *Huff* and *Ellison* (*144*) present an easier and simpler method which is very favourable for computational purposes; it also depends on truncating the potential after the harmonic term: the 3 N mass-weighted (m_i) Cartesian displacement co-ordinates Δx_i are called q_i, and are given by

$$q_i = \sqrt{m_i}\, x_i. \qquad (6.1)$$

This means, as usual, that

$$2\,V = \sum_k \sum_l f_{kl}\, q_k\, q_l \qquad (6.2)$$

$$2\,T = \sum_k \dot{q}_k^2. \qquad (6.3)$$

The orthogonal transformation to external (mass-weighted) symmetry co-ordinates is given by (7)

$$S_i = \sum_k a_{ik} q_k, \tag{6.4}$$

where

$$2\,V = \sum_i \sum_j F_{ij} S_i S_j \tag{6.5}$$

$$2\,T = \sum_i \dot{S}_i^2 \tag{6.6}$$

and

$$F_{ij} = \sum_k \sum_l a_{ik} a_{jl} f_{kl} \tag{6.7}$$

is zero if S_i and S_j belong to different irreducible representations of the point group of the calculated equilibrium molecule, or to different orthogonal components of a given degenerate species. The secular equations

$$\sum_i (F_{ij} - \delta_{ij} \lambda)\, A_i = 0;\; i\, j = 1, 2, \cdots 3\,N; \tag{6.8}$$

through the secular determinants then yields the roots λ_k (five or six are equal to zero because rotation was not factored off!). From

$$l_{ik} = A_{ik} / \sqrt{(\sum A_{ik}^2)} \tag{6.9}$$

the normal co-ordinates can be obtained

$$S_i = \sum_k l_{ik} Q_k \tag{6.10}$$

or from its transpose. The vibrational eigenfuctions and the normal frequencies are obtained as usual *via* the Hermite polynomials. *Huff* and *Ellison* apply it to the H_3^+ case (*144*); this compares favourably with the four approximations used by *Schwartz* and *Schaad* (*143*) and *Christofferson* (*145*). This type of vibrational analysis should be only valid where all atoms move in the same plane.

Phillipson (*146*) has given an extension of his diatomic molecule force-constant theory to polyatomic molecules, using the scaling procedure and the virial theorem; no numerical calculations have appeared.

The calculations of all these potential-energy hypersurfaces are all within the Born-Oppenheimer framework in which the electronic motion is partitioned off from the internuclear motion. This partitioning of two independent types of motion is only possible if the frequencies of the nuclear vibrations are small compared to the differences in electronic energy levels in the molecule. The Born-Oppenheimer approach breaks down if the vibration splits degenerate electronic energy levels (*147*), just like the Λ-doubling in diatomic molecules where the angular momentum of the rotating molecule couples with the electronic angular momentum (*65*). The case of linear CO_2 is exhaustively treated by *Renner* (*148*) for the bending vibration which is doubly degenerate (and in which the three atoms move along circular paths around the internuclear axis) so that the molecule is non-linear. This perturbs the Π and Δ-electronic terms; the splitting of the Π, Δ-terms can be of the same order as the potential energy of the bending vibration, so that the Born-Oppenheimer approach fails (this does not occur in the cases of the symmetric stretching and antisymmetric stretching vibrations!).

The failure of the Born-Oppenheimer approximation for degenerate vibrational modes which can lift electronic degeneracies can thus account for the poor results obtained in frequency calculations which *assume* the validity of the Born-Oppenheimer approximation in the energy hypersurface calculation. A case in point is *Grein's* comparison (*149*) of the results on BH_4^-, CH_4^- and NH_4^+ (see also above, as well as (*44—48, 50, 52—57*)) which shows the discrepancies between the experimental and calculated values of $\omega_2(E)$, $\omega_3(F_2)$ and $\omega_4(F_2)$ of the XY_4 molecules; the agreement with $\omega_1(A_1)$ is far better.

The failure of the Born-Oppenheimer approach is also discussed by *Lefebvre* and *Sucre* (*13*) in connection with dimer spectra; they discern weak- and strong-coupling cases of the zeroth-order Born-Oppenheimer approximation, and derive appropriate equations. *Kollmann* and *Allen* (*154*) developed an *ab-initio* calculation of the hydrogen bond, and calculated various hydrogen-bond force constants. *Dewar et al.* (*155*) considered the ground states of a large number of σ-bonded molecules, and calculated force constants (e.g. simple organic compounds), obtaining reasonable values. *Body, McClure* and *Clementi* (*156*) made an accurate SCF study of ammonia, and calculated the potential energy surface and vibrational force constants, comparing their values with other calculations; the angle-bending force constants are in reasonable agreement, while the bond-stretching values are to high, and the mixed stretching-angle force constants are too small. *Peyerimhoff* and *Buenker* have been very active in the ab-initio field, and have calculated *inter alia* the allylic system (*157*), HCOOH (*158*), triatomic molecules (*159*). *Bak, Clementi* and *Kortzeborn* (*160*) determined the structure, vibrational spectra and dipole moment of gaseous LiCN and LiNC using an *ab-initio* Hartree Fock calculation; this in a very interesting calculation.

There is thus considerable activity in this field, and it will no doubt continue to be improved in the future as the understanding of the binding forces in molecules becomes better.

Since this review was written, many SCF-calculations of force constants and frequencies of polyatomic molecules have appeared, as well as advances of the basic theory. *Parr* and *Brown* (*149, 150a*) extended their work on diatomic molecules (*114*) to polyatomic molecules in the Born-Oppenheimer framework and the quantum mechanical virial theorem; it is applied to CO_2, and the results are encouraging. *Smyser* and *Wilson* (*150b*) published a paper on the quantum dynamics of triatomic molecules with Morse bond-stretching potentials; they make the very important point that in treating multidimensional problems as direct products of one-dimensional problems, it is very important to remove all possible off-diagonal terms in the Hamiltonian before forming the direct product representation. *Poshusta, Haugen* and *Zetik* (*151*) developed an approximate *ab-initio* method for predicting *inter alia* the vibrational frequencies of small (fewer than six binding electrons) unipositive polyatomic ions (e.g. BeH_2^+, $LiHeH^+$, etc.), but no experimental frequencies are available for comparison. *Newton et al.* (*152*) used a previously derived SCF method (*153*) to calculate the quadratic force constants of a number of first-row polyatomic molecules ($CH_3.CH_3$, CH_2CH_2, CH_3CCH, NH_3, OH_2, CH_3F, NF_3, OF_2, H_2CO, HCN, CO_2); the results are reasonable although the method overestimates the force constants.

7. References

1. *Bjerrum, N.:* Nernst Festschrift, Halle, 1912.
2. *Kratzer, A.:* Z. Physik *3,* 289 (1920).
3. a) See any (elementary) text on quantum mechanics, as well as: b) *Dunham, J. L.:* Phys. Rev. *41,* 721 (1932), who discusses the Schrödinger rotating-vibrator for the first time, using the Wentzel-Brillouin-Kramers method; the internuclear potential V is developed in a power series about $\xi = (r - r_e)/r_e = 0$. Dunham ascribes the fact that his treatment of the rotating-vibrator eight years after the formulation of quantum mechanics is the first paper in this field, to the success of Morse's empirical function.
4. *Morse, P. M.:* Phys. Rev. *34,* 57 (1929).
5. *Nielsen, H. H.:* The Vibration-rotation Energies of Molecules and their Spectra in the Infra-red. In: *S. Flügge* (Editor), Handbuch der Physik, Bd. XXXVII/1, Atome III-Moleküle I, p. 172—313. Berlin–Göttingen–Heidelberg: Springer 1959.
6. *Mizushima, S.:* Raman Effect. In: *S. Flügge* (Editor), Handbuch der Physik, Bd. XXVI, Licht und Materie II, p. 170—243. Berlin–Göttingen–Heidelberg: Springer 1958.
7. *Wilson, E. B., Jr., Decius, J. C., Cross, P. C.:* Molecular Vibrations. New York: McGraw Hill 1955.
8. *Nakamoto, K.:* Infra-red Spectra of Inorganic and Co-ordination Compounds. New York: Wiley Second Edition 1970.
9. *Born, M., Oppenheimer, J. R.:* Ann. Physik *84,* 457 (1927).
10. — Festschrift. Gött. Nachr. Math. Phys., Kl. 1 (1951).

11. a) *Hobey, W. D., McLachlan, A. D.:* J. Chem. Phys. *33*, 1695 (1960);
b) *Dalgarno, A., McCarroll, R.:* Proc. Roy. Soc. (London) *237A*, 383 (1956);
c) *Liehr, A. D.:* Ann. Physik *1*, 221 (1957).
12. *Slater, J. C.:* Quantum Theory of Molecules and Solids, Vol. I. New York: McGraw-Hill 1963.
13. *Lefebvre, R., Sucre, M. G.:* Intern. J. Quantum Chem. *1S*, 339 (1967).
14. *Hirschfelder, J. O., Curtis, C. F., Bird, R. B.:* Molecular Theory of Gases and Liquids. New York: Wiley 1954. London: Chapman and Hall 1954.
15. *Eyring, H., Walter, J., Kimball, G. E.:* Quantum Chemistry. New York–London: Wiley 1944.
16. Ref. 7, Chapter 11 and Appendix.
17. *Mathews, J., Walker, R. L.:* Mathematical Methods of Physics, p. 376f. New York: Benjamin 1964.
18. a) *Diehl, H., Fluegge, S., Schroeder, U., Voelkel, A., Weiguny, A.:* Z. Physik *162*, 1 (1961);
b) *Fluegge, S., Weiguny, A.:* Z. Physik *162*, 15 (1961);
c) *Diehl, H., Fluegge, S.:* Z. Physik *162*, 21 (1961).
19. *Burrau, Ø.:* Kgl. Danske Videnskab. Selskab, Mat. Fys. Medd. *7*, 14 (1927).
20. *Margenau, H., Murphy, G. M.:* The Mathematics of Physics and Chemistry, Vol. I, 2nd Edition, p. 411ff. Princeton N. J.: D. van Nostrand Company, Inc. 1956.
21. *Eyring, H., Polanyi, M.:* Z. physik. Chem. *B12*, 279 (1931).
22. a) *Curtiss, C. F., Hirschfelder, J. O., Adler, F. T.:* J. Chem. Phys. *18*, 1638 (1950);
b) *Curtiss, C. F.:* J. Chem. Phys. *21*, 1199 (1953);
c) *Hirschfelder, J. O., Wigner, E. P.:* Proc. Natl. Acad. Sci. *21*, 113 (1935).
23. *Hellmann, H.:* Quantenchemie, p. 285. Leipzig: Deuticke & Co. 1937.
24. *Feynman, R. P.:* Phys. Rev. *56*, 340 (1939).
25. Ref. 14, p. 932f.; see also p. 838ff.
26. Ref. 20, p. 344.
27. See Ref. 14, p. 916ff.
28. *Platt, J. R.:* Chemical Bond and Distribution of Electrons in Molecules. In: *S. Flügge* (Editor), Handbuch der Physik, Bd. XXXVII/2, Moleküle II. Berlin–Göttingen–Heidelberg: Springer 1961.
29. *Berlin, T.:* J. Chem. Phys. *19*, 208 (1951); the "Coulson-confusion" referred to in this paper has been cleared up by the work reported in Refs. (30—32).
30. *Hirschfelder, J. O., Coulson, C. A.:* J. Chem. Phys. *36*, 941 (1962).
31. *Coulson, C. A., Hurley, A. C.:* J. Chem. Phys. *37*, 448 (1962).
32. *Hurley, A. C.:* J. Chem. Phys. *37*, 449 (1962).
33. Ref. 14, p. 42f. and p. 930ff.
34. *Marion, J. B.:* Classical Dynamics, p. 256. New York: Academic Press 1965.
35. *Goldstein, H.:* Classical Mechanics, p. 69. Cambridge (Mass.): Addison-Wesley 1950.
36. Ref. 15, p. 355.
37. *Slater, J. C.:* J. Chem. Phys. *1*, 687 (1933).
38. Ref. 12, p. 254 (Appendix 3).
39. Ref. 14, p. 68.
40. Ref. 15, p. 355.
41. *Clinton, W. L.:* J. Chem. Phys. *33*, 1603 (1960).
42. *Finkelstein, B. N.:* Z. Physik *50*, 293 (1928).
43. See, e.g. *Schutte, C. J. H.:* The Wave Mechanics of Atoms and Molecules, p. 63ff. London: Edward Arnold 1968

44. *Menna, E., Moccia, R., Randaccio, L.:* Theoret. Chim. Acta (Berl.) *4*, 408 (1966).
45. *Moccia, R.:* J. Chem. Phys. *40*, 2164 (1964).
46. — J. Chem. Phys. *40*, 2176 (1964).
47. — J. Chem. Phys. *40*, 2186 (1964).
48. *Bishop, D. M.:* Mol. Phys. *6*, 305 (1962).
49. *McLean, A. D., Weiss, A., Yoshimine, M.:* Rev. Mod. Phys. *32*, 211 (1960).
50. *Bishop, D. M., Hoyland, J. R.:* Mol. Phys. *7*, 161 (1963).
51. *Herzberg, G.:* Molecular Spectra and Molecular Structure, Vol. II, Infra-red and Raman Spectra of Polyatomic Molecules. Princeton, N. J.: D. van Nostrand Company, Inc. 1945.
52. *Allavena, M.:* Theoret. Chim. Acta (Berl.) *5*, 21 (1966).
53. a) *Albasiny, E. L., Cooper, J. R. A.:* Proc. Phys. Soc. (London) *85*, 1133 (1965); b) see also *Albasiny, E. L., Cooper, J. R. A.:* Mol. Phys. *4*, 353 (1961) and Proc. Phys. Soc. (London) *82*, 289 (1963) for BH_4^-, CH_4, NH_4^+.
54. *Sutton, A., Banyard, K. E.:* Mol. Phys. *12*, 377 (1967).
55. *Hartman, H., Gliemann, G.:* Z. Phys. Chem. N. F. *15*, 108 (1958).
56. *Shimanouchi, T., Nakagawa, I., Hiraishi, J., Ishii, M.:* J. Mol. Spectr. *19*, 78 (1966).
57. *Bishop, D. M., Hoyland, J. R., Parr, R. G.:* Mol. Phys. *6*, 466 (1962).
58. *McLean, A. D., Yoshimine, M.:* I. B. M. Journal of Research and Development, Nov. 1967.
59. — — Tables of Linear Molecule Wave Functions. I. B. M. Publication, obtainable upon request.
60. *Sugiura, Y.:* Z. Physik *45*, 484 (1927).
61. *Heitler, W., London, F.:* Z. Physik *44*, 455 (1927).
62. *Parr, R. G.:* The Quantum Theory of Molecular Electronic Structure, p. 13ff. W. A. Benjamin 1963.
63. a) Ref. 12, p. 18; Ref. 43, p. 67.
b) A rapid method to construct "true" potential curves of diatomic molecules using the Morse function was given by *Vaidyan, V. K., Santaram, C.:* J. Chem. Phys. *52*, 3068 (1970); see also Ref. (139).
64. *Morse, P. M.:* Phys. Rev. *34*, 57 (1929).
65. *Herzberg, G.:* Molecular Spectra and Molecular Structure, Vol. I, Spectra of Diatomic Molecules, 2nd Edition, p. 101f. Princeton, N. J.: D. van Nostrand Co. 1950.
66. a) *Wind, H.:* J. Chem. Phys. *43*, 2956 (1965);
b) — J. Chem. Phys. *42*, 2371 (1965).
67. *Slater, J. C.:* Quantum Theory of Atomic Structure, Vol. I, p. 25. New York: McGraw-Hill 1960.
68. *Bates, D. R., Ledsham, K., Stewart, A. L.:* Phil. Trans. Roy. Soc. London *246*, 215 (1953).
69. *Pauling, L. C., Wilson, E. B., Jr.:* Introduction to Quantum Mechanics, p. 326ff. New York: McGraw-Hill 1935.
70. *Hurley, A. C.:* Proc. Roy. Soc. (London) *A 226*, 179 (1954).
71. a) *Varshni, Y. P., Shukla, R. C.:* Rev. Mod. Phys. *35*, 130 (1963);
b) — Rev. Mod. Phys. *29*, 664 (1957); *31*, 839 (1959);
c) *Steele, D., Lippincott, E. R., Van der Slice, J. T.:* Rev. Mod. Phys. *34*, 239 (1962).
72. *Matcha, R. L.:* J. Chem. Phys. *47*, 5295 (1967). — *Matcha, R. L.:* J. Chem. Phys. *48*, 335 (1968). — *Cashion, J. K.:* J. Chem. Phys. *48*, 94 (1968).
73. *Karo, A. M., Olson, A. R.:* J. Chem. Phys. *30*, 1232, 1241 (1959).
74. *Hylleraas, E. A.:* Z. Physik *71*, 739 (1931).

75. *Richardson, O. W.:* Nuovo Cimento *15*, 232 (1938).
76. *Cohen, S., Hiskes, J. R., Riddell, R. J.:* Phys. Rev. *119*, 384 (1960).
77. Ref. 65, p. 95 and p. 440.
78. Ref. 43, p. 64, 423f.; see also Ref. 43, p. 62ff., p. 423.
79. *Bingel, W. A.:* J. Chem. Phys. *30*, 1250, 1254 (1959); see also Ref. 28, pp. 254—262, as well as Ref. 79.
80. *Preuss, H.:* Mol. Phys. 440 (1963); see also Ref. 124.
81. *McLean, A. D.:* J. Chem. Phys. *40*, 2774 (1964); *38*, 343 (1963).
82. *Phillipson, P.:* J. Chem. Phys. *39*, 3010 (1963).
83. *Löwdin, P.-O.:* J. Mol. Spectry. *3*, 46 (1959).
84. *Kolos, W., Roothaan, C. C. J.:* Rev. Mod. Phys. *32*, 219 (1960); see also *Baran, J., Kolos, W.:* J. Mol. Spectry. *8*, 121 (1962).
85. *McLean, A. D.:* J. Chem. Phys. *39*, 2653 (1963).
86. *Meizner, J.:* Mathieusche Funktionen und Sphäroidfunktionen, p. 14ff. Berlin–Göttingen–Heidelberg: Springer 1954.
87. *Madelung, E.:* Die Mathematischen Hilfsmittel des Physikers, 2nd Edn., p. 223ff. Berlin–Göttingen–Heidelberg: Springer 1953.
88. *Hall, G.:* Phil. Mag. *6*, 249 (1961).
89. *Goodisman, J.:* J. Chem. Phys. *39*, 2397 (1963).
90. *Stanton, R. E.:* J. Chem. Phys. *36*, 1298 (1960).
91. *Ross, W. R., Phillipson, P.:* J. Chem. Phys. *44*, 844 (1966).
92. *Heitler, W., London, F.:* Z. Physik *44*, 455 (1927).
93. *Hurley, A. C.:* Proc. Roy. Soc. (London) *A 226*, 179 (1954).
94. *Weinbaum, S.:* J. Chem. Phys. *1*, 593 (1933).
95. *Wang, S.:* Phys. Rev. *31*, 579 (1928).
96. *Coulson, C. A.:* Trans. Faraday Soc. *33*, 1479 (1937).
97. *Davidson, E.:* J. Chem. Phys. *36*, 2527 (1962).
98. *Platt, J. R.:* J. Chem. Phys. *18*, 932 (1950).
99. *Clinton, W. L.:* J. Chem. Phys. *33*, 1603 (1960).
100. *Bratoz, S., Daudel, R., Roux, M., Allavena, M.:* Rev. Mod. Phys. *32*, 412 (1960), and references therein.
101. *Empedocles, P.:* J. Chem. Phys. *46*, 4474 (1967).
102. *Hoare, M. F., Linnett, J. W.:* Trans Faraday Soc. *46*, 885 (1950).
103. *Bader, R. F. W.:* Mol. Phys. *3*, 137 (1960).
104. *Murrell, J. N.:* J. Mol. Spectr. *4*, 446 (1960).
105. a) *Beyers Brown, A.:* Proc. Cambridge Phil. Soc. *54*, 251 (1958);
 b) *Beyers Brown, W., Steiner, E.:* J. Chem. Phys. *37*, 461 (1962).
106. *Salem, L.:* J. Chem. Phys. *38*, 1227 (1963).
107. *Hall, G. G., Rees, D.:* Theoret. Chim. Acta (Berl.) *1*, 448 (1963).
108. *Schwendeman, R. H.:* J. Chem. Phys. *44*, 556 (1966).
109. a) *Gerratt, J., Mills, I. M.:* J. Chem. Phys. *49*, 1719 (1968);
 — — J. Chem. Phys. *49*, 1730 (1968).
110. *Roothaan, C. C. J.:* Rev. Mod. Phys. *23*, 69 (1951).
111. *Freed, K. F.:* J. Chem. Phys. *52*, 253 (1970).
112. a) *Bishop, D. M., Randič, M.:* J. Chem. Phys. *44*, 2480 (1966);
 b) — *Macias, A.:* J. Chem. Phys. *51*, 4997 (1969), applied to H_2^+.
113. *Bratoz, S.:* Colloq. Intern. Centre Natl. Rech. Sci. (Paris) *82*, 287 (1958).
114. *Parr, R. G., Borkman, R. F.:* J. Chem. Phys. *49*, 1055 (1968);
 b) *Borkman, R. F., Simons, G., Parr, R. G.:* J. Chem. Phys. *50*, 58 (1969).
115. *Fues, E.:* Ann. Physik *80*, 367 (1926).
116. *Allen, L. C., Karo, A. M.:* Rev. Mod. Phys. *32*, 275 (1960).

117. *Kotani, M., Mizuno, Y., Kayama, K., Yoshizumi, H.:* Ann. Rev. phys. Chem. *9*, 245 (1958).
118. a) *Ransil, B. J.:* Rev. Mod. Phys. *32*, 239 (1960);
b) — Rev. Mod. Phys. *32*, 245 (1960).
119. *Wahl, A. C., Bertoconcini, P. J., Das, G., Gilbert, T. C.:* Intern. J. Quantum Chem. *1 S*, 123—152 (1967) (Slater Symposium; this symposium contains many important papers on these topics).
120. *Pople, J. A., Beveridge, D. L., Ostlund, N. S.:* Intern. J. Quantum Chem. *1 S*, 293 (1967).
121. *Preuss, H.:* In: *P. O. Löwdin* (Ed.), Quantum Theory of Atoms, Molecules and the Solid State, p. 281; see also Ref. 80. New York: Academic Press 1966.
122. *Nesbet, R. K.:* Advan. Quantum Chem. (*P. O. Löwdin* (Ed.)) *3*, 1—23 (1967).
123. *Bishop, D. M.:* Advan. Quantum Chem. (*P. O. Löwdin* (Ed.)) *3*, 25—59 (1967).
124. *Bratoz, S., Allavena, M.:* J. Chem. Phys. *37*, 2138 (1962).
125. — Advan. Quantum Chem. (*P. O. Löwdin* (Ed.)) *3*, 209—237 (1967).
126. *Joshi, B.:* J. Chem. Phys. *43*, S40 (1965).
127. *Preuss, H.:* Fortschr. Chem. Forsch. *9*, 325 (1968).
128. *Kouba, J., Öhrn, Y.:* J. Chem. Phys. *52*, 5387 (1970).
129. *Ellison, F. O., Slezak, J. A.:* J. Chem. Phys. *50*, 3942 (1969) and references therein.
130. *Kolos, W., Wolniewicz, L.:* J. Chem. Phys. *49*, 404 (1968), and references therein; *Rush, D. G.:* Trans. Faraday Soc. *64*, 2013 (1968) devised a non-iterative method to numerically determine the eigenvalues of any vibrational potential, and applied it to the *Kolos* and *Wolniewicz* results.
131. *Sahni, R. C., Sawhney, B. C., Hanley, M. J.:* Trans. Faraday Soc. *65*, 3120 (1969) and references therein.
132. *Salotto, A. W., Burnelle, L.:* J. Chem. Phys. *52*, 2936 (1970).
133. *Pack, R. T., Hirschfelder. J. O.:* J. Chem. Phys. *52*, 521 (1970).
134. *Bunker, P. R.:* J. Mol. Spectry. *28*, 422 (1968).
135. *Tuan, D. F.:* J. Chem. Phys. *51*, 607 (1969).
136. *Loeb, R. J., Rasiel, Y.:* J. Chem. Phys. *52*, 4995 (1970).
137. *King, W. T.:* J. Chem. Phys. *49*, 2866 (1968). — *King, W. T., Zelano, A. J.:* J. Chem. Phys. *47*, 3197 (1967).
138. *Simonetta, M., Gianinetti, E.:* In: *P. O. Löwdin* and *B. Pullman* (Editors), Molecular Orbitals in Chemistry, Physics and Biology. New York: Academic Press 1964; see also *Kotani, M., Ohno, K., Kayama, K.:* Quantum Mechanics of Electronic Structure of Simple Molecules. In: *S. Fluegge* (Editor), Handbuch der Physik, Vol. XXXVII/2, Moleküle II. Berlin–Göttingen–Heidelberg: Springer 1961.
139. *Greenawalt, E. M., Dickinson, H. S.:* J. Mol. Spectry. *30*, 427 (1969).
140. *Le Roy, R. J., Bernstein, R. B.:* J. Chem. Phys. *49*, 4312 (1968).
141. *Cashion, J. K.:* J. Chem. Phys. *48*, 94 (1968).
142. a) *Ellison, F. O.:* J. Am. Chem. Soc. *85*, 3540 (1963);
b) —, *Huff, N. R., Patel, J. C.:* J. Am. Chem. Soc. *85*, 3544 (1963).
143. *Schwartz, M. E., Schaad, J. C.:* J. Chem. Phys. *47*, 5325 (1967).
144. *Huff, N. R., Ellison, F. O.:* J. Chem. Phys. *42*, 364 (1965); calculations on the H_3^+ system (see (142)).
145. *Christofferson, R. E.:* J. Chem. Phys. *41*, 960 (1964). Also deduces the formulae discussed in text; see also *Conroy, H.:* J. Chem. Phys. *40*, 603 (1964), *erratum, ibid. 40*, 3121 (1964).
146. *Phillipson, P.:* J. Chem. Phys. *44*, 633 (1966).
147. *Herzberg, G., Teller, E.:* Z. phys. Chem. *21 B*, 410 (1933).

148. *Renner, R.:* Z. Physik 172 (1934).
149. *Grein, F.:* Theoret. Chim. Acta (Berl.) *1*, 52 (1962).
150. a) *Parr, R. G., Brown, J. E.:* J. Chem. Phys. *49*, 4849 (1968);
b) *Smyser, W. E., Wilson, D. J.:* J. Chem. Phys. *50*, 182 (1969).
151. *Poshusta, R. D., Haugen, J. A., Zetik, D. F.:* J. Chem. Phys. *51*, 3343 (1969).
152. *Newton, M. D., Latham, W. A., Hehre, W. J., Pople, J. A.:* J. Chem. Phys. *52*, 4064 (1970).
153. *Hehre, W. J., Stewart, F. R., Pople, J. A.:* J. Chem. Phys. *51*, 2657 (1969).
154. *Kollmann, P. A., Allen, L. C.:* J. Chem. Phys. *52*, 5085 (1970).
155. *Bodor, N., Dewar, M. J. S., Harget, A., Haselbach, E.:* J. Am. Chem. Soc. *92*, 3854 (1970).
156. *Brody, R. G., McClure, D. S., Clementi, E.:* J. Chem. Phys. *49*, 4916 (1968).
157. *Peyerimhoff, S. D., Buenker, R. J.:* J. Chem. Phys. *51*, 2528 (1969).
158. — — J. Chem. Phys. *50*, 1846 (1969).
159. — — J. Chem. Phys. *49*, 2473 (1968).
160. *Bak, B., Clementi, E., Kortzborn, R. N.:* J. Chem. Phys. *52*, 764 (1970).

Received September 21, 1970

Structure and Bonding: Contents Vol. 1-9

GPSR Compliance

The European Union's (EU) General Product Safety Regulation (GPSR) is a set of rules that requires consumer products to be safe and our obligations to ensure this.

If you have any concerns about our products, you can contact us on ProductSafety@springernature.com

In case Publisher is established outside the EU, the EU authorized representative is:

Springer Nature Customer Service Center GmbH
Europaplatz 3
69115 Heidelberg, Germany

Batch number: 10682411

Printed by Printforce, the Netherlands